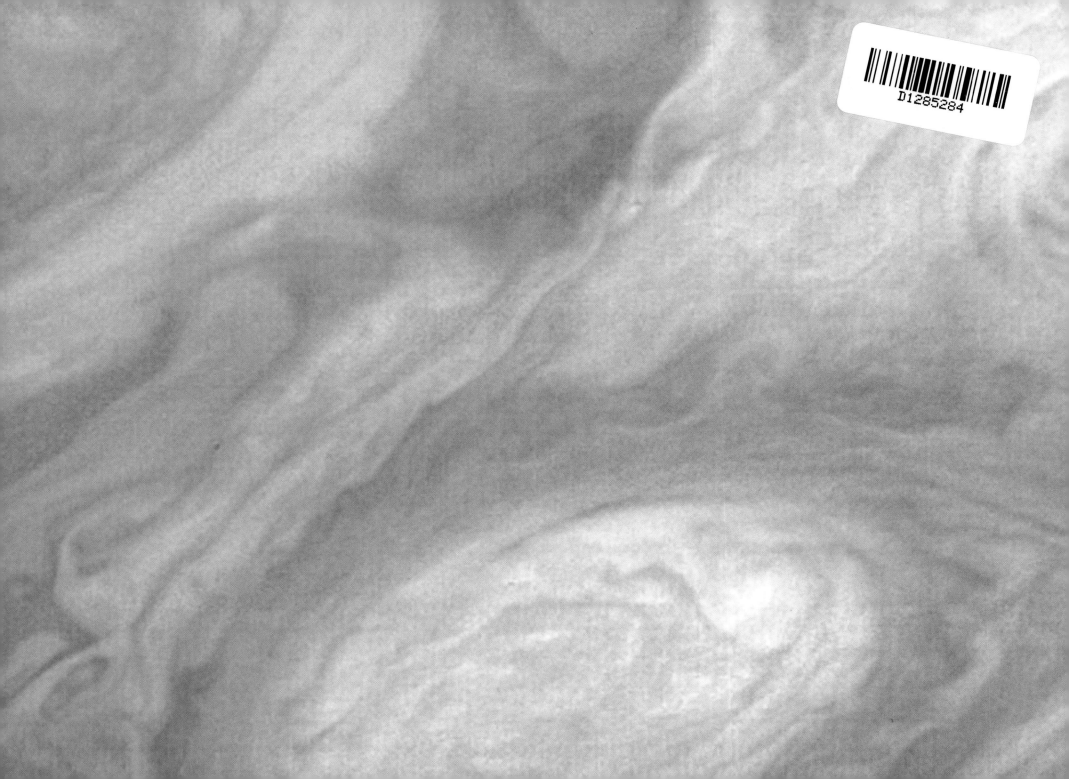

THE SOLAR SYSTEM

THE SOLAR SYSTEM

Exploring the Planets and Their Moons
from Mercury to Pluto and Beyond

Giles Sparrow, Editor

THUNDER BAY
P·R·E·S·S

San Diego, California

Thunder Bay Press
An imprint of the Advantage Publishers Group
5880 Oberlin Drive, San Diego, CA 92121-4794
www.thunderbaybooks.com

All notations of errors or omissions should be addressed to Thunder Bay Press, Editorial Department, at the above address. All other correspondence (author inquiries, permissions) concerning the content of this book should be addressed to Amber Books Ltd., Bradley's Close, 74–77 White Lion Street, London N1 9PF, England, www.amberbooks.co.uk.

ISBN-13: 978-1-59223-579-7
ISBN-10: 1-59223-579-4

Library of Congress Cataloging-in-Publication Data available upon request.

Printed in Singapore

1 2 3 4 5 10 09 08 07 06

CONTENTS

INTRODUCTION

such as Jupiter, the objects in the Solar System are our immediate celestial neighbors, and the only ones we can photograph in detail, send space probes to, and perhaps even visit in the foreseeable future.

An image of the surface of Mars, returned by the U.S. Pathfinder probe in 1997. It was the first spacecraft to land on the Red Planet since Viking in 1976.

This book offers a tour of the Solar System through a series of fact files, each concentrating on a specific topic, be it a specific object, a general process, or a specific aspect of one of the larger and more complex worlds. It begins with a general guide to the system as a whole—its scale, content, origins, and the specific processes that have shaped it through the past five billion years. The second chapter focuses on the Sun—the great energy source whose sphere of

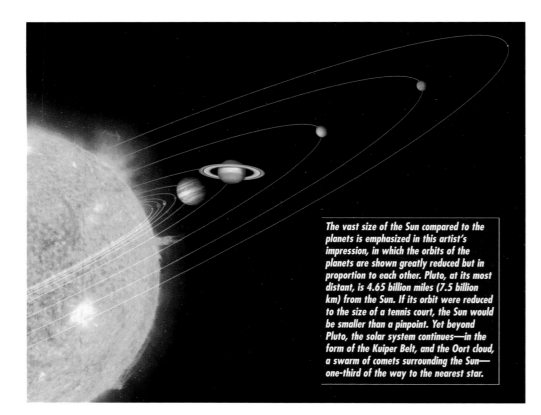

The vast size of the Sun compared to the planets is emphasized in this artist's impression, in which the orbits of the planets are shown greatly reduced but in proportion to each other. Pluto, at its most distant, is 4.65 billion miles (7.5 billion km) from the Sun. If its orbit were reduced to the size of a tennis court, the Sun would be smaller than a pinpoint. Yet beyond Pluto, the solar system continues—in the form of the Kuiper Belt, and the Oort cloud, a swarm of comets surrounding the Sun—one-third of the way to the nearest star.

Right: Our home planet—the Earth. The presence of liquid water makes Earth unique among the terrestrial planets, and makes it the only known body in the Solar System capable of supporting life.

Around five billion years ago, a dense cloud of gas and dust in the outer reaches of the Milky Way galaxy collapsed and gave birth to a young star of fairly average mass—our Sun. Like most young stars, this one was surrounded by a halo of debris—material too far away to be pulled in by its gravity, but too close to escape its influence altogether. Through aeons of collision, cohesion, and growth, this material was transformed into our present-day Solar System. Ranging in size from microscopic dust fragments, through rocky moons and planets like our own Earth, to giant balls of gas

Through aeons of collision, cohesion, and growth, this material was transformed into our present-day Solar System.

influence defines the limits of the Solar System, and which also offers astronomers their only opportunity to study a star close up. The Sun is a complex object—the nuclear reactions deep within its core generate a huge amount of activity at the surface, with effects that can reach across billions of miles of surrounding space.

From here, we move on to the planets, beginning in Chapter 3 with tiny Mercury, the rocky, moonlike world that skirts closest to the Sun, and whose near airless surface is alternately baked and frozen as it moves from searing daylight to darkest night. Chapter 4 looks at Earth's nearest neighbor, Venus—a non-identical twin of similar size and mass to our planet, but with a choking and corrosive atmosphere as a result of a runaway greenhouse effect.

Chapter 5 considers Earth itself as a planet and member of the Solar System, while Chapter 6 looks at Earth's constant companion the Moon—a satellite so large that some astronomers consider Earth a "double planet." From here we move on to Mars, the last of the solar system's inner rocky planets, but arguably the most fascinating, with giant volcanoes, icy polar caps, and evidence for a watery past during which life might have begun to evolve.

With Chapter 8 we move into the realm of the outer gas giant

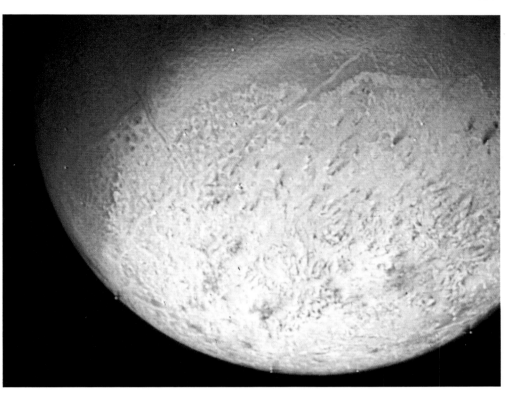

A false-color image of Neptune's largest satellite, Triton, taken by Voyager 2 in 1989. In reality, Triton would appear mostly pink, probably due to a covering of frozen water, methane, and nitrogen mixed with dust.

This infra-red image of Jupiter's Great Red Spot was captured by the Galileo probe in June 1996. The pink clouds are high and thin, while the blue areas are deeper within Jupiter's atmosphere.

planets, starting with Jupiter, a world larger than all the other planets of the Solar System put together, Jupiter, like all the outer giants, has a huge family of moons, several of which are complex enough to be considered as worlds in their own right. Chapter 9 looks at Saturn, the second largest planet and the giant with the most prominent, spectacular system of rings. Chapter 10 considers the two outer giants—turquoise Uranus and blue Neptune—both of which are markedly different from their inner neighbours. Chapter 11 takes us into the outer limits—a region of icy bodies such as Pluto, the small worlds of the Kuiper Belt, and the great comet reservoir of the Oort Cloud. Finally, Chapter 12 considers the debris of the Solar System—asteroids, comets, meteors, and meteorites.

INTRODUCING THE SOLAR SYSTEM

The Solar System consists of everything that comes within the Sun's region of influence—a volume of space extending out to approximately one light-year (six trillion miles/10 trillion km) from the Sun itself. The vast majority of the Solar System's material lies much closer to the Sun, however—even Pluto orbits just a few thousand million miles away from the centre. This inner region is the realm of the planets, eight of which follow more-or-less circular orbits around the Sun, with Pluto, the ninth and smallest, in a more elongated orbit. In fact Pluto's designation as a planet is little more than an accident of history, as we shall see. The major planets divide neatly into two types: rocky or terrestrial planets orbiting close to the Sun (of which Earth is the largest), and giant gas planets such as Jupiter, orbiting further out. Many of the planets have their own natural satellites or moons, and the gas giants are also encircled by ring systems, of which Saturn's is the most spectacular. A belt of rocky asteroids divides the terrestrials from the giants between Mars and Jupiter, while beginning around Neptune's orbit is the Kuiper Belt, a region of small icy worlds of which Pluto is the largest known member. Some comets originate here, while others orbit further out, at the very limits of the Sun's gravitational reach, in the Oort Cloud.

A montage of images taken by the Voyager spacecraft of the planets and four of Jupiter's moons, with the surface of Earth's moon in the foreground. Studying and mapping Jupiter, Saturn, Uranus, Neptune, and many of their moons, Voyager provided scientists with better images and data than they had ever expected.

OVERVIEW OF THE SOLAR SYSTEM

Once, the solar system seemed a simple place, its members forming a neat hierarchy: a central star, nine planets in stately orbit, a dozen or so dead moons, the occasional comet, and a collection of asteroids. At the orbit of Pluto, the most distant planet, our system appeared to end. In the last 30 years, modern technology has helped to change this simple image. Moons can be larger than planets, and comets swarm by the trillion halfway to the nearest star. We live in a dynamic and complex neighborhood.

WHAT IF...

...ALIENS WERE WATCHING US?

One of the most intriguing projects in astronomy is the search for life in other solar systems. The stars of the Alpha Centauri system, our closest solar neighbors, are promising candidates for life-supporting suns. Suppose there are, or were, inhabited planets around Alpha Centauri, with intelligent life similar to 20th-century humans. Would they know that we were here?

The Alpha Centaurans are especially interested in the skies above them. They know that their sun is the largest of three stars, all of them bound together by gravity. The Alpha Centauran calendar is based not only on their planet's orbit around their governing sun, but also on the gradual motion of the second sun, Alpha Centauri B. Probes and deep space telescopes have so far provided no evidence of life elsewhere in their solar system.

Recent technological advances have allowed astronomers to measure the distances to many of the stars around them. They discover that the nearest star to their own system is just 4.35 light-years away. Perhaps this star could support life-bearing planets.

Detailed research on the star begins. It turns out to be a middle-aged, stable, yellow dwarf, a good candidate for a life provider. Closer examination reveals that the star displays a series of telltale wobbles. Alpha Centauran astronomers soon calculate that there are likely to be at least four planets orbiting their closest neighbor. If one of them supports life, it is probably orbiting between 100 and 300 million miles (160–480 million km) from the star.

But there is bad news, too. The star itself is about a billion years younger than the Alpha Centaurans' own—if life did exist, it would probably be unadvanced. This would explain why they haven't picked up any radio activity from the area. But radio transmissions aimed at the system are continued—just in case.

As it happens, the Alpha Centaurans are right. Life does indeed exist in their neighboring solar system, but only in a very primitive form. Their radio signals have been falling on deaf ears—in fact, not on ears at all. Life on Earth, the third planet from the central star, is yet to emerge from the simple cellular stage of its evolution. A few billion years later, long after the Alpha Centaurans have been wiped out by a massive asteroid impact, descendants of these cells will in turn gaze up at Alpha Centauri—and wonder whether life exists there.

SOLAR SYSTEM PROFILE

LIFE-SCALE	4.5 BILLION YEARS
DIAMETER	3 LIGHT-YEARS
DIAMETER OF PLANETARY ZONE	7.4 BILLION MILES (11.9 BILLION KM)
KNOWN PLANETS	9 COMMONLY ACCEPTED
SATELLITES	AT LEAST 63
KNOWN LIFE-BEARING PLANETS	1
LARGEST PLANET	JUPITER (318 TIMES THE MASS OF THE EARTH)
SMALLEST PLANET	PLUTO (⅕ THE DIAMETER OF THE EARTH)
PLANET WITH FASTEST SPIN	JUPITER (9 HOURS 55.5 MINUTES)

SYSTEM TOUR

As we approach the solar system, the first thing that strikes us is that it is a single-star system. The galaxy is full of multiple star systems, which can lead to very eccentric planet orbits. The Sun's single state may result from its birth in a modest-size gas cloud, leaving it to develop a system of planets that all orbit in the same direction and in the same plane.

This is not to say that everything in the solar system is well organized. In the Oort Cloud, the outermost shell of the solar system, icy bodies orbit slowly in all directions, taking millions of years to do so. But as we move inward, the orbits become more regular. At the fringe of the planetary area lies the Kuiper Belt, home to numerous small ice worlds, including Pluto, the outermost accepted planet, and perhaps several other Kuiper Belt Objects of similar size.

Closer in lie the giant planets, with extended gaseous or liquid atmospheres. The largest, Jupiter, dominates. No planet can exist without feeling the pull of Jupiter, which carries with it most of the solar system's momentum around the Sun.

Closest in are the rocky planets, no more than a tenth the diameter of Jupiter. At their outer fringes are the asteroids, small bodies that were prevented by Jupiter's gravity from clumping into a planet. One of the rocky planets, Earth, has the good fortune to be in an almost circular orbit at a distance from the Sun where water can exist as a liquid on its surface. This has led to a remarkable development—life.

JOURNEY TO THE CENTER

The solar system is shrouded in the Oort Cloud, which consists of the icy residue left over from the formation of the outer planets. Deep inside this chilly region, the planets orbit the system's central star.

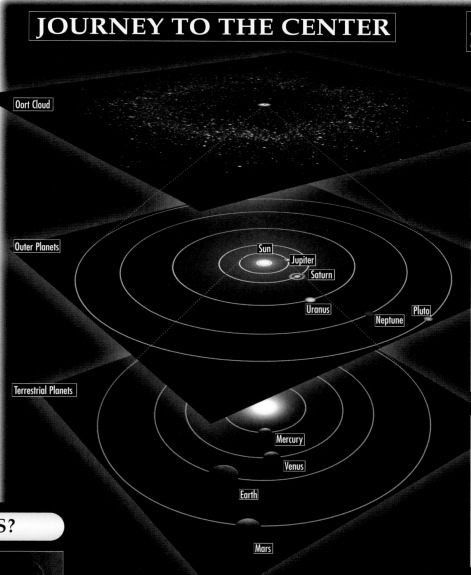

Oort Cloud

Outer Planets

Terrestrial Planets

Sun
Jupiter
Saturn
Uranus
Neptune
Pluto
Mercury
Venus
Earth
Mars

THE HALO
In the Oort Cloud, trillions of comets slowly swarm around the solar system. At this immense distance from the Sun—a light-year or more—the comets are inert icy cores with chaotic orbits. They only become active—and visible from Earth—when collisions or gravitational encounters drive them toward the inner system.

OUTER PLANETS
Two of the widely accepted nine planets vie for the distinction of being the most distant from the Sun, Neptune and Pluto. Pluto (left) is a tiny planet in an elongated orbit that sometimes brings it inside the orbit of Neptune. Even though its has a moon, Charon, it is often described as merely the largest Kuiper Belt object.

HEARTLAND
The inner solar system is the home of the rocky planets, Mars, Earth, Venus and finally Mercury. Earth and Mars (left) orbit within the so-called "life zone"—the region in which liquid water might exist. On Mars, though, atmospheric conditions make liquid water impossible. So far, life is known to have evolved only on Earth.

ONE MORE—OR LESS?

IN 2005, ASTRONOMERS ANNOUNCED THE DISCOVERY OF THE FIRST NEW KUIPER BELT OBJECT TO RIVAL PLUTO IN SIZE. THIS WORLD, NICKNAMED "SANTA" FOR ITS DISCOVERY AT CHRISTMAS 2004, ORBITS TWO AND A HALF TIMES FURTHER FROM THE SUN THAN PLUTO. THE DISCOVERY HAS REIGNITED THE DEBATE ABOUT THE NUMBER OF PLANETS — SOME SAY THE NEW OBJECT SHOULD BE CLASSED AS A TENTH PLANET, WHILE OTHERS ARGUE THAT THIS SHOWS PLUTO IS ONLY A PLANET BY HISTORICAL ACCIDENT, AND IT SHOULD NOW BE DEMOTED TO A MERE KUIPER BELT OBJECT.

SCALE OF THE SOLAR SYSTEM

The Sun is enormous, packing within its huge globe 99% of the solar system's matter. Yet measured on the scale of the entire solar system, the Sun is almost as insignificant as its nine planets. Pluto, at its most distant, is 4.65 billion miles (7.5 billion km) from the Sun. If its orbit were reduced to the size of a tennis court, the Sun would be smaller than a pinpoint. Yet beyond Pluto, the solar system continues—in the form of the Kuiper Belt, and the Oort cloud, a swarm of comets surrounding the Sun—one-third of the way to the nearest star.

WHAT IF...

...WE COMPARE THE SOLAR SYSTEM WITH OTHER OBJECTS?

The solar system is enormous in comparison with everyday objects that we find on Earth, and even next to the Sun itself. But how does it measure up to some other astronomical objects?

The giant molecular clouds from which stars form can contain millions of solar masses' worth of gas and dust, and some of them stretch for hundreds of light-years. The Orion Nebula is a famous example. These complexes are a thousand times larger than the Sun's tiny family: If we imagine that the solar system, including the Oort cloud—the sparse but gigantic cloud of trillions of icy, inert comets that orbit up to 1.6 light-years from the Sun—is the size of a grapefruit, then the largest molecular clouds would be 20–30 feet (6–9 m) across. In other words, giant molecular clouds are as large in relation to the Solar System as the orbits of the inner planets are compared with the Sun.

On an even larger scale are the giant islands of stars called galaxies. The Milky Way galaxy in which our solar system resides, for example, is about 120,000 light-years from one side to the other. This is more than 100 times bigger than even the largest molecular clouds, and 90,000 times more extensive than the solar system. Compared with our grapefruit-size solar system, the Milky Way would extend for about seven miles. The Sun on this scale would be barely larger than a subatomic particle, while the Earth would be far smaller even than quarks, the tiniest known fragments of matter.

Galaxies themselves group together into even larger structures called clusters. And clusters of galaxies are assembled into still larger groups called superclusters. So although the solar system is already incomprehensibly large, there are other, much bigger structures in existence—and the largest scale of all is that of the universe itself.

The entire observable universe is estimated to be about 30 billion light-years from one side to the other. If we could shrink it to the same scale as the grapefruit-size solar system, it would have the same diameter as the real orbit of the innermost planet, Mercury.

SOME STEPS IN THE SCALE

Object	Diameter	Size compared with previous object
Jupiter ring particle	0.00004 inches (0.001 mm)	N/A
Saturn ring particle	Up to 15 feet (4.5 m)	5,000,000 times
Typical asteroid	300 feet (90 m)	20 times
Largest known crater (on Mars)	1,550 miles (2,500 km)	5.2 times
Largest known moon (Ganymede)	3,270 miles (5,250 km)	2.1 times
Earth	12,800 miles (20,600 km)	3.9 times
Largest known planet (Jupiter)	88,800 miles (143,000 km)	6.9 times
Saturn's ring span	210,000 miles (337,962 km)	2.4 times
Sun	865,000 miles (1.4 million km)	4 times
Orbit of inner planet (Mercury)	36 million miles (58 million km)	42 times

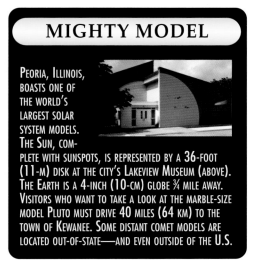

MIGHTY MODEL

PEORIA, ILLINOIS, BOASTS ONE OF THE WORLD'S LARGEST SOLAR SYSTEM MODELS. THE SUN, COMPLETE WITH SUNSPOTS, IS REPRESENTED BY A 36-FOOT (11-M) DISK AT THE CITY'S LAKEVIEW MUSEUM (ABOVE). THE EARTH IS A 4-INCH (10-CM) GLOBE ¾ MILE AWAY. VISITORS WHO WANT TO TAKE A LOOK AT THE MARBLE-SIZE MODEL PLUTO MUST DRIVE 40 MILES (64 KM) TO THE TOWN OF KEWANEE. SOME DISTANT COMET MODELS ARE LOCATED OUT-OF-STATE—AND EVEN OUTSIDE OF THE U.S.

THE SUN AND ITS OFFSPRING

When the Sun and its nine planets are shown together to the same scale, tiny Pluto is so small that it could easily be mistaken for a background star. And there is no room to show more than a section of the Sun's 865,000-mile (1.4 million km) disk.

Pluto

Neptune

Uranus

Saturn

Mercury

Earth

Mars

Venus

Jupiter

PALE BLUE DOT
As Voyager 1 passed beyond Neptune and headed on into the depths of space, it turned to face the Sun one last time to capture a grainy but inspiring image of a blue dot suspended in a sunbeam (the beam is actually an illusion caused by spacecraft reflections). The insignificant speck is the planet Earth and all it contains. The image inspired the late Carl Sagan to write his last book, appropriately entitled *The Pale Blue Dot*.

MOON MISCELLANY
In an image captured by Voyager 2 in 1979, the moon Io passes above Jupiter's stormy southern hemisphere. The solar system's satellites come in all sizes. The smallest are little more than lumps of rocks a few miles across. But the two largest moons—Jupiter's Ganymede and Saturn's Titan—are both bigger than the planet Mercury.

SIZE MATTERS

The human brain evolved to deal with a small-scale environment. Whether hunting prey or avoiding predators, near was more important than far. The horizon was as distant a perspective as early Homo sapiens needed to care about, and even today it is difficult for humans to comprehend the truly large-scale. The Earth seems vast to us. But all the planets and even the Sun itself, a glowing ball of plasma some 865,000 miles (1.4 million km) in diameter, are insignificant specks in the immensity of the whole solar system.

To reduce such vastness to sizes that humans can more readily grasp, imagine the entire solar system scaled down by a factor of 13 billion. That's the number needed to reduce the Sun to the size of a grapefruit. On such a scale, Earth becomes a pinhead about 38 feet (11.5 m) away, while the Moon, smaller than a grain of sand, swings around its parent planet at a distance of just 1.1 inches (28 mm). Farther away, some 200 feet (61 m) from the grapefruit Sun, lies the giant planet Jupiter, which on this scale is about the size of a pea. Jupiter's largest moons could easily pass through the eye of a needle—and the smallest are too tiny to be seen with the naked eye. Far beyond Jupiter, a barely discernible

fleck of dust orbits more than a quarter of a mile from the grapefruit Sun. This is the outermost planet, Pluto, whose elliptical orbit carries it 50 times farther away from the Sun than the Earth is. To drive there at a steady 60 miles an hour (96 km/h) without stopping would take almost 9,000 years.

THE OUTER REACHES

Pluto marks the commonly accepted boundary of the realm of the planets—as far as we know. But the Sun's family actually extends much farther afield than that. Surrounding the Sun at a distance of perhaps 9 trillion miles (15 trillion km)—one-and-a-half light-years and one-third of the way to the nearest stars—is the Oort cloud. Named after the Dutch astrophysicist Jan

Hendrik Oort, this structure is a vast, spherical swarm of icy debris, left over from the formation of the solar system more than 4.5 billion years ago. Oort cloud objects are so far from the Sun that the tether of gravity is feeble. From time to time, a passing star will nudge a few of them from their fragile orbits and send them falling into the inner system, where their icy material boils off and they cross our sky as comets. If we took our grapefruit Sun and placed it in Salt Lake City, Utah, the Oort cloud would stretch from the Mexican border to Canada—a distance of about 1,400 miles (2,250 km).

Within the immense span of the solar system are objects that range from dust grains to planets. Between them, they have a remarkable range of sizes. Among the smallest orbiting objects are the flecks of dust that make up Jupiter's rings—each

particle is only about 0.00004 inch (0.001 mm) across. At the other extreme is the entire Oort cloud—in scientific parlance, 22 orders of magnitude larger. An order of magnitude simply means a factor of 10, and in scientific notation 22 orders would be written as 10^{22}—that is, 1 followed by 22 zeroes, or the mind-numbing figure of 10 billion trillion.

The middle of this scale corresponds to a length of about 100 miles (160 km), the diameter of a fairly large asteroid. So an asteroid is as big compared with a Jovian ring particle as the Oort cloud is compared with the asteroid. Our own planet Earth, roughly two orders of magnitude larger than the asteroid, is still a billion times smaller than the Oort cloud. Yet even the vastness of the Oort cloud cannot compare with the expanse of the universe itself—which is 20 billion times bigger still.

PLANETARY ORBITS

Dominating the sky, a rogue planet the size of Jupiter looms above a flooded, stormy Earth. The gravity of such a giant would produce tides at least 100 times higher than those raised by the Moon. It would also buckle the Earth's crust and drag our planet into a new orbit. The Earth might even become one of the giant's own moons.

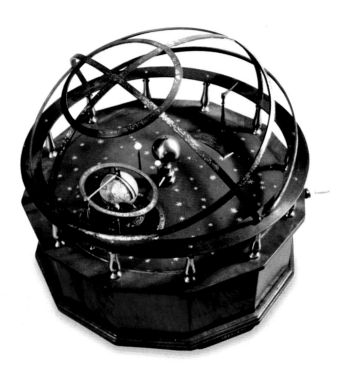

WHAT IF...

...EARTH CHANGED ITS ORBIT?

The Earth sits roughly in the middle of the solar system's habitable band—an area about 40 million miles (64 million km) wide where liquid water can exist and life should theoretically be able to thrive. So it might be possible for our planet to move a considerable distance without killing off its vast population of living things.

Beyond the edges of the habitable band, the planet would either scorch or freeze, and life as we know it would cease. But as all orbits are elliptical, even a big change in the Earth's orbit would not necessarily scorch us or freeze us all year round. The Earth could stay mainly within the habitable band, but in a more elongated orbit—one that increased the maximum distance from the Sun and

reduced the minimum. That way, we would scorch for a small part of the year and freeze for much of the rest—if we survived at all. The cause of such an orbital change might be as catastrophic as its effects, since the Earth is not easy to shift. An asteroid impact big enough to move us off-course would probably shatter the planet and would certainly strip it of any life.

The close approach of a rogue giant planet, as it fell toward the Sun from interstellar space, could move the Earth without completely destroying it. But the gravitational pull of such a hypothetical body would not only alter the Earth's orbit. At close range, it would trigger massive earthquakes and volcanic eruptions and raise vast, destructive tides. Even before a changed orbit inflicted possibly fatal climatic changes, life on Earth would be in serious trouble.

The movements of the planets in the sky are not easy to account for. Some follow fairly predictable paths; others roam without any obvious pattern and appear to change direction almost at random. Over the centuries, many of humanity's best brains tried to find an explanation. But until the 17th century and the genius of Johannes Kepler, they failed. Kepler was the first to realize that the planetary orbits are ellipses, not circles, and that the Earth moves around the Sun in just the same way as the other planets do.

ORBITS OF THE PLANETS

PLANET	CLOSEST TO SUN (MILLIONS OF MILES/KM)	FARTHEST FROM SUN (MILLIONS OF MILES/KM)	ORBITAL VELOCITY (MILES/KM PER SECOND)	TIME TAKEN FOR ONE ORBIT
MERCURY	28.5/45.9	43.3/69.7	29.80/47.96	87.97 DAYS
VENUS	66.7/107.3	67.7/109.0	21.77/35.04	224.70 DAYS
EARTH	91.3/146.9	94.4/151.9	18.50/29.77	365.26 DAYS
MARS	129.0/207.6	155.0/249.4	15.00/24.14	686.98 DAYS
JUPITER	460.4/740.9	506.9/815.8	8.11/13.05	11.86 YEARS
SATURN	837.0/1,347.0	936.0/1,506.3	6.00/9.66	29.46 YEARS
URANUS	1,699.0/2,734.3	1,867.0/3,004.6	4.23/6.81	84.07 YEARS
NEPTUNE	2,769.0/4,456.3	2,819.0/4,536.7	3.37/5.42	164.82 YEARS
PLUTO	2,939.0/4,729.9	4,583.0/7,375.6	2.90/4.67	248.60 YEARS

CIRCLING THE SUN

When Polish astronomer Nicolas Copernicus died in 1543, he left an explosive legacy to his fellow scholars. In a book that he had quietly prepared for publication after his death, he declared that the Earth was not at the center of the universe, as most scholars and the Church still firmly believed. Instead, the Earth orbited the Sun—just like all the other planets.

But Copernicus was unable to explain just how the planets revolved around the Sun. He believed that they moved in perfect circles or, sometimes, in circles within circles. But observations of the night sky did not match his theory. The main problem was Mars, which seemed to wander back and forth almost as it pleased. Why did Mars move in this way? It was another half century before the German astronomer Johannes Kepler (1551–1630) provided the answer.

Kepler broadly agreed with Copernicus' theory, but saw the need for fine-tuning and went to work with Danish astronomer Tycho Brahe (1546–1601). Over the course of several decades, Brahe and his team had logged a huge

number of very accurate measurements of Mars' position in the sky. Given Brahe's data—obtained without telescopes—Kepler soon realized that Mars could not orbit the Sun in a perfect circle. By trial and error, he calculated that its course could only be explained if it moved in an ellipse, a type of elongated circle. All points on a circle are the same distance from the center, but an ellipse has two "centers," or foci. And from any point on an ellipse, the sum of the distances to each of the foci remains constant.

KEPLER'S LAWS

If one planet orbited in an ellipse, why not all the others? On that assumption, backed up by careful observation, Kepler proposed three laws of planetary motion. First, planetary orbits are ellipses, with the Sun at one of the foci. Second, a line drawn from the Sun to a moving planet sweeps through equal areas in equal times. Third, the square of the time each planet takes to orbit the Sun is proportional to the cube of its

mean distance from the Sun. The second and third laws mean that a planet moves fastest when it is closest to the Sun, and slowest when it is most distant.

The governing force, to which Kepler had discovered the key, was gravity. After his death, another great scientist—Isaac Newton—would reveal how it worked. But Kepler's laws still hold true, and combined with Newton's own laws, they explain the movements of any object in space—planet, satellite or spacecraft.

JOHANNES KEPLER

JOHANNES (OR JOHANN) KEPLER BECAME ASSISTANT TO DANISH ASTRONOMER TYCHO BRAHE IN 1600 AND COMPLETED BRAHE'S TABLES OF PLANETARY MOTION. THIS DATA AND KEPLER'S OWN GENIUS LED TO HIS LAWS OF PLANETARY MOTION, WHICH HAVE GIVEN HIM A PLACE AMONG THE GREAT SCIENTISTS OF HIS AGE. HE PUBLISHED A DESCRIPTION OF A SUPERNOVA—NOW KNOWN AS KEPLER'S STAR—THAT HE HAD OBSERVED IN 1604 IN THE CONSTELLATION OPHIUCHUS. HE LATER BUILT ONE OF THE FIRST TELESCOPES AND IN 1611 PUBLISHED A BOOK ON OPTICS.

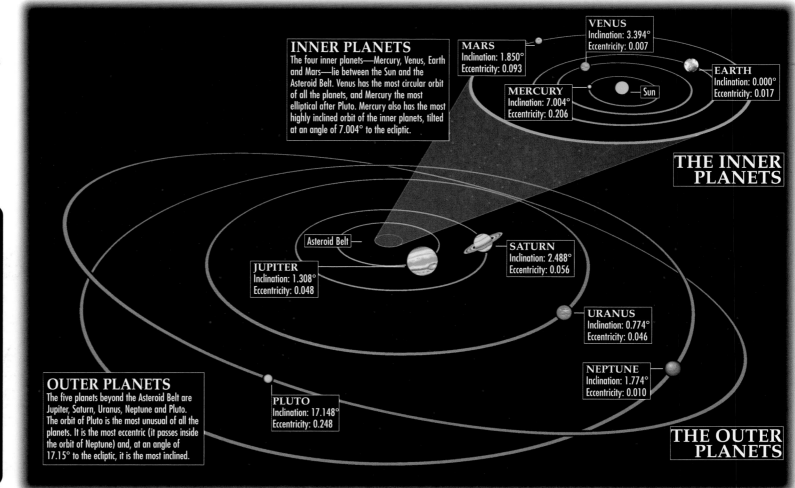

INNER PLANETS
The four inner planets—Mercury, Venus, Earth and Mars—lie between the Sun and the Asteroid Belt. Venus has the most circular orbit of all the planets, and Mercury the most elliptical after Pluto. Mercury also has the most highly inclined orbit of the inner planets, tilted at an angle of 7.004° to the ecliptic.

VENUS
Inclination: 3.394°
Eccentricity: 0.007

MARS
Inclination: 1.850°
Eccentricity: 0.093

EARTH
Inclination: 0.000°
Eccentricity: 0.017

MERCURY
Inclination: 7.004°
Eccentricity: 0.206

Sun

THE INNER PLANETS

Asteroid Belt

JUPITER
Inclination: 1.308°
Eccentricity: 0.048

SATURN
Inclination: 2.488°
Eccentricity: 0.056

URANUS
Inclination: 0.774°
Eccentricity: 0.046

NEPTUNE
Inclination: 1.774°
Eccentricity: 0.010

OUTER PLANETS
The five planets beyond the Asteroid Belt are Jupiter, Saturn, Uranus, Neptune and Pluto. The orbit of Pluto is the most unusual of all the planets. It is the most eccentric (it passes inside the orbit of Neptune) and, at an angle of 17.15° to the ecliptic, it is the most inclined.

PLUTO
Inclination: 17.148°
Eccentricity: 0.248

THE OUTER PLANETS

BIRTH OF THE INNER PLANETS

The planets formed 4.6 billion years ago from a vast disk of gas and dust that surrounded the newly forming Sun. The material in the disk began to cool and condense, initially forming grain-sized bodies and then coalescing into planets. Only rocky and metallic materials could survive the heat close to the Sun, and the inner planets—Mercury, Venus, Earth and Mars—still have compositions that reflect this. These planets probably took 100 million years to grow—and another 700 million to mature into the planets we know today.

WHAT IF...

...WE VISITED A NEWLY FORMING PLANETARY SYSTEM?

Planetary systems are very dangerous and hectic places during their formation. The Moon, for example, has a surface scarred by the impacts of asteroids and comets, some of which were probably dozens of miles across. Most of these impacts happened fairly early in the solar system's history, within a few hundred million years of the creation of the planets themselves. But the most dangerous period of all would be at the very beginning—the first thousand years.

In its earliest stages, shortly after the disk starts to condense, a typical protoplanetary disk quickly fills to capacity with tiny particles of rock and metal less than a millionth of an inch across. In the innermost regions of the disk, temperatures probably reach 3,000°F (1,650°C), so this material would be molten. Densities, too, are high, with just inches separating the closest rocky and metallic fragments. And as if these conditions were not bad enough, the particles zoom around the growing central star at speeds of dozens of miles per second.

A visit to such an environment in a spaceship would be a suicidal venture without advanced shield technology to prevent the craft and its occupants from being completely vaporized by the intense heat. Even quite far from the central star, conditions would be equally hazardous. Although the abundance of rock and metal fragments reduces with distance, in the cooler outer regions there comes a point where ice grains start to condense—and these can outnumber the rock and metal by 10 to one.

The distance at which this happens is called the "snow line," by analogy with the height up a mountain where conditions are always cold enough for snow to lie on the ground. Beyond the planetary snow line, though the temperature is much lower, densities are still very high—and the speeds, though lower also, are still measured in miles per second. Any foolhardy passengers on such a spaceship would first be sandblasted, and then their fragments would be frozen solid—one day ending up inside a planet or moon.

Protoplanetary disks, in their earliest stages, are best scratched off the list of interstellar tourist resorts. But to scientists they are fascinating places, and even now plans are being laid to study nearby examples, surrounding stars such as Beta Pictoris. Within the next few years, astronomers are hoping to launch spacecraft that will indeed explore the inner reaches of such regions—but from the safe vantage point of our own solar system.

PLANET-BUILDING TIMES

EVENT	DURATION
COLLAPSE OF GAS GLOBULE TO FORM SOLAR NEBULA	1—2 MILLION YEARS
CONDENSATION FORMS FIRST GRAINS	2,000 YEARS
GROWTH OF ASTEROID-LIKE PLANETESIMALS	A FEW THOUSAND YEARS
APPEARANCE OF MOON-SIZE PROTOPLANETS	10,000 TO 100,000 YEARS
INNER FOUR PLANETS REACH HALF THEIR EVENTUAL MASS; ACCRETION SLOWS	10 MILLION YEARS
INNER PLANETS REACH MODERN MASS; CRUSTS SOLIDIFY	100 MILLION YEARS
BOMBARDMENT MODIFIES PLANETS' CRUSTS	100 TO 800 MILLION YEARS
EVOLUTION OF PLANETS TO PRESENT DAY	3.8 BILLION YEARS

SIGNS OF CREATION

SOME METEORITES, KNOWN AS CHONDRITES, HAVE INTERIORS THAT CONSIST OF COLLECTIONS OF DIFFERENT MINERALS LUMPED TOGETHER (RIGHT). THE INDIVIDUAL PARTICLES WITHIN THEM ARE THOUGHT TO BE LARGELY UNCHANGED SINCE THE EARLY DAYS OF THE SOLAR SYSTEM, AND SHOW THE SORT OF PEBBLE-SIZE BODIES THAT ACCRETED TO FORM LARGER BODIES. THE ELEMENTS IN THEM SEEM TO BE IN THE SAME PROPORTIONS AS ASTRONOMERS BELIEVE EXISTED IN THE SOLAR NEBULA AT THE BIRTH OF THE SOLAR SYSTEM.

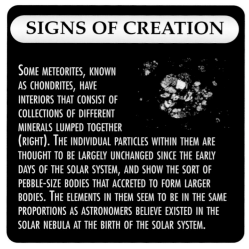

HOT START

Some 4.6 billion years ago, the Earth and all the other planets existed as little more than a thin scattering of gas and grains of dust. The raw material from which the planets sprang probably took the form of a vast disk known to astronomers as the solar nebula.

Like the newly forming Sun, or protosun, that it surrounded, the solar nebula was born when a much larger cloud of gas and dust particles contracted under gravity. Close to the protosun in the center of the solar nebula, temperatures may have been greater than 3,000°F (1,650°C). Eventually, with material in the disk spiraling in to the protosun, the disk grew sparser and its heat was able to escape into space. Then the disk began to cool and its material started to condense, with single atoms grouping together one at a time until they had grown into tiny grains less than a ten-millionth of a inch across.

This process, condensation, was the first step in the planet-building process. Far from the protosun, cooler conditions allowed water, ammonia and methane to condense into their ice form. But closer in, only rocky and metallic materials could condense.

As the condensed particles orbited the protosun, all swirling in the same direction, some of them began to stick to their neighbors.

IN THE BEGINNING

The scene in the solar system about 4.6 billion years ago, as individual planets are beginning to form from the solar nebula. At this stage, the planets are much smaller than those we know today, but a few bodies are starting to become sufficiently large that they dominate the scene.

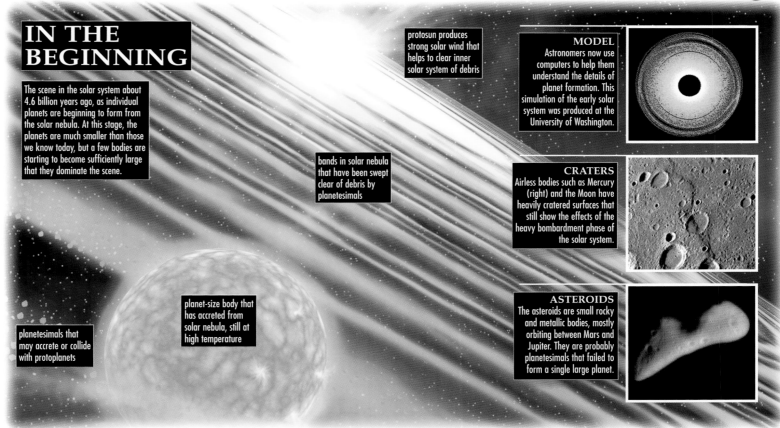

protosun produces strong solar wind that helps to clear inner solar system of debris

bands in solar nebula that have been swept clear of debris by planetesimals

planet-size body that has accreted from solar nebula, still at high temperature

planetesimals that may accrete or collide with protoplanets

MODEL
Astronomers now use computers to help them understand the details of planet formation. This simulation of the early solar system was produced at the University of Washington.

CRATERS
Airless bodies such as Mercury (right) and the Moon have heavily cratered surfaces that still show the effects of the heavy bombardment phase of the solar system.

ASTEROIDS
The asteroids are small rocky and metallic bodies, mostly orbiting between Mars and Jupiter. They are probably planetesimals that failed to form a single large planet.

Astronomers are uncertain what caused the grains to stick together, but it might have been electrostatic forces. Because of this "agglomeration," individual grains grew steadily larger as they merged with adjacent particles.

BOMBARDMENT BEGINS

Within perhaps 2,000 years, the innermost regions of the solar nebula were swarming with countless pebble-size particles. After a few more thousand years, these pebbles had grown to the dimensions of asteroids, with the biggest being miles in diameter. Known as planetesimals, these fragments were by now so large that they grew not only by chance collisions with others, but because they could actually attract their neighbors by virtue of their gravity—a process known as

accretion. It was at this point that the planet-building process stepped up a gear.

After about 10 million years, the innermost regions of the disk were populated by four dominant protoplanets that would later become Mercury, Venus, Earth and Mars, plus maybe one or two others. But by now these objects had mopped up much of the available debris, so their growth rate diminished. It took perhaps another 100 million years for these protoplanets to double in mass to their modern values.

Violent times lay ahead. For some 800 million years—a period known to astronomers as the heavy bombardment phase—the primitive planets continued to sweep up smaller pieces of debris as they orbited around the Sun. Only after this period ended, about 3.8 billion years ago, did the inner planets as we know them truly emerge.

MOON CRASH

THE MOON IS THOUGHT TO HAVE BEEN CREATED SHORTLY AFTER THE EARTH STARTED TO SOLIDIFY. ASTRONOMERS SUSPECT THAT A LARGE PROTOPLANET—PERHAPS UP TO THREE TIMES AS MASSIVE AS MARS—COLLIDED OBLIQUELY WITH THE EARTH, AND THE IMPACT TOTALLY VAPORIZED THE PROTOPLANET, ALONG WITH A SUBSTANTIAL PART OF THE EARTH'S CRUST. THE DEBRIS WENT INTO ORBIT AROUND THE EARTH, WHERE MUCH OF IT LATER FORMED THE MOON.

BIRTH OF THE GAS GIANTS

All planets form by accretion—the lumping together of material in a vast spinning disk of dust and gas. But our solar system has planets of two very different kinds. Near the Sun, the four inner planets are small and rocky. Farther out, where conditions were cooler, planets grew from accumulations of snowflakes. In time, they became large enough to attract hydrogen and helium. But the four giant planets—Jupiter, Saturn, Uranus and Neptune—have differences that are not so easy to explain.

WHAT IF...

...GIANT PLANETS FORMED CLOSE TO A STAR?

The first planets orbiting Sun-like stars were discovered in 1996, detected because the gravity of their considerable masses—many of them are bigger than Jupiter—makes their parent stars wobble in space. By now, several dozen of these exoplanets are known. As the number has grown, astronomers have found that many of these planetary systems differ substantially from ours. For example, the planet orbiting the Sun-like star called 70 Virginis takes just 116.6 days to complete an orbit and does so at remarkable proximity to its star—only a bit farther out than Mercury is from our Sun. Other exoplanets are in even closer orbits.

In our own solar system, by contrast, only planets made primarily of rock and metal are found so near the Sun. The more massive planets, made of gas and ice, are much farther away. Such a distribution fits current theories of planet building. Close to the infant Sun in the protoplanetary disk that spawned the planets, the high temperature would have prevented water and other ices from existing as solids. The planets that formed there were rocky, and never grew large enough to scoop up hydrogen or helium gas. In the lower temperatures of the outer system, ice played a major role in planet-building. Icy protoplanets were large enough to accumulate

layer upon layer of gas, and so grew larger still.

But the theory cannot explain systems like 51 Pegasi and 70 Virginis. The question is, how did these form? Nobody is really sure, but one idea that has been put forward is inward migration. In this theory, giant planets always form far from the star, but something makes them spiral gradually inward until they are in very tight orbits. Friction between the remaining gases and the planet may be to blame.

But why do the planets stop so close to their stars instead of falling all the way to impact? Perhaps we see only those that remain when the inner gases evaporated. Astronomers will have to investigate many more planetary systems before the theory can be tested.

The backlit ring system of a far-off giant planet makes an impressive sight from its outermost moon, which has a polar orbit. Most of the planet's other moons orbit in the plane of the rings.

OTHER STARS WITH GIANTS

STAR NAME	LOCATION	PLANET MASS (JUPITER=1)
51 Pegasi	Pegasus	0.4
Upsilon Andromedae	Andromeda	0.7
55 Cancri	Cancer	0.8
Rho Coronae Borealis	Corona Borealis	1.1
16 Cygni B	Cygnus	1.6
Iota Horologii	Horologium	2.2
47 Ursae Majoris	Ursa Major	2.3
Tau Bootis	Bootes	3.6
14 Herculis	Hercules	4.7
70 Virginis	Virgo	7.4

ANOMALIES

URANUS AND NEPTUNE ARE FAR LARGER THAN THEORY SAYS THEY OUGHT TO BE. IN THEIR FAR-FLUNG ORBITS, THERE WOULD HAVE BEEN TOO LITTLE MATERIAL FOR THESE PLANETS TO HAVE GROWN TO THEIR PRESENT SIZES. RECENT WORK BY EDWARD THOMMES (ABOVE) AND MARTIN DUNCAN OF QUEENS UNIVERSITY, TORONTO, CANADA, SUGGESTS THAT URANUS AND NEPTUNE ACTUALLY FORMED MUCH CLOSER TO THE SUN, NEAR JUPITER. THERE THEY WOULD HAVE BEEN ABLE TO GROW UNTIL JUPITER'S GRAVITY EJECTED THEM AND THEY SETTLED INTO THEIR CURRENT, DISTANT ORBITS.

BIG BABIES

The Solar System contains two distinct classes of planet. Huddled in close to the Sun are small, dense planets made of rock and metal: Earth and Mars, for example. But farther out, the planets are very much more massive, composed primarily of hydrogen and helium. These outer giants are so large that they contain about 99% of the combined mass of all the planets, satellites and asteroids we know of. How they grew so large, and how the solar system developed these general characteristics, is a natural consequence of the way the system was born.

About 4.6 billion years ago, a cloud composed mainly of hydrogen and helium, with water and other ices and particles of carbon dust, began to collapse under its own gravity. As it contracted, the cloud rotated steadily faster. Its material spread into a disk, with the Sun slowly taking shape at its center. At this stage the Sun was not a true star: It had not yet grown massive enough to ignite thermonuclear reactions in its core. It shone only dimly by gravitational contraction—the heat generated when a body shrinks under its own weight. In the disk, gas and dust particles were colliding to form progressively larger objects: grains, boulders, asteroids and then little protoplanets. Although the Sun was relatively cool, the inner system was still too warm for ice to

FROM PROTOPLANET TO GIANT

GAS GIANTS
This far from the Sun, protoplanets begin forming around cores of ice. Soon, they are so large that their gravity traps light, fast-moving atoms of hydrogen and helium. They grow until they sweep a clear space along their orbit.

HYDROGEN
A cloud of hydrogen gas in the Swan Nebula gives some indication of what our solar system may have looked like before it formed. Hydrogen is by far the most common material in the universe.

INNER TERRESTRIALS
Close to the protosun, ice cannot form. But rocky and metallic fragments begin to coalesce to form the inner, terrestrial planets.

JUPITER
At more than twice the combined mass of all the other planets, and 1,300 times as large as the Earth, Jupiter is a vast ball of hydrogen and helium around a small core of rock and ice.

At the center of the protoplanetary disk, the Sun is forming. At this stage in the process, it is only a protostar, very much larger than its final size and still collapsing under gravity. It shines dimly, not by nuclear fusion but by the heat released during gravitational contraction.

70 VIRGINIS
This star has a planet more than seven times the mass of Jupiter. Astronomers suspect that this and other planets found around other stars—all of them very large—are gas giants of the Jovian type.

form. Only rock and metal could contribute to the planet-building process this far in. When these materials were all used up, the innermost protoplanets stopped growing.

Much farther out in the disk, the planetary nursery was still in full production: The giants were forming. The relatively cool environment meant that outer protoplanets could grow by the accumulation not only of rocky material but also of the plentiful ice—orbiting snowflakes squeezed by gravity into ever-larger snowballs. With so much material on hand, these distant protoplanets grew more massive—several times larger than Earth. By the time the protoplanets that were to become Jupiter and Saturn had grown to 15 to 20 Earth-masses of material, their gravity was strong enough to haul in the light

gases hydrogen and helium.

After 1–10 million years, raw material in the disk became scarce. Far beyond Jupiter and Saturn, in the vicinity of Uranus and Neptune, the combination of the rapidly thinning material and the longer orbital timescales meant that these planets could not accumulate as much mass as Jupiter or Saturn had. As a result, they remained as largely water ice, without the extensive hydrogen and helium atmospheres of the closer giants.

It was also about this time that the Sun first grew hot enough to become a true star. Very soon, its increasing heat blew any remaining gaseous material in the planetary disk into interstellar space, and the planets effectively stopped growing altogether. The result was the solar system we see today.

WATERY LINK

THE PLANET NEPTUNE WAS GIVEN ITS NAME IN 1846 AFTER ITS DISCOVERY BY PROFESSOR GALLE OF BERLIN. THIS CONTINUED THE TRADITION OF NAMING PLANETS AFTER MYTHOLOGICAL ROMAN GODS. NEPTUNE WAS THE GOD OF THE SEA, WHOSE GREEK EQUIVALENT WAS POSEIDON. THE SYMBOL GIVEN TO THE PLANET IS THE TRIDENT CARRIED BY NEPTUNE. BY COINCIDENCE, IT IS NOW KNOWN THAT NEPTUNE DOES INDEED CONTAIN SIGNIFICANT AMOUNTS OF WATER AND WATER ICE—A FACT WHICH COULD NOT HAVE BEEN KNOWN WHEN THE PLANET RECEIVED ITS NAME.

ORIGINS OF ATMOSPHERES

Only some of the planets in the solar system have a recognizable atmosphere—a film of gases that clings to the surface of the planet. Of the rocky planets, Venus, Earth and Mars have substantial atmospheres—mostly carbon dioxide, oxygen, water and nitrogen gas. An atmosphere can make all the difference in a planet's history. On Earth, it has helped protect and provide for life, while on Venus, it has covered the surface with an acidic overcoat. But where did such different atmospheres come from?

WHAT IF...

...THE EARTH'S ATMOSPHERE WERE BLOWN AWAY?

As eternal as our blue skies may seem, they will eventually be destroyed by the Sun. Even today, the Sun continually strips away particles from our upper atmosphere. Solar radiation—most importantly, short-wave radiation such as ultraviolet and X-rays—beats down with its full force on upper-atmosphere particles. It imparts energy that excites the particles and accelerates their motion. And the radiation also breaks up these active molecules into lighter fragments that are more likely to speed off into space.

The heaviest losses occur during solar flares— eruptions of the Sun's surface that are heavy on X-ray radiation. Luckily, solar flares are too weak to cause much trouble. And it will remain that way for a very long time—but not forever.

As the Sun slowly ages, it will grow brighter. Stronger radiation and greater solar activity will strip off more and more of the Earth's atmosphere. Over millions of years, the Sun will slowly erode our skies away.

But that may not be the end of the story. The Earth's atmosphere could actually be replenished by its own crust. As crustal rock is melted by plate tectonics, and molten magma surfaces through volcanoes, chemicals bonded in the rocks are released. And sedimentary rocks often capture atmospheric or surface compounds as they form— compounds that are released as gas when the rocks are heated. In a hotter future, increased emissions from the surface may compensate for increased losses into space.

Life, too, may help save the atmosphere. Plant and animal remains eventually form natural gas (mainly methane) and oil deposits—all of which could prove gas-rich when their parent rock is heated.

In spite of the barrage of solar radiation, it is possible that a new atmosphere of gases released from the crust might stabilize. After all, Venus receives about 1.5 times as much radiation as the Earth, and its atmosphere is still thick. But Venus— encased in hot, high-pressure and corrosive gases— also warns us that a new atmosphere may be less hospitable than the present version.

In the longer term, though, around 7 billion years from now, the Sun will swell to almost engulf the Earth. Then, its radiation will overwhelm any remaining atmosphere.

Volcanic vents or fumaroles in Iceland are caught in the act of releasing gases into the atmosphere. Rocky planets may rely on emissions like this to replenish their atmospheres.

ATMOSPHERIC BREAKDOWN

	PERCENTAGE OF TOTAL ATMOSPHERE			PRESSURE CREATED BY EACH GAS (FULL PRESSURE OF THE EARTH'S ATMOSPHERE = 100%)		
	VENUS	EARTH	MARS	VENUS	EARTH	MARS
CARBON DIOXIDE	96%	0.035%	95%	8,640%	0.035%	0.62%
NITROGEN	3.5%	77%	2.7%	320%	78%	0.018%
OXYGEN	–	21%	–	–	21%	–
WATER VAPOR	0.01%	1%	0.006%	0.9%	1%	3.9×10^{-5} %
ARGON	0.007%	0.93%	1.6%	0.63%	0.94%	0.01%

TRAPPED GAS

An atmosphere first coalesces around a planet as the planet is forming. Some 4.6 billion years ago, the solar system formed out of a giant spinning pancake of dust and gas particles. Asteroid-size lumps of rocky material called planetesimals picked up thin shrouds of gas—hydrogen and helium. But in the inner solar system, this first atmosphere dissipated easily, because the planetesimals' gravity was too weak to keep a grip on the gas.

Planet-building continued. Planetesimals grew by colliding and clumping together—and the force of impact heated the rock to release gas. This secondary gas formed atmospheres around the inner planets.

But how do rocks make gas? They are a bit like mushrooms: They do not drip when you squeeze them, but they give off lots of steam when fried. Even solid rock—rock without holes or pores—can contain molecules such as water or carbon dioxide, chemically bound up in its mineral structure. When rock is heated—by impacts or by volcanic activity—its chemical bonds are broken, and it releases molecules as vapor.

The secondary gases were heavier than hydrogen and helium, so they were attracted more strongly to the surface. But by the time the solar system came of age, not every planet had held on to an atmosphere. At a given temperature, there is a minimum mass for a planet with an atmosphere—and the Moon and Mercury are below it. Gases can escape the relatively weak gravity of these small worlds. Unlike Venus, Earth and Mars, neither the Moon nor Mercury is large enough to hold on to gases.

MERCURY
Mercury is probably too small and hot to retain any of the gases emitted by its rocks. Some gas hangs above the surface—but too little to be considered a real atmosphere.

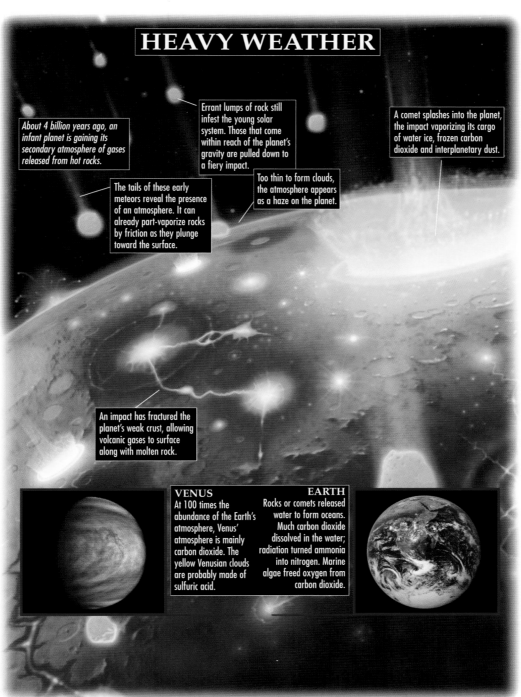

HEAVY WEATHER

About 4 billion years ago, an infant planet is gaining its secondary atmosphere of gases released from hot rocks.

Errant lumps of rock still infest the young solar system. Those that come within reach of the planet's gravity are pulled down to a fiery impact.

A comet splashes into the planet, the impact vaporizing its cargo of water ice, frozen carbon dioxide and interplanetary dust.

The tails of these early meteors reveal the presence of an atmosphere. It can already part-vaporize rocks by friction as they plunge toward the surface.

Too thin to form clouds, the atmosphere appears as a haze on the planet.

An impact has fractured the planet's weak crust, allowing volcanic gases to surface along with molten rock.

VENUS
At 100 times the abundance of the Earth's atmosphere, Venus' atmosphere is mainly carbon dioxide. The yellow Venusian clouds are probably made of sulfuric acid.

EARTH
Rocks or comets released water to form oceans. Much carbon dioxide dissolved in the water; radiation turned ammonia into nitrogen. Marine algae freed oxygen from carbon dioxide.

LOST ATMOSPHERES

At double the mass of Mercury, but only a tenth that of the Earth, Mars may have an intermediate history. Although its eroded surface suggests it once flowed with water and had a thick atmosphere, today its atmosphere is thin and its surface dry.

Not all the gas in the atmospheres of the rocky planets has come from within. The dirty balls of water ice and dry ice called comets have also made contributions. Unlike planetesimals, which formed near the hot center of the solar nebula, comets coalesced near the edge of the nebula, at temperatures low enough to allow ice. Comets may have been tipped off their wide orbits by the gravity of large chunks of passing debris and flown down toward the center of the forming solar system to collide with planetesimals. On impact, they would have vaporized, releasing gas that originated at the edge of the solar system into a planetary atmosphere. Some scientists think that the Earth's oceans could not have been filled without help from ancient comets.

An alien visitor to our solar system would quickly realize that the atmosphere of the third planet cannot be explained by volcanic outgassing and comet impact alone. On our world—uniquely—the atmosphere is modeled and managed by the engine of life.

MARS
Surface features suggest that Mars once had a dense atmosphere and liquid water. Now, though, the atmosphere is only 1/100 as abundant as the Earth's. Weak gravity may have let Mars' gases slip away.

IMPACTS IN THE SOLAR SYSTEM

Nothing alters a landscape more dramatically than an impact from space. On some worlds, the resulting craters are filled in by erosion or geological processes. But the surfaces of most moons and planets are sterile, and still bear the scars of billions of years of bombardment. Impacts are far less common today than they were in the young solar system. But the vast reservoirs of comets and asteroids still supply plenty of ammunition, and impacts will continue to pound the solar system for billions of years.

WHAT IF...

...THE IMPACTS STOPPED?

Compared with the heavy artillery of asteroid-like bodies that bombarded the solar system 4 billion years ago, the rate of impacts today amounts to only an occasional unlucky smash. But these collisions have not stopped altogether. Although Venus' surface is just 500 or 600 million years old, it has already accumulated at least 1,000 craters. There would be many more if it were not for the planet's dense atmosphere, which burns up the smaller would-be impactors before they hit the ground. But there are reasons to suspect that the collision rates will diminish—for the inner planets at least.

There are differences between the bodies that pummel the inner planets and those that crash into Jupiter and the planets beyond it. The inner planets are mainly besieged by asteroidal fragments— lumps of rock or iron. Crater counts show that these bodies have impacted at a constant rate for 3 billion years—ever since the heavy bombardment stopped.

With every strike on the inner planets, the population of asteroidal debris loses one member. The fact that the impact rate has not dropped off means that these asteroid reserves must be renewed from somewhere—most likely from the main asteroid belt between Jupiter and Mars. Collisions between these objects send many into the inner solar system, where they keep asteroid numbers up and the impact rate steady. But no reserve is inexhaustible, and one day the asteroid belt will be too thin to send fresh supplies toward the inner solar system. As the asteroid population dwindles, the inner planets will be impacted less and less.

There is no such quiet future in store for the outer planets. Beyond Jupiter, it is cometary nuclei rather than asteroids that dominate the bombardment. Comets hail from the Oort cloud—a vast swarm of trillions of icy fragments that stretches from far beyond Pluto's orbit halfway to the nearest star. At this distance from the Sun's gravity, the Oort cloud is easily perturbed by the gravity of stars in the Milky Way's spiral arms, and by the movement of nearby stars, which might pass by every couple dozen million years or so. When this happens, new comets plunge into the planetary regions, where Jupiter's gravity nudges them into orbits that will sooner or later end in collision with one of the outer planets. These bits of debris are unlikely to diminish in substantial numbers for billions of years to come. In the meantime, the outer solar system will remain a danger zone.

BIG HITS

Name	Location	Diameter
Caloris	Mercury	830 miles (1,340 km)
Mead	Venus	180 miles (290 km)
Aitken	Moon	1,560 miles (2,510 km)
Vredefort	Earth	190 miles (310 km)
Hellas Planitia	Mars	1,550 miles (2,500 km)
Valhalla	Callisto	370 miles (600 km)
Herschel	Mimas	80 miles (130 km)
Odysseus	Tethys	250 miles (400 km)
Mazomba	Triton	17 miles (28 km)

STRIKE OUT

At the beginning of the solar system, 4.6 billion years ago, there was little going on except impacts. The planets grew out of the mass of dust that swirled around the young Sun. These particles were welded together by collisions. The impact rate tailed off as more debris was locked away, but on asteroids, moons and planets, there are still crater scars everywhere.

All impacts are basically explosions. As one body hits another, its velocity is converted into energy. But there are big differences between the craters on various bodies. Impacts on the Moon show a central bowl surrounded by debris or ejecta. In larger craters, the rock is not strong enough to hold a bowl shape, and the crater slumps into a central peak. The craters on Mercury formed in the same way. But because bodies orbit faster the closer they are to the Sun, these impacts occurred at a higher

velocity. Still, the impact ejecta on the Moon sprayed over a wider area—the stronger gravity of Mercury brought it back to the ground before it could travel far.

GRAND SLAMS

From planets to asteroids—impacts are everywhere in the solar system. Ancient, heavily cratered terrains like the surface of Mercury are the best record of impacts. But even young surfaces have scars—proving that the threat of receiving a direct hit never goes away.

SCAR STORIES
Craters on Mercury (above) look similar to those of the Moon at first glance—but the secondary craters, caused by the debris from the big explosions, fall closer to their parent crater on Mercury.

PLUTO AND CHARON
unknown surfaces

NEPTUNE
heavily cratered moons except for the largest, Triton, which has ice volcanoes

URANUS
cratered icy moons

SATURN
icy surfaces of moons are heavily cratered; crater Herschel on Mimas is one-third of the moon's diameter

MARS
some areas heavily cratered

ORBIT CHANGE
Pluto's orbit is elongated compared with the orbits of most other objects in the solar system, and crosses the path of Neptune (right). An impact may have knocked Pluto into this eccentric course.

JUPITER STRIKE
The July 1994 collision of comet Shoemaker-Levy 9 and Jupiter (right) provided a graphic insight to the fall-out of a large impact. The strike left Jupiter's gaseous atmosphere with huge impact scars: some larger than the Earth.

MOON BUILDER
A giant impact may have been responsible for forming our own Moon (above). After the Earth formed, it was probably struck a glancing blow by a rogue protoplanet. The collision blasted debris into Earth orbit, which coalesced to form the Moon.

VENUS
about 1,000 craters

MERCURY
heavily cratered

EARTH
about 150 large craters; Earth's Moon is heavily cratered

JUPITER
Jupiter's moon Callisto is the most heavily cratered body in the solar system

DEATH BY IMPACT

ASTEROID OR COMET IMPACTS HAVE BEEN BLAMED FOR SEVERAL MASS EXTINCTIONS IN THE HISTORY OF LIFE ON EARTH—MOST FAMOUSLY FOR THE DEATH OF THE DINOSAURS, 65 MILLION YEARS AGO. THE IMPACT OCCURRED IN WHAT IS NOW MEXICO, AND LEFT A CRATER OVER 100 MILES (160 KM) ACROSS. BESIDES REGIONAL ANNIHILATION, A COLLISION OF THIS SIZE COULD PUT ENOUGH DUST INTO THE ATMOSPHERE TO BLOCK OUT THE SUN, LEADING TO A GLOBAL FOOD CRISIS.

COMPARING SCARS

On Mars there is another difference. Some of the larger craters sit on splatters that look like the white of a fried egg. These were created as the force of the impact melted subsurface ice. Farther out in the solar system, ice plays an even larger part in shaping impact craters. The moons of Jupiter are mainly ice, and strange craters called palimpsests occur, most dramatically on Ganymede. These are probably ghost craters—impact scars that formed early in the moon's history and have been all but erased by the ice melted through later collisions.

Smaller bodies have not escaped impacts either. The size of the largest crater on an asteroid is a good index of its strength—some withstand a strike that blasts a hole nearly as large as their diameter without splitting up.

Crater counts in a region of a planet or moon are good indicators of the area's age. For example, craters are less common on the Moon's young lava seas than they are on the old highlands. And Venus' low crater score indicates that the planet's surface is only about 500 million years old. The plentiful craters gathered on older surfaces are a sobering reminder of what is to come. Impacts may be less frequent than they were, but the numerous asteroids that roam the solar system promise that one day, the Earth will be back in the crosshairs.

PLANETARY RINGS

...THE EARTH DEVELOPED A RING SYSTEM?

Whatever your view of planetary ring formation, size matters. Only the Jovian planets, Jupiter, Saturn, Uranus and Neptune, have ring systems, and even the smallest of these giants is four times the radius of the Earth.

The smaller size of the Earth and the other terrestrial planets means that they simply do not have the gravity to hold onto a ring system. These planets may have lost their ring potential very early in the solar system, at a time when disks of debris began to circle the Jovian giants. The small terrestrial planets were not able to grip the debris long enough for it to form rings. Their position did not help either. As the closest planets to the Sun, they occupied the hottest part of the solar nebula, and were never surrounded by the icy materials that make up Saturn's rings.

Some scientists argue that ring systems formed much later, from the remains of shattered asteroids and comets. If that is the case, the terrestrials would have lost out again. Giant planets capture comets easily, and the puny gravitational pull of the rocky planets is ineffective by comparison.

None of this means that a rocky planet is incapable of sustaining a ring system—just that this is statistically unlikely. If a large enough comet or asteroid ranged close enough to the Earth with the right speed and trajectory, it is quite possible that its destruction would lead to the formation of a planetary ring. Rings can also form from broken-up moons. Mars' moon Phobos is currently spiraling

inward, and the planet's tidal forces may tear it apart to form a faint ring system.

But the Earth's best chance of acquiring a ring is by far less traditional means. Already, the Space Age has littered near-Earth orbit with thousands of pieces of human-made junk. At present, there is nowhere near enough space debris to form visible rings, but in a few thousand years, there could be enough orbiting satellites, habitats and plain scrap metal to transform the Earth's night sky.

From the Earth's surface, a ring system would be a beautiful sight. At the equator the rings would arch across 180° of the sky from due east to due west as a narrow line illuminated by the Sun. Farther north or south, the rings would cover less of the sky, but instead of a narrow line, they would appear as a broad band.

In the far future, a ring arches above the Earth—one consequence of the Space Age. Made of metallic debris, the ring reflects light well and is visible even by day.

The first set of planetary rings was spotted around Saturn in 1610, but it took almost four more centuries for observers to put the rings of Uranus, Jupiter and Neptune in place. For giant planets, it seems that rings are the rule rather than the exception. Each is a thin pancake of particles—perhaps the remains of moons that failed to form, or those that have shattered. None of the inner planets have ring systems—at least, not yet. Human activity in space could eventually provide Earth with the strangest rings of all.

PLANETARY RINGS SPECS

Planet	Year of Ring Discovery	Outermost Diameter miles/km	Thickness miles/km	Particle Size inches/mm
Saturn	1610	300,000/480,000	0.006–0.06/0.01–0.1	0.5–200/13–6000
Uranus	1977	32,000/52,000	0.006–0.06/0.01–0.1	4–400/100–10,000
Jupiter	1979	143,000/23,000	20/32	0.00004/0.01
Neptune	1984	39,000/63,000	20/32	0.5–50/13–1,250

CIRCLE STORY

Saturn, Jupiter, Uranus and Neptune: Every giant planet in our solar system is surrounded by rings. In photographs, these structures appear solid, but in fact each one consists of millions of particles doing their own thing. Although they all circle in the same thin plane, each one follows its own individual orbit.

No two ring systems are the same. Saturn's are the boldest—broad bands of icy fragments that reflect light. The rings of Uranus are much narrower and jet black. Jupiter's rings are also dark, but their particles are finer. And the rings of Neptune—again, dark and narrow—have an unusual brightening along certain stretches that gives the appearance of arcs rather than whole rings.

There are several plausible models of ring formation, but the definitive version has yet to be agreed. All of these ideas involve the Roche limit—the minimum distance from a planet at which a satellite can remain stable. Inside this limit, tidal forces from the nearby planet induce enough stresses to tear the body apart.

Most of the rings in the Solar System orbit well inside their respective Roche limits—which suggests that they are the debris of something that strayed too close. But what? Some scientists argue that the rings came from the planet's own moons, others that they are the shattered remains of comets or asteroids that came from afar. In either case, the composition of the destroyed body would determine the makeup of the rings—and that might explain some of the differences between the ring systems.

EARLY STARTERS

Another school of thought dates the rings back to the time when the planets themselves formed. Some scientists believe that the planets condensed out of a vast doughnut of gas and dust that surrounded the young Sun. As these particles collided, they stuck together and eventually grew into planets. The giant planets were surrounded by excess material, some of which coalesced to form their moons. But this was not the fate of material inside the Roche limits. Instead, it gradually collapsed into a plane that encircled the planets and then spread out to form the rings as we see them today.

But there are aspects of ring formation that do not fit into any current model. Astronomers think that ring systems persist for less than a few hundred million years. So the fact that we see rings today is not consistent with their creation alongside the planets 4.6 billion years ago. But the breakup of a moon or comet suggested by the alternative model asks for too much of a coincidence—it implies that all of the catastrophes that formed ring systems happened in the recent past. Also, the solar system does not carry as much debris as it did billions of years ago. This means that impacts capable of forming ring systems grow less likely as time goes on—and unless we see one in the making soon, we may never know how the giant planets got their rings.

GROWING RINGS

A cloud of fragments—from the early solar system or a shattered moon—surrounds the planet

Although every giant planet in our solar system has rings, their origins are unclear. Scientists know broadly how rings form, but are less certain about where their material comes from, or how the rings will eventually disperse.

Collisions between particles reduce their size and push them toward a common plane, which lies perpendicular to the planet's spin axis

Collisions push the innermost particle farther in and the outer farther away. This spreads the ring radially and the a true ring system takes shape

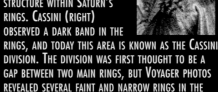

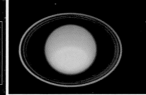

JUPITER'S HALO
Particles in Jupiter's rings are charged, so they are affected by the planet's magnetic field. For very small particles, this magnetism overpowers gravity. This Galileo image (above) shows how particles are pulled out of the main ring (yellow) into a halo (blue).

SHARP EDGES
From the Earth, the structure of ring systems is best revealed by the planet's movement in front of a star. Uranus' rings were discovered in this way in 1977. Uranus' ring boundaries are very crisp (above)—evidence that their edges are "supervised" by nearby moons.

FILLING IN THE GAPS

IN 1675, FRENCH ASTRONOMER GIOVANNI CASSINI (1625–1712) BECAME THE FIRST PERSON TO OBSERVE STRUCTURE WITHIN SATURN'S RINGS. CASSINI (RIGHT) OBSERVED A DARK BAND IN THE RINGS, AND TODAY THIS AREA IS KNOWN AS THE CASSINI DIVISION. THE DIVISION WAS FIRST THOUGHT TO BE A GAP BETWEEN TWO MAIN RINGS, BUT VOYAGER PHOTOS REVEALED SEVERAL FAINT AND NARROW RINGS IN THE REGION.

RINGS TRUE

SATURN'S RINGS WERE DISCOVERED IN THE 1600S, BUT IT TOOK TWO MORE CENTURIES AND THE GENIUS OF SCOTTISH PHYSICIST JAMES CLERK MAXWELL (1831–79) TO FINALLY UNRAVEL THEIR TRUE NATURE. MAXWELL (RIGHT) DEMONSTRATED THAT THE RINGS MUST BE COMPOSED OF COUNTLESS INDIVIDUAL PARTICLES, EACH IN AN INDEPENDENT ORBIT AROUND THE PLANET—NOT SOLID SHEETS OF MATTER. MAXWELL PROVED THAT SOLID OR FLUID RINGS WOULD BE PHYSICALLY UNSTABLE.

WATER IN THE SOLAR SYSTEM

Of all the molecules in the solar system, the one that holds the greatest interest for the human race is water. Not only do we depend on it for our survival, but it is easily broken down into its components—hydrogen and oxygen—which together make a very useful fuel. Although water made up only about 1% of the gas cloud from which the solar system formed, it is common throughout the solar system in the form of ice—though Earth remains the only planet with large amounts of surface water.

WHAT IF...

...WE COULD DRINK EXTRA-PLANETARY WATER?

When the first trans-solar-system missions started in the 22nd century, scientists had no idea that water supplies would become a problem. With ice in the vicinity of every planet, the experts thought that water would flow as smoothly as it did on the Mars bases—where solar heat focusers melted terrestrial quantities from the polar caps.

To begin with, the human colonies in the realm of the gas giants were not short of drinking water, either. Jupiter, Neptune, Saturn and Uranus all have icebound satellites. Although solar panels are inefficient heaters this far from the Sun, the colonists found that there's nothing a radioisotope thermal generator can't melt these days. The water seekers simply took a supply of plutonium and used the energy from its radioactive decay to heat and melt the ice on any convenient moon.

The Outer Space Water Network (OSWN) also found these murky liquids easy to purify. Tethys, Mimas and Enceladus are all great water reservoirs orbiting Saturn. Their water content is mixed up with methane and ammonia ices, but it was no problem to separate out drinking water. The ice impurities all boil at lower temperatures than water, and after they had evaporated into space, all that was left in the satellite distilleries was pure water.

So far, so good—until the Earth-based company Liquid Planet put in a bid for the exclusive rights to space water supply. Liquid Planet immediately shut down all of its moon distilleries, and found a new one on Oberon—a far-flung satellite of Uranus. Robot probes had already established that Oberon's craters were brimming over with dirty water ice. Advance teams installed melt-pans, plutonium heat plants, and a series of magnetic accelerators that pointed spaceward. These giant tubes were lined with electromagnets that accelerated steel barrels full of purified water to Oberon's escape velocity.

Water Stations were carefully positioned in the orbits of Uranus and Saturn to receive and distribute the water payloads.

For the directors of Liquid Planet back on the Earth, the strategy worked brilliantly. But space dwellers were far from happy. Oberon water took months, or sometimes years, to arrive—and the price was exorbitant. As everyone now knows, within a few years the Liquid Planet monopoly led to the Declaration of Colonial Independence and the first war in space. But that is another story.

After they are caught in the nets of the Water Stations, barrels of Oberon water are fired away to their human customers. People on Saturn have a long wait—it takes months for fresh water to arrive.

DISCOVERING ICE

1957 Gerard Kuiper detects ice on Europa and Ganymede	**1986** Passing Uranus, Voyager 2 records ice on five moons
1971 Mariner 9 confirms water ice at poles of Mars	**1988** Charon, Pluto's moon, shown to be icy
1979 Passing Jupiter, Voyagers 1 and 2 reveal first detailed views of Callisto, Ganymede, Europa and Io	**1991** Radar images of Mercury suggest presence of polar ice
1980 Voyager 1 provides first images of Saturn's icy moons	**1995** Galileo spacecraft provides close-ups of ice-covered Europa
1981 Bypassing Saturn, Voyager 2 gives more details of Tethys, Phoebe and Iapetus: all contain large amounts of water ice	**1996** Clementine mission finds polar crater on Moon, where water may exist
	1998 Lunar Prospector finds further evidence for the existence of ice at lunar poles

THE TEMPERATURE OF THE SUN'S SURFACE IS AROUND 10,000°F (5,530°C) —MUCH TOO HOT FOR WATER TO EXIST. BUT SUNSPOTS (RIGHT) ARE MUCH COOLER, AT 5,300°F (2,900°C), AND THE INTENSE MAGNETIC ACTIVITY THAT GENERATES THEM ALSO SUPPRESSES THE FLUX OF HOT MATERIAL FROM THE STAR'S INTERIOR. OXYGEN AND HYDROGEN ARE ABLE TO COMBINE, AND SMALL AMOUNTS OF SOLAR WATER APPEAR AS SUPERHOT STEAM.

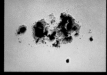

WATER WORLDS

Water is everywhere in the solar system. It has been found from Pluto's tiny moon Charon to the Sun's neighbor Mercury—and on the seven other planets and their moons that fill the billions of miles of space in between. Water only makes up 1% of the solar system's total mass, but most of this bulk is contained in the Sun—so a much greater proportion of the planets and their moons is water.

Much of this water exists as ice, which occurs anywhere that temperatures are cool enough: either far away from the Sun in the outer solar system, or shielded from the fierce heat on the planets closest to the Sun.

Most of the water has always been there. The raw ingredients of water—hydrogen and oxygen—were abundant in the solar nebula that condensed into our Sun and planets. The inner zone of the nebula was too hot for water to freeze into ice, but farther away from the infant Sun, water ice was stable. The Jovian planets may have formed around icy nuclei whose gravity attracted more and more material. The cores of Jupiter and Saturn attracted nebular gases, so their composition is similar to that of the Sun. Farther out, both Uranus and Neptune consist of little else but water and carbon ices. In the outermost zone of the nebula, temperatures at the birth of the solar system were lower still.

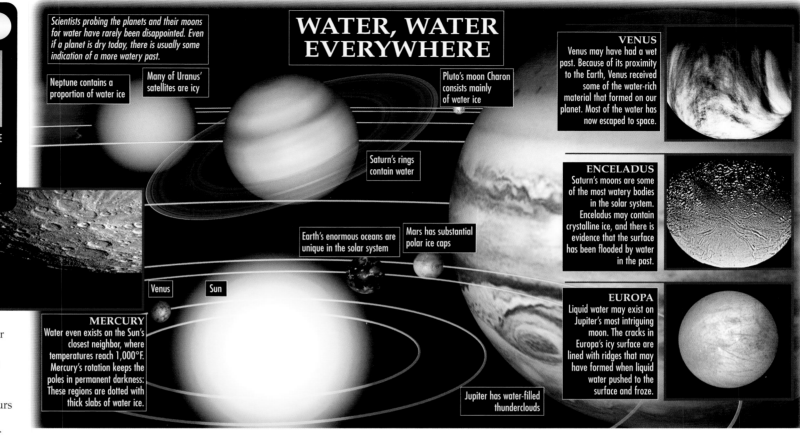

WATER, WATER EVERYWHERE

Scientists probing the planets and their moons for water have rarely been disappointed. Even if a planet is dry today, there is usually some indication of a more watery past.

Neptune contains a proportion of water ice

Many of Uranus' satellites are icy

Pluto's moon Charon consists mainly of water ice

VENUS
Venus may have had a wet past. Because of its proximity to the Earth, Venus received some of the water-rich material that formed on our planet. Most of the water has now escaped to space.

Saturn's rings contain water

Earth's enormous oceans are unique in the solar system

Mars has substantial polar ice caps

ENCELADUS
Saturn's moons are some of the most watery bodies in the solar system. Enceladus may contain crystalline ice, and there is evidence that the surface has been flooded by water in the past.

Venus

Sun

MERCURY
Water even exists on the Sun's closest neighbor, where temperatures reach 1,000°F. Mercury's rotation keeps the poles in permanent darkness: These regions are dotted with thick slabs of water ice.

Jupiter has water-filled thunderclouds

EUROPA
Liquid water may exist on Jupiter's most intriguing moon. The cracks in Europa's icy surface are lined with ridges that may have formed when liquid water pushed to the surface and froze.

Here, water ice mixed with dust to form the comets in the Kuiper Belt.

HIDDEN OCEANS?

One of the most watery worlds in the solar system is Jupiter's satellite Europa. This moon is covered with a crust of water ice 100 miles (160 km) thick, and some of the surface fissures are impossible to explain unless the ice floats on a liquid layer. Scientists now believe that there is an ocean on Europa, kept liquid by the heat of tidal stresses from Jupiter's gravity.

Vast stores of water ice were expected, and found, in the environs of all the planets in the outer solar system. Closer to the Sun, water is scarcer—but the mystery is why the inner solar system contains any water at all. The standard model for the appearance of water on the Earth's surface is outgassing. Soon after the planet formed, volcanoes pumped out water vapor trapped deep within the planet though fractures in the surface. This model explains how water arrived in the first place, but it does not explain why it has stuck around: In its early days, the Earth was hot enough to boil any water off into space. Now, some scientists believe that bombardment from the distant comet and asteroid belts was at least as important as volcanoes as a source of the water on Earth, Venus and Mercury. These impactors released their icy load into atmospheres and surfaces — and delivered a little of the water in the outer solar system to the inner worlds.

FINDING WATER

ASTRONOMERS DETECT SPACE WATER BY ANALYZING THE SPECTRUM OF LIGHT FROM DISTANT OBJECTS. DARK LINES IN THE SPECTRUM REVEAL THE ABSENCE OF CERTAIN WAVELENGTHS. DIFFERENT MOLECULES ABSORB DIFFERENT WAVELENGTHS, SO THE SPECTRUM IS, IN EFFECT, A BAR CODE THAT LISTS THE ELEMENTS AND COMPOUNDS AT THE LIGHT'S ORIGIN—INCLUDING WATER. EARTH-BASED TELESCOPES—THE MCMATH-PIERCE TELESCOPE AT KITT PEAK, ARIZONA (ABOVE) CAN NOW "SEE" WATER AS FAR AWAY AS PLUTO.

INTERPLANETARY MEDIUM

The list of ingredients reads like a noxious concoction: searing hot winds, blowing dust and enough energy to fry a rattlesnake. In fact, it describes the interplanetary medium—the extremely thin but ever-present mixture of energized gas, dust and subatomic particles that occupy the space between the planets in the solar system. Earth's thick atmosphere and magnetic field protect us from this deadly cocktail. Space travelers, though, need elaborate shielding to survive the rigors of such a harsh environment.

WHAT IF...

...THE INTERPLANETARY MEDIUM BECAME THICKER?

The solar wind that dominates the interplanetary medium creates a giant bubble—the heliosphere—in the vast galactic cloud of gas and dust that lies in our region of the Milky Way. At the moment, this bubble forms a protective cocoon around the Earth and the other planets that effectively shields them from the even harsher environment of interstellar space. But in around 50,000 years, the solar system may enter a galactic cloud that is a million times thicker than the present one. The material in that cloud could overpower the solar wind, exposing the Earth to intense bombardment by cosmic rays, disrupting its magnetic field and perhaps even changing the climate.

Like the interplanetary medium, the space between the stars is filled with a thin mixture of gas, dust and energetic particles. The difference is that this mixture is less uniform; there are dense clumps where new stars begin to form, and relatively thin regions that have been scoured clean by exploding supernovae. Our solar system is currently in one of the thinner regions near the edge of one of the Milky Way's spiral arms. But we are moving toward a much denser cloud—although astronomers cannot be sure if we will pass through it, because they have yet to plot its motion.

If we do encounter the cloud, the effects could be unpleasant—or catastrophic. The material in the cloud would push the heliosphere back to within 100–200 million miles (150–300 million km) of the Sun—only one to two times the Earth's distance. The gas and dust in the cloud, which is much thicker than the material inside the heliosphere today, would then dominate the interplanetary medium.

An immediate visible effect would be more intense meteor showers and brighter aurorae than we experience today. But with less protection from the solar wind, the Earth would also receive much higher, and potentially life-threatening, doses of cosmic rays. Indeed, some scientists have suggested that intense bombardment by cosmic rays may have caused or contributed to the planet's previous mass extinctions.

Higher doses of cosmic rays would also further break down the ozone layer in the atmosphere, letting through more of the Sun's harmful ultraviolet radiation. At the same time, the number of ice particles in the atmosphere would increase, reflecting more sunlight and causing surface temperatures to plummet. It seems unlikely that life on Earth would perish altogether—but it would certainly become a good deal less comfortable.

WHAT'S IN THE MEDIUM?

	SOLAR WIND	DUST	COSMIC RAYS
ORIGIN	CHARGED GAS PARTICLES (PLASMA) EJECTED FROM THE SUN	DEBRIS FROM COMETS AND ASTEROID COLLISIONS	SUBATOMIC PARTICLES FROM FAR BEYOND THE SOLAR SYSTEM
EXTENT	EXTENDS AT LEAST 7 BILLION MILES (11 BILLION KM) BEYOND THE SUN	THIN CLOUD THAT EXTENDS ROUGHLY 200 MILLION MILES (300 MILLION KM) AROUND THE SUN	FOUND THROUGHOUT THE UNIVERSE
DISCOVERY	1959, BY THE SOVIET LUNA 2 PROBE	1683, BY ITALIAN ASTRONOMER JEAN-DOMINIQUE CASSINI	1912, BY GERMAN SCIENTIST VICTOR HESS USING A BALLOON-MOUNTED ELECTROSCOPE
MOTION	TRAVELS OUTWARD FROM THE SUN, FOLLOWING THE SUN'S MAGNETIC FIELD	ORBITS THE SUN, LIKE THE PLANETS	ENTER THE SOLAR SYSTEM FROM ANY DIRECTION, TRAVELING AT NEAR LIGHT-SPEED

FILLING THE SPACE

The space between the planets is a near-perfect vacuum, far emptier than any that could be created on Earth. But it is not totally empty. Swirling in the void are streams of hot, energized matter from the Sun, along with the dusty remnants of ancient collisions and deadly cosmic rays from deep in space.

The interplanetary medium is dominated by the solar wind, the stream of electrically charged gas particles that is ejected from the surface of the Sun at a rate of several million tons every second. These particles are highly energetic, which heats the wind to around 180,000°F (100,000°C). The particles spiral out along the lines of the Sun's magnetic field, which its rotation twists and turns into a bubble called the heliosphere that extends far beyond the planets.

The solar wind races outward from the Sun's equator at around 900,000 mph (1.5 million km/h), and recent measurements suggest that near the poles it may travel at twice this speed. Magnetic storms on the surface of the Sun also cause the wind to gust. During the solar maximum, the peak of the Sun's 11-year magnetic cycle, these sudden blasts of energy can damage satellites, disrupt radio communications and knock out power grids on the Earth. They could be even more hazardous to crewed spacecraft traveling to Mars or beyond, where the length of the voyage would make it impossible to avoid them.

GHOSTLY LIGHT

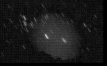

A GHOSTLY PYRAMID OF LIGHT THAT POINTS UPWARD FROM THE HORIZON (RIGHT) IS SOMETIMES VISIBLE JUST BEFORE DAWN IN THE FALL, OR AT TWILIGHT IN THE SPRING. CALLED THE ZODIACAL LIGHT, IT IS CAUSED BY LIGHT REFLECTED FROM THE CLOUD OF DUST PARTICLES THAT ENVELOPS THE SUN. THIS CLOUD SPANS A DISTANCE OF AROUND ABOUT 400 MILLION MILES (650 MILLION KM)—ROUGHLY FOUR TIMES THE DISTANCE FROM THE SUN TO THE EARTH.

SOLAR WIND
Each day, the Sun shoots millions of tons of charged gas particles—plasma—into space. This steady stream pushes back the contents of interstellar space to form a relatively empty bubble known as the heliosphere. The heliosphere may extend as far as 7 billion miles (11 billion km) beyond the Sun.

DUST CLOUD
The solar system is filled with dust particles that orbit the Sun in the same way as the planets. The density of this dust cloud increases toward the Sun, but is still incredibly low—a near-perfect vacuum by Earth standards.

COSMIC RAYS
Some cosmic rays originate in exploding stars, while others are thought to come from quasars and black holes that lie billions of light-years away. Most cosmic rays cannot be detected directly from the Earth's surface. But the rays split apart molecules in the upper atmosphere, triggering subatomic particle showers called secondary cosmic rays that can be detected.

DUST AND DANGER

Mixed in with the solar wind are vast quantities of dust particles. Scientists have estimated that there may be enough of them scattered around the inner solar system to make a small asteroid, but they are spread incredibly thinly. Some of these particles date back to the formation of the solar system; others are fragments from comet tails. Others still are the result of collisions and impacts between comets, asteroids and planetary bodies.

Most of the particles are no bigger than grains of sand and have been drawn into a cloud around the Sun by the pull of the Sun's gravitational field. But some of the dust consists of larger fragments that occasionally make their way to

Earth: Scientists have found at least 13 pieces of Mars that were chipped off in ancient asteroid impacts, only to land on Earth as meteorites after journeying around the solar system for millions of years.

The third component of the interplanetary medium consists of highly energized subatomic particles whose origins lie far beyond the solar system. Scientists discovered these particles in 1912, when an electric charge detector carried aboard an atmospheric research balloon registered an increasing number of hits at higher altitudes.

Since the particles clearly came from beyond Earth, they were named "cosmic rays."

Most cosmic rays are protons—the positively charged particles in the nuclei of atoms. Other rays consist of entire atomic nuclei, and a few are electrons—the negatively charged particles that surround the nucleus.

Cosmic rays are classified by energy level, which also gives clues to their origins. Low-energy cosmic rays are thought to be the fallout from stars that have died an explosive death within our own galaxy. The more powerful rays, by contrast, are more likely to have originated far beyond the Milky Way, possibly in the massively high-energy environment of quasars or the black holes that lie at the heart of giant galaxies.

THE FUTURE OF THE SOLAR SYSTEM

Six and a half billion years from now, the Sun will swell up into a red giant, engulfing the orbit of Mercury and melting the surfaces of Venus and the Earth. Two billion years later, the Sun will shrink and cool into a dim white dwarf. Computer simulations suggest that much of the rest of the solar system will survive these cataclysms. Shifted into new orbits, the fiercely altered planets will still revolve around the dying Sun. And on a few favored moons, there is a chance that life could spring forth anew.

WHAT IF...

...THE SOLAR SYSTEM IS STILL INHABITED WHEN THE SUN DIES?

For human life to survive in the solar system, people will have to abandon the Earth long before the Sun enters its red giant phase. NASA scientist James Kasting has revealed that the Sun is due to give Earth a "midlife crisis" just 1.1 billion years from now. The Sun's luminosity is steadily increasing, so by that time it will be shining 10% more brightly than it is now. This apparently modest increase could be enough to induce catastrophic temperature rises on Earth.

Even if our descendants manage to mitigate its worst effects with feats of planetary engineering, they will be fighting a losing battle. The Sun's luminosity will continue to rise, and evacuation will be the only option. Earth's population would probably move to Mars or settle on other space colonies.

Once they have safely abandoned the Earth, though, future humans might actually look forward to the time the Sun enters its red giant stage. This will take place in about six and a half billion years. As the Sun begins to burn helium instead of hydrogen, its luminosity will start to increase even further. For a time, the solar system will become a much warmer place. Saturn and Jupiter will change in appearance. Saturn's rings will be long gone, and high temperatures will have dramatic effects on the planets' atmospheres. But our descendants would be more interested in the gas giants' moons.

The ice-moons of Ganymede, Europa and Callisto, in orbit around Jupiter, would be transformed into more hospitable waterworlds. And research at NASA's Ames Research Center in California shows that Saturn's moon Titan would receive enough energy to thaw out the frozen ammonia and water that makes up its surface. This could be enough to kick-start the evolution of life on the gas giant's moon.

Life that may already exist in the hidden ice-covered oceans of Europa and Callisto would also thrive. Unfortunately, these small planetary moons may not have strong enough gravitational fields to retain their atmospheres. Also, the planetary nebula expelled from the dying Sun could block out some of the solar energy received by these new worlds, causing climatic variations. And when the Sun begins to shrink Europa and Callisto will refreeze, never to be thawed again. But for two billion years at least, the Sun's red giant phase might bring about a new age of discovery and colonization for the human race.

FUTURE ORBITS

THE SUN WILL HAVE LOST HALF ITS MASS BY THE TIME IT EXITS THE RED GIANT PHASE TO BECOME A WHITE DWARF. THE ORBITS OF THE SURVIVING PLANETS WILL SHIFT ACCORDINGLY, HEADING FARTHER OUT INTO SPACE.

SOLAR SYSTEM TODAY MILES/KM	SOLAR SYSTEM CIRCA 8000000000 A.D. MILES/KM
MERCURY: 35,983,000/57,909,000	MERCURY: DESTROYED
VENUS: 67,232,000/108,196,000	VENUS: 124,379,000/200,163,000
EARTH: 92,957,000/149,596,000	EARTH: 171,970,000/276,751,000
MARS: 141,635,000/227,933,000	MARS: 262,025,000/421,677,000
JUPITER: 483,633,000/778,310,000	JUPITER: 894,721,000/1,439,875,000
SATURN: 886,661,000/1,426,904,000	SATURN: 1,640,323,000/2,639,772,000
URANUS: 1,783,954,000/2,870,917,000	URANUS: 3,300,315,000/5,311,197,000
NEPTUNE: 2,794,356,000/4,496,957,000	NEPTUNE: 5,169,559,000/8,319,371,000
PLUTO: 3,674,500,000/5,913,373,000	PLUTO: 6,797,825,000/10,939,740,000

END OF AN ERA

The Sun and the solar system were born together around 5 billion years ago, forming from a huge contracting cloud of interstellar dust and gas. As the cloud contracted and became denser, the temperature at its core increased until it became hot enough to start a self-sustaining nuclear reaction. The Sun burst into life, and for four and a half billion years it has been fusing together hydrogen atoms to generate heat and light. There is enough hydrogen left for this process to continue for billions of years to come—but not forever. In six and a half billion years the Sun's hydrogen will be exhausted, and this 11-billion-year nuclear reaction will cease.

The only element left in the Sun's core will be helium, which reacts at a higher temperature than hydrogen. Without the pressure generated by nuclear reactions, the core will shrink under the force of its own gravity. But as it becomes denser, it will heat up until the nuclear reaction temperature of helium is reached. The sudden ignition of helium atoms will cause the Sun to grow bigger and brighter than ever before—200 times its present size and 5,000 times more luminous.

The solar wind, a stream of charged subatomic particles emitted by the Sun, will increase in strength as the star expands. And in a series of pulsations every 10,000 years or so, the outer layers of the Sun will be hurled out into space, creating a series of shells of gas known as a planetary nebula. Some astronomers believe that the Sun will lose almost half its mass through a combination of increased solar wind and planetary nebula generation.

So what effects will the Sun's expansion have on the planets of the solar system? Mercury has no chance of survival and will be swallowed up by the expanding star. The diminished mass of the Sun also means that its gravitational pull will decrease. As a result, the orbits of Venus and Earth will expand outward until both planets are almost double their present-day distances from the Sun. This should save both planets from being engulfed, although the Sun's increased luminosity could still be sufficient to boil away

both planets' atmospheres, and reduce Earth to red-hot liquid rock. Venus is probably doomed to fall into the Sun's fiery atmosphere, and some recent work suggests that Earth may suffer the same fate.

After two billion years, the Sun will run out of helium. During that time, though, the outer solar system may have a happy future. Warm, comfortable environments on some of the moons of Jupiter and Saturn could see life emerge. If so, the organisms of Europa or Titan will have to evolve in a hurry. On Earth, it took 4.5 billion years for life to develop from bacteria to a high-tech civilization. Any Titanians will have less than half that time to acquire the skills of deep space travel—and possible escape—before the Sun finally dims to a white dwarf and their environments freeze around them.

THE SOLAR SYSTEM'S LATE LATE SHOW

MARS
With its climate warmed by the enlarged Sun, the Martian environment will improve. Liquid water will flow once more across the planet's surface, making Mars more hospitable to life—for a brief summer. Soon, rising temperatures will spell doom for life.

THE SUN
By 7000000000 A.D., the Sun will have expanded out as far as the current orbit of the Earth, engulfing Mercury in the process. The loss of about half of the star's mass through the ejection of its outer layers will also have diminished it gravitational pull. As a result, the remaining planets will have drifted further out into space.

VENUS
Scorched hot enough to melt rock, Venus may be susceptible to tidal effects that could cause it to spiral into the swollen Sun.

JUPITER
Increased solar output will thaw out the giant planet's icy moons. For a short time at least, they may become new worlds with oceans and atmospheres.

THE EARTH
Our planet's atmosphere will have boiled away into space. A home to life no longer, the planetary surface will be a wasteland of molten rock at temperatures of up to 2,400°F (1,300°C).

SATURN
Saturn's moon Titan, now at a temperature of −290°F (−550°C), will be warm enough for organic chemicals to react. The moon is the late solar system's best bet for a second chance at the evolution of life.

MISNAMED NEBULAE

THE PLANETARY NEBULA THAT WILL MARK THE END OF THE SUN HAS NOTHING WHATSOEVER TO DO WITH PLANETS. SUCH NEBULAE OWE THEIR NAME TO ENGLISH ASTRONOMER SIR WILLIAM HERSCHEL (1738–1822). WHEN HE VIEWED THESE OBJECTS THROUGH HIS TELESCOPE, HERSCHEL NOTED THAT MANY NEBULAE APPEARED AS DEFINITE DISKS, LIKE PLANETS. SO HE CALLED THEM "PLANETARY" NEBULAE, AND THE NAME STUCK. THE "TURTLE IN SPACE"—NGC 6210 (ABOVE, WITH DETAIL OF CENTER INSET)— IS A GOOD EXAMPLE.

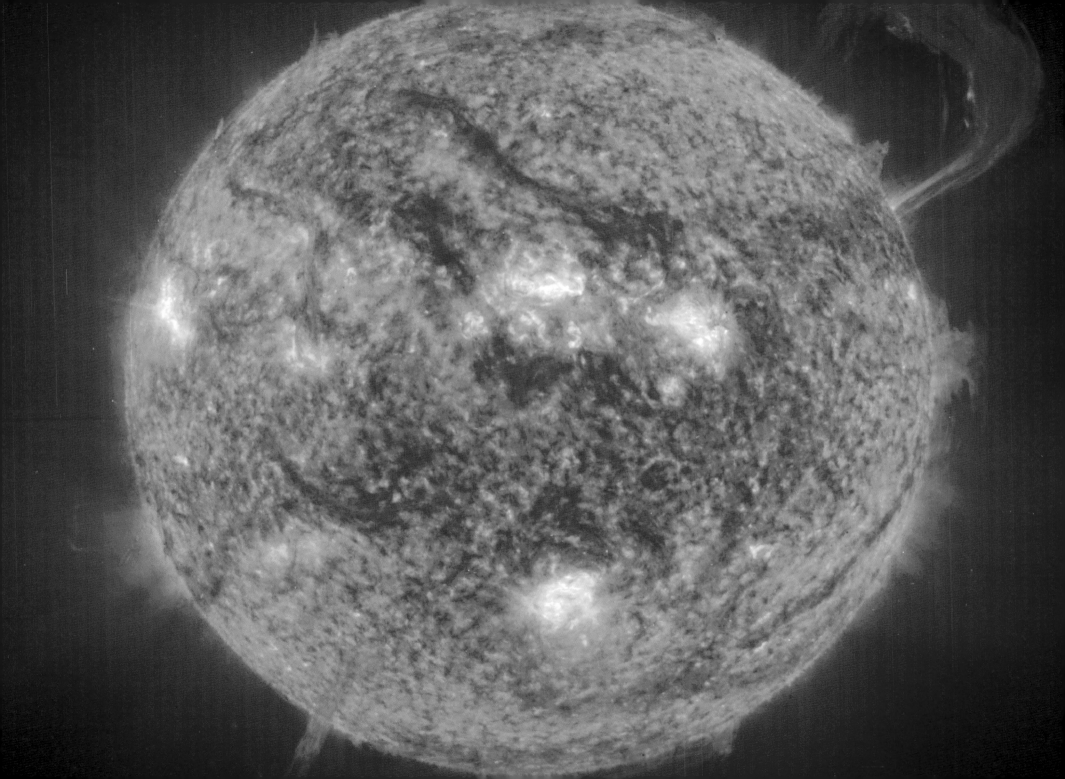

THE SUN

Our local star is the dominant force in the Solar System. The Sun's vast mass holds everything in its gravitational grasp, and the streams of material blown out in its solar wind create "space weather" that affects the planets as far out as Jupiter and beyond. As stars go, the Sun is a fairly average middle-aged example. It creates energy by nuclear fusion in its core, but the amount of energy emitted is small compared to that of more massive stars. Fortunately this means that the Sun will have a long lifespan—it has existed for five billion years, and will not undergo any major changes until it runs out of fuel in another five billion. In its present life stage, the Sun produces energy at a steady rate, although it is constantly changing on a small scale. Sunspots (cooler regions on the surface) and solar flares (vast outbursts of gas and particles into space) are both signs of the solar cycle, in which the Sun breaks down its magnetic field and regenerates it roughly once every eleven years.

The feature at the top right of this image of the Sun is a solar flare, a huge cloud of relatively cool dense plasma erupting from the Sun's hot, thin corona. In this image taken by the Extreme Ultraviolet Imaging Telescope, the hottest areas appear almost white, while the darker red areas indicate cooler temperatures.

THE SUN

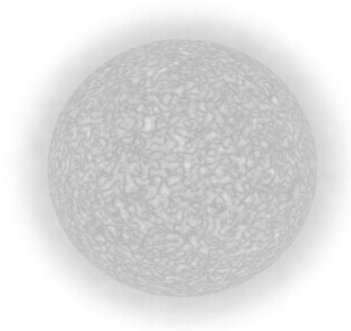

T he warmth that you feel on your skin on a sunny day originated in one of the most extreme environments in the entire universe—the nuclear cauldron that lies at the heart of a star. The star in question is the Sun, the body that supplies heat and light to the small planet we call Earth. The way the Sun generates energy in its fiery furnace, and the means by which that energy moves outward from the core to radiate as sunshine on the surface, are the keys to the evolution of life on this planet.

WHAT IF...

...THE SUN IS GETTING HOTTER?

F ortunately for life on Earth, the Sun's output of heat is remarkably consistent. But even a small change would have grave consequences, triggering either a new ice age, or runaway global warming. What evidence is there that this has happened in the past—or could happen in the future?

Scientists have been measuring the Sun's output for many years, initially from ground level, and now from space. Its level is almost constant, varying less than 1% over many years. But even a small change like this can have a marked effect on our climate. Between the 16th and 18th centuries the Earth was gripped by a "Little Ice Age." The climate became colder and caused rivers such as the Thames in London to freeze over in winter, allowing "frost fairs" to be held on the ice. Since then, the Thames has never frozen.

Looking farther into the past, studies of the fossilized remains of plants and animals suggest that the temperature on Earth has hardly varied. However, flare-ups in the Sun's output lasting only 200 or 300 years would not be expected to show up in the fossil records.

In 1972, Sun expert Douglas Gough of Cambridge University, England, worked out that the processes going on in the center of the Sun are,

by their nature, unstable. Before this it had always been assumed that the Sun was reliably stable.

If Gough is right, the Sun could change in brightness by as much as 5% over a period of millions of years—a big change in astronomic terms. "The work created quite a stir," says Gough. "No one has yet been able to show that the idea is wrong, but people have rather been ignoring it in the hope that it will go away."

Now, other scientists are beginning to take more of an interest in Gough's work. Some believe that the global warming we are currently experiencing is not caused by greenhouse gases in the atmosphere, but by the Sun heating up.

It is true that the Sun will grow steadily hotter as it reaches the end of its life about 5 billion years from now, although the change will be so slight that it may be millions of years before we notice it. And by then, humans will probably possess the technology to roam the galaxy in search of new Suns to warm them.

Life on Earth relies on heat and light from the Sun. Once scientists believed that the Sun's energy output was constant. Now there are indications that it may be unstable, with potentially disastrous consequences for the Earth.

SOLAR STATISTICS

Average distance from Earth	92.957 million miles (149.595 million km)
Diameter	864,950 miles (1.392 million km)
Age	4.5 billion years
Mass	2×10^{33} lb (2×10^{30} kg)
Average density	1.4 times the density of water
Surface temperature	10,900°F (6,040°C)
Core temperature	27 million°F (15 million°C)
Composition	At least 90% hydrogen, the rest mostly helium with traces of other elements

THE SHINING

The Sun is made of gas, mainly hydrogen. But because this body of gas is so massive, the Sun's own gravity creates enormous heat (27 million°F/15 million°C) and pressure (500 billion pounds per square inch/3.5 billion megapascals) at the core. Under these conditions, the hydrogen atoms cannot exist in their usual form; they become stripped of their orbiting electrons, leaving just the naked nuclei, called protons.

The heat and pressure agitates these protons to a point where they continually collide with each other. This causes some pairs of hydrogen atoms to fuse together, creating a single atom of a new element—helium. Each time this happens, a minute amount of matter is converted into energy. For each ounce of matter annihilated, enough energy is produced to power a 100W light bulb for about 750,000 years. And in the Sun, some 5 million tons of matter is annihilated every second.

The energy released in these nuclear reactions heats up the core of the Sun still further, and produces high levels of radiation. Photons—tiny "packets" of this radiation—slowly make their way outward from the core through a superdense region called the radiative zone. After that, the photons reach the convective zone where the Sun is less dense and where giant pockets of super-hot gas bubble to the surface.

Eventually, the energy-carrying photons reach the surface and radiate out into space. Much of the energy takes the form of visible light, but there are also infrared light, x-rays and harmful ultraviolet rays. The light comes from a region known as the photosphere which, effectively, is all we see of the Sun.

The photosphere appears grainy and is constantly moving. The grains seem to come and go, each one lasting some 25 minutes. These 600- to 1,000-mile-wide (1,000- to 1,600-km-wide) "granules" are actually the surface bubbling as energy is carried up from below. The surface of the Sun also shows a larger pattern, called supergranulation. Supergranules are each about 20,000 miles (32,000 km) across, and are related to the massive convection bubbles in the convective zone below the photosphere.

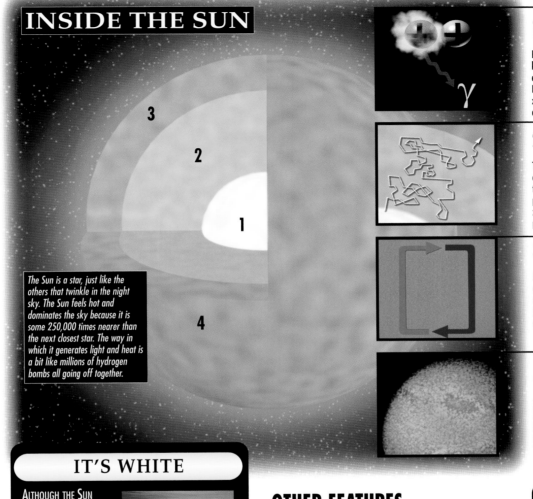

INSIDE THE SUN

The Sun is a star, just like the others that twinkle in the night sky. The Sun feels hot and dominates the sky because it is some 250,000 times nearer than the next closest star. The way in which it generates light and heat is a bit like millions of hydrogen bombs all going off together.

1 IN THE CORE

Pairs of hydrogen atoms combine to form helium atoms in a process known as nuclear fusion. During the process, matter is destroyed and energy given off. The amount of energy generated is given by Einstein's famous equation $E=mc^2$ (energy = mass x the speed of light x the speed of light). The numbers are enormous, as is the amount of energy generated!

2 RADIATIVE ZONE

The matter near the center of the Sun is so densely packed that energy-carrying photons produced during the nuclear reactions have trouble finding their way through. They bounce from particle to particle in a so-called "random walk pattern" through the radiative zone. Their path is so slow that it can take over a million years for a photon to find its way out.

3 CONVECTION ZONE

Energy is carried from the radiative zone outward through the convective zone. Here, the hot gases boil up in giant convection cells rather like soup boiling in a pot—except that these "pots" are up to 20,000 miles (32,000 km) across. The gases radiate the energy to the surface, then cool and sink down again, ready to pick up more energy.

4 PHOTOSPHERE

Here, smaller convection cells—up to 600 miles (1,000 km) across—bubble up to the surface with more energy, giving the surface of the Sun a grainy appearance. The sunshine we see on Earth comes from the photosphere, which is the only part of the Sun we can see directly. The corona and flares are only visible during an eclipse.

IT'S WHITE

ALTHOUGH THE SUN APPEARS YELLOW WHEN SEEN FROM EARTH, IT IS ACTUALLY WHITE. WE SEE THE SUNLIGHT AFTER IT HAS BEEN FILTERED THROUGH THE EARTH'S ATMOSPHERE. AIR SCATTERS THE BLUE COMPONENT, MAKING THE SKY APPEAR BLUE AND THE SUNLIGHT YELLOW.

OTHER FEATURES

Above the photosphere is the chromosphere, which can be seen as a pink ring around the Sun during an eclipse. Above this is the corona, from which a hot, thin stream of particles—the solar wind—blows outward into space. Other visible features of the Sun include darker regions called sunspots and bright flares.

The internal workings of the Sun may be complex, but they are nevertheless crucial to our existence. Without them, we would have no light, no energy, and, of course, no life on Earth.

GETTING TOGETHER

SCIENTISTS HAVE CALCULATED THAT THE AVERAGE TIME ANY ONE PROTON IN THE SUN'S CORE WILL SPEND WAITING TO COLLIDE WITH ANOTHER PROTON IS GREATER THAN THE AGE OF THE UNIVERSE! EVEN SO, THERE ARE SO MANY PROTONS IN THE CORE OF THE SUN THAT THERE ARE COUNTLESS COLLISIONS EVERY SPLIT SECOND.

CHANGING VIEWS OF THE SUN

The Sun has cast its light and heat upon the Earth for 4.5 billion years, and it will continue to do so for at least 5 billion years more. Human history spans just over 5,000 years—only a tiny fraction of the Sun's lifetime. Yet in this time our concept of the Sun has changed many times, from benevolent god to ball of fire to thermonuclear reactor visible across many light-years of space. But through it all, one idea has remained unchanged: Without the Sun's light and warmth, humankind would not exist.

WHAT IF...

...OUR CURRENT IDEAS ABOUT THE SUN ARE WRONG?

Scientists agree on the basic facts about our Sun. It is a ball of gas about 865,000 miles (1.4 million km) in diameter. Its core is a nuclear reactor, combining hydrogen atoms to make helium. And its surface temperature is about 10,000°F (5,500°C), making it appear yellow. Few would argue that this basic picture of the Sun is incorrect.

Yet after more than a century of studying the Sun with advanced astronomical instruments, scientists still have not solved many of its mysteries. As they decipher each mystery in turn, the Sun may reveal yet more new faces.

One mystery we have yet to solve is why the Sun has "spots"—the dark blotches that sometimes mar the Sun's surface. Astronomers understand that sunspots are cooler patches on the Sun's surface caused by the magnetic field breaking through, and that their numbers rise and fall in an 11-year cycle. Astronomers even know basically how sunspots form. But they are not entirely sure exactly why sunspots form, or why they have an 11-year cycle.

Another long-running mystery has been the missing neutrinos. Each fusion reaction in the Sun's core yields the nucleus of one helium atom, a photon (a "packet" of energy), and two subatomic particles called neutrinos. These particles have little or no mass, and almost never interact with other matter, so they are difficult to detect; about 700 billion neutrinos zip through every square inch (6.45 square cm) of our bodies each second, and we never know the difference.

Yet the Sun produces so many neutrinos that scientists can detect a few of them as they pass through Earth. However, detectors have found only about one-third as many as predicted, and this difference in number at first suggested that theories of the Sun's interior might need some work.

There are several possible explanations for the missing neutrinos. One theory says that the Sun's core is much cooler than is generally believed. But if that were so, then the Sun's size and surface temperature would also be different than observed.

However, there are three types of neutrino, and another theory is that neutrinos change from one type to another after leaving the Sun's core. Since detectors on Earth can "see" only one of the types generated by the Sun, the numbers detected are lower than expected. Recent evidence from neutrino "observatories" suggests this theory is right, but in turn this has raised questions about the nature of neutrinos.

These are just some of the mysteries about our Sun yet to be solved. It appears we have yet to fill in all the features in the changing face of our Sun.

PIONEERS OF SUN THEORIES

NAME	DATE	THEORY
GREEK PHILOSOPHERS	500 B.C.	THOUGHT THAT THE SUN MIGHT HAVE BEEN SMALL AND CLOSE TO EARTH, OR FLAT AND SUPPORTED BY AIR
ANAXAGORAS	C. 434 B.C.	SAID THE SUN WAS A "FIERY STONE"
ARISTOTLE	4TH CENTURY B.C.	SAID THE SUN WAS A FLAWLESS BALL OF "PURE FIRE"
ARISTARCHUS	C. 270 B.C.	SUGGESTED THAT THE PLANETS CIRCLE THE SUN
GALILEO GALILEI	1610	OBSERVED THE SUN WITH A TELESCOPE AND ANNOUNCED THAT HE SAW DARK SUNSPOTS ON ITS SURFACE
JOSEF VON FRAUNHOFER	1815	DISCOVERED DARK LINES—THE "FINGERPRINTS" OF DIFFERENT ELEMENTS—IN THE SUN'S SPECTRUM
GEORGE ELLERY HALE	1908	DISCOVERED THAT SUNSPOTS ARE MAGNETIC STORMS ON THE SUN'S SURFACE
ARTHUR EDDINGTON	1920	DEDUCED THAT THE SUN IS POWERED BY NUCLEAR FUSION

SUN SENSE

Khufu was one of the most powerful pharaohs of ancient Egypt—and he was taking no chances. When he died, he expected to ascend into the heavens and join with the Sun god, Re, the god of the Old Kingdom pharaohs. To make the climb a little easier, Khufu built the Great Pyramid to serve as a stairway. And to make his divine status clear to his subjects and his successors, he proclaimed himself the incarnation of Re; in essence, Khufu told the world that he was the Sun, bringing life to the valley of the Nile.

Khufu was not the last king to claim kinship with the Sun. Through most of human history, the Sun was regarded as a god of life and order that chased away the demons of darkness and chaos at each sunrise. Since most kings ruled by divine right, it was only natural for them to link themselves to this supremely important god.

One example of this divine link comes from the Inca of South America. In their cosmology, the Inca people were created by the Sun god, Inti. Their ruler, who was known simply as "the Inca," was considered a direct descendant of Inti, making him a god in his own right. Like the pharaohs of Egypt, the Inca expected to join the Sun god after his mortal body died.

LIFE GIVER

In many cultures worldwide, the Sun was worshiped not only for its life-sustaining qualities, but because of its continual cycle of death and rebirth. Each sunrise was a renewal not only of the Sun and Sun god, but of the whole world. Each sunset required the faith of the people that the Sun would meet the challenges awaiting it in the underworld and survive to rise another day.

Some cultures thought it was their duty to help sustain the Sun to meet these challenges. The Toltecs and Aztecs of central Mexico, for example, sacrificed human captives to the Sun god. They believed that the blood of these victims would renew the Sun and keep it strong.

By the time the Spanish conquered the Aztecs in the 16th century, Western scientists understood that the Sun is a hot, glowing ball—although it

took them centuries more to learn what makes it shine. Some suggested it was a big lump of burning coal, while others said it was hot because it was constantly shrinking, and any object that is squeezed gets hotter.

Today, the Sun maintains little of its mystic hold on humanity. We know that it is a star, like the thousands of twinkling lights visible in the night sky, and that it shines because it "fuses" hydrogen atoms in its core into heavier helium atoms, releasing enormous energy in the process.

Yet the Sun's importance to life on our planet is clearer than ever. Its warmth prevents the Earth from freezing, while its light provides energy for the plants that form the first link in the global food chain. Ancient civilizations from Egypt to Japan to Mexico were right after all: The Sun sustains our living planet.

THE SUN'S EVOLVING FACE

SUN GOD
Ancient Egyptians believed the Sun was the god Re, or Ra, who sailed across the sky-ocean by day in a boat. Boats found buried near the Great Pyramid may have been intended to carry the 4th-dynasty pharaoh Khufu (left) on this journey after his death.

FIERY STONE
In about 434 b.c., Greek philosopher Anaxagoras (left) said the Sun was a "fiery stone" larger than the southern half of Greece. This was perhaps the first scientific description of the Sun.

COAL FURNACE
Many 19th-century scientists saw the Sun as a burning coal. Lord Kelvin (left) said it could shine for 100 million years by gravitational contraction. But he was wrong—it is slowly expanding.

From depictions of the Sun god by the ancient Egyptians (far left) to an image of the Sun taken with a color filter by SoHO (Solar Heliospheric Observatory) in 1999 (below), our perceptions of the Sun have altered considerably. But the one thing that has remained the same is the vital role of the Sun in sustaining life on Earth.

NUCLEAR REACTOR
In 1920, Arthur Eddington (left) suggested that the Sun shines from nuclear fusion reactions in its core. The Sun has enough hydrogen gas to power this nuclear reactor for another 5 billion years.

SUN KINGS

In Shinto, the principal religion of Japan, the Japanese emperor is a direct descendant of the Sun Goddess, Amaterasu (right). Shinto theology says the Sun Goddess sent her grandson, Jimmu, to Earth to establish a line of kings more than 2,000 years ago.

ROTATION OF THE SUN

...WE COULD MONITOR THE SUN CONTINUOUSLY?

The Earth's rotation means that the Sun can only be seen for half a day at most from any one observatory. For helioseismologists—astronomers who study earthquake-like waves inside the Sun—such daily interruptions cause annoying gaps in their data. These observational gaps make it hard to keep track of the "sunquakes" that most interest helioseismologists. Fortunately, there are ways to deal with Earth's troublesome night and day cycles.

One method is to observe from the South Pole during the southern hemisphere's summer, when the Sun doesn't set for months. Astronomers have already done this. But even during the South Pole's so-called summer, the weather makes it almost impossible to observe for more than a few days at a time.

A better solution is provided by the GONG project. GONG, short for Global Oscillations Network Group, is a network of six small telescopes situated at stations in Australia, the Canary Islands, California, Chile, Hawaii and India. They are strategically located so that as the Sun sets in one location, it rises in another, enabling our star to be monitored continuously. GONG is a fully automated project, requiring little human intervention. As soon as the Sun rises at one of the tracking stations, the telescope there begins taking measurements. When the Sun sets, another tracking station takes over the operation. The project is designed to run continuously for at least three years, and is expected to amass enough data to fill more than 2,000 CD-ROMs. GONG will give astronomers observations of the Sun's sound waves that are at least 10 times more accurate than any currently available, offering unprecedented glimpses of the Sun's interior.

A final and obvious way to eliminate the problems of observing the Sun from Earth is to observe it from space, and that is exactly what the SoHO spacecraft is doing. Launched in December 1995 by the European Space Agency and NASA, SoHO has 12 instruments on board, three of which are dedicated to examining different aspects of the Sun's oscillations. When the data from SoHO and GONG has all been analyzed, astronomers will know much more about the Sun's internal structure and rotation than they do today.

Like the planets, the Sun spins around its axis. But because the Sun is a ball of plasma, not solid like the Earth, its rotation period varies with latitude and depth. The area near the equator moves the fastest, completing a revolution in about 25 days—compared with 35 days at the poles. This latitude variation is detected 30% of the way into the Sun, and then there is a change: The core of the Sun seems to rotate more like a rigid body. The Sun's rotation is also linked to the magnetic knots we see as sunspots.

ROTATION PERIODS

OBJECT	ORBITAL PERIOD	ROTATION PERIOD
SUN	N/A	25–35 EARTH DAYS
MERCURY	0.24 EARTH YEARS	58.65 EARTH DAYS
VENUS	0.62 EARTH YEARS	243.01 EARTH DAYS
EARTH	1.00 EARTH YEARS	1.00 EARTH DAYS
MARS	1.88 EARTH YEARS	1.03 EARTH DAYS
JUPITER	11.86 EARTH YEARS	0.41 EARTH DAYS
SATURN	29.46 EARTH YEARS	0.43 EARTH DAYS
URANUS	84.01 EARTH YEARS	0.75 EARTH DAYS
NEPTUNE	164.79 EARTH YEARS	0.80 EARTH DAYS
PLUTO	248.54 EARTH YEARS	6.39 EARTH DAYS

TWISTER SUN

The first recorded observations of the Sun's rotation were made about 400 years ago. Among the handful of Europeans at the forefront of this new science was Italian astronomer Galileo Galilei (1564–1642). With his newly invented telescope, he observed dark spots superimposed on the solar disk and suggested that they were physically associated with the Sun itself—they were not, as was previously believed, dark clouds or planets situated between us and the Sun. It was a simple step to measure the time it took for these spots—now called sunspots—to move across the solar disk. Galileo found that the Sun's "day" was a little under one month long.

But there were problems. As more astronomers carried out the same experiment, it became clear that the Sun's rotation rate was difficult to calculate exactly. Sometimes the spots appeared to move quickly across the Sun, and at other times they moved quite slowly. It was not until the mid-19th century that English astronomer Richard Carrington (1826–75) found the answer to the puzzle. In 1863, he observed that the solar equator was spinning once every 27 days as seen from Earth, but that at a latitude roughly halfway to the poles, the period was closer to 30 days. The Sun's rotation rate does indeed vary with latitude, and shows a smooth variation in spin period from 25 days at the equator to 35 days or even more at the poles. This rotation at different speeds is known as differential rotation. It is also seen in the gas planets of our solar system and in spiral galaxies.

THE SUN'S HUM

More recently, with the advent of helioseismology, astronomers have been able to see how the Sun rotates internally. Just as the study of earthquakes, or seismology, reveals properties about our planet's interior, so the study of vibrations on the Sun—helioseismology—offers clues to the interior of the solar furnace. The Sun is a violent place,

MURDER MYSTERY

RICHARD CARRINGTON, THE WEALTHY AMATEUR ASTRONOMER WHO DISCOVERED THE SUN'S DIFFERENTIAL ROTATION, WAS INVOLVED IN ONE OF THE SHADIEST EPISODES IN THE HISTORY OF ASTRONOMY. IN 1875, TWO WEEKS AFTER HIS UNFAITHFUL WIFE WAS FOUND SUFFOCATED IN HER BED, CARRINGTON HIMSELF WAS FOUND DEAD IN HIS OBSERVATORY HOUSE IN SUSSEX, ENGLAND (ABOVE) WITH EMPTY SEDATIVE BOTTLES NEARBY.

and so it is also very noisy. The sound waves that carry this noise move through the Sun and change direction when they encounter regions of different density. This process is similar to the bending of light rays when they cross the boundary between water and air, an effect that makes swimming pools appear shallower than they really are. When the sound waves reach the Sun's surface, they cause it to pulsate. By observing these pulsations, astronomers can make accurate deductions about the interior of the Sun.

The results suggest that the convective zone—the outer layer of the Sun—has the same rotation pattern as the surface. Near the boundary region between the convective zone and the deeper radiative zone, the Sun starts to show a difference in rotation rate with depth. The equatorial rotation speed decreases while the polar rotation rate increases. The rates equalize around 40% of the way into the Sun. From here on in, the Sun rotates as a rigid body. The exact period is uncertain, but it appears to spin roughly once every 25 days.

What happens at even greater depths, though, is a mystery. Because the deepest sound waves suffer the most modifications in their journey to the surface, the information that they carry about the deep interior is easily

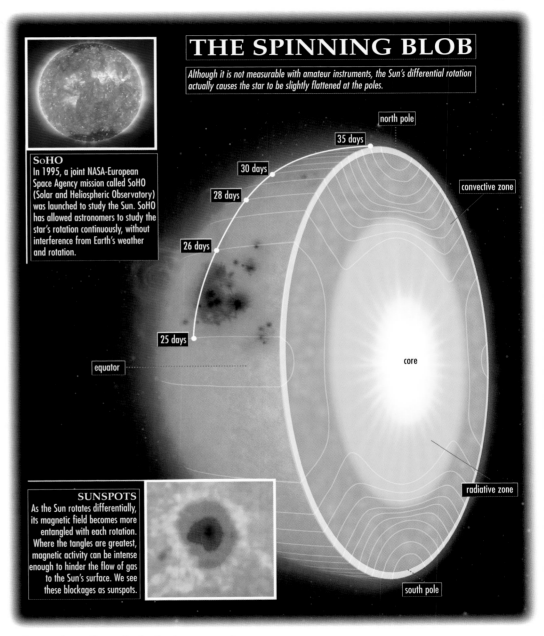

THE SPINNING BLOB

Although it is not measurable with amateur instruments, the Sun's differential rotation actually causes the star to be slightly flattened at the poles.

SoHO
In 1995, a joint NASA-European Space Agency mission called SoHO (Solar and Heliospheric Observatory) was launched to study the Sun. SoHO has allowed astronomers to study the star's rotation continuously, without interference from Earth's weather and rotation.

north pole
35 days
30 days
28 days
26 days
25 days
convective zone
equator
core
radiative zone
south pole

SUNSPOTS
As the Sun rotates differentially, its magnetic field becomes more entangled with each rotation. Where the tangles are greatest, magnetic activity can be intense enough to hinder the flow of gas to the Sun's surface. We see these blockages as sunspots.

drowned out. To study the Sun's core requires much more sensitive equipment than that currently available, and is a puzzle that future generations of astronomers will have to face.

SOLAR ECLIPSES

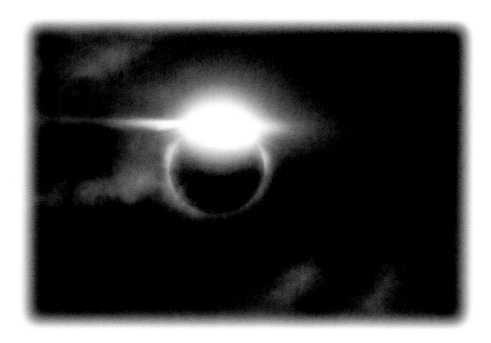

A total eclipse occurs when the Moon passes between the Earth and the Sun, blotting out the Sun's disk and turning day into night. Throughout history, few natural phenomena have filled the human race with such awe. Even for professional astronomers, the sudden darkening of the Sun, accompanied by a spectacular range of optical effects, is an experience never to be forgotten. But eclipses are not just entertaining—they also enable us to examine features of the Sun that are invisible at other times.

WHAT IF...

...ECLIPSES STOP HAPPENING?

One day, millions of years from now, total solar eclipses will cease to occur. In fact, we are lucky to see them at all: In cosmic terms, it is pure coincidence that the human race evolved at a time when the disks of the Moon and the Sun in the sky happen to be almost the same size.

Even today, in around 20% of solar eclipses, the Moon's disk is smaller than the Sun's. This is because the Moon's orbit around the Earth is elliptical, not circular: At the times when the Moon is farther from us, its disk appears smaller.

Annular eclipses are not as spectacular as total eclipses, but they can still be dramatic, especially at sunset, as seen here. Over the coming millennia they will become more common.

The result is an annular eclipse, in which there is a ring of light around the Moon. There is no full shadow (umbra) and the Sun's corona is hard to see. As the Moon gradually moves away from the Earth, spiraling into space at the rate of about 1.4 inches (36 mm) per year, annular eclipses will become the rule rather than the exception. In about 100 million years, total eclipses will cease altogether.

...YOU SEE THE NEXT ECLIPSE?

The next total solar eclipse in the U.S. is due on August 21, 2017. It will cover a belt of land about 70 miles (110 km) wide and will last for a maximum of around 2½ minutes, traversing the country from northwest to southeast, starting in Oregon and ending in South Carolina. It is expected that millions of people will be able to observe the event, and that they will be joined by crowds of eclipse-chasers from around the world.

TOTALLY ECLIPSED...

Date	Duration (minutes)	Max. Umbra Width (miles/km)	Where visible
2006, March 29	4:06	114/183	Africa, Turkey, Azerbaijan, Kazakhstan, Russia
2008, August 1	2:27	147/237	Greenland, Russia, China
2009, July 22	6:38	160/257	India, China, Pacific Ocean
2010, July 11	5:20	160/257	S. Pacific Ocean, southern tip of S. America
2012, November 13	4:02	111/179	Australia, Pacific Ocean
2015, March 20	2:46	287/461	N. Atlantic Ocean, Norwegian Sea
2016, March 9	4:09	96/154	Indonesia, N. Pacific Ocean
2017, August 21	2:40	71/114	United States of America
2019, July 2	4:32	124/200	S. Pacific Ocean, Chile, Argentina
2020, December 14	2:09	56/90	Chile, Argentina
2021, December 4	1:54	260/418	Antarctica, S. africa, S. atlantic
2023, April 20	1:16	30/48	S.E. Asia, Australasia
2024, April 8	4:28	122/196	N. & C. America
2026, August 12	2:18	369/594	N. America, W. Africa, Europe

DARKNESS IN THE DAYTIME

As the Moon orbits the Earth every month or so, it reaches a point at which it is in roughly the same direction as the Sun when seen from Earth. From this you might imagine that a total eclipse of the Sun would be a monthly event—but as we know, it is actually quite rare.

Partial eclipses, where just part of the Sun is obscured by the Moon, are infrequent enough. In most years there are between two and four, although in 1935 there were five. Total eclipses are rarer still: There are seldom more than about 70 per century.

The infrequency of eclipses is largely explained by the angle of the Moon's orbit around the Earth, and by the size of the Moon relative to the Sun. The Moon's orbit is at about 5° relative to the orbit of the Earth around the Sun. This means that the Moon does not pass over the face of the Sun every time it orbits the Earth. The Moon is also relatively small, which means that its shadow often misses the Earth altogether.

When the Moon does cast its shadow on the Earth, the path of the shadow crosses the Earth in a general west-to-east direction. Those fortunate enough to be lining the route are in for a spectacular show.

It takes several hours for the eclipse to unfold—a sequence of events that astronomers call contacts. Each point along the path sees exactly the same thing, but at a later time of day the farther east you go. For this reason some dedicated eclipse watchers take to aircraft, in an effort to chase the eclipse around the globe.

BAD LUCK

IN INDIA IT IS CONSIDERED UNLUCKY FOR PREGNANT WOMEN TO WATCH AN ECLIPSE, AND A CHILD BORN DURING AN ECLIPSE IS A BAD OMEN. IN SOME PLACES IT IS ALSO BELIEVED THAT TO STOP THE SUN FROM BEING "CONSUMED BY THE CELESTIAL BEAST," PEOPLE SHOULD PLUNGE THEIR HEADS INTO WATER.

ANATOMY OF AN ECLIPSE

The first stage of an eclipse, called first contact, occurs when the Moon appears to touch the edge of the Sun. The eclipse is now in its partial phase. Second contact, or totality, starts at the instant the Moon is completely in front of the Sun and usually lasts no more than a few minutes. Third contact is when the Sun is just about to be revealed again. From then on the eclipse reverts to being partial until, at fourth contact, the Moon

clears the Sun's disk completely. Because the Sun is so bright, there is no noticeable reduction in daylight until some 80% of it is covered. But toward second contact, an eerie darkness begins to descend with increasing rapidity.

In ancient times, people reported that the ground became filled with "writhing snakes." We now know that these mysterious serpents are, in fact, moving bands of shadow—the result of the Sun's narrowing crescent being focused by turbulence in the air.

SHINING ON
When the brilliant disk of the Sun is obscured, there is a good chance of seeing the normally invisible red tongues of hydrogen known as prominences. Usually these are tiny, but if the Sun is in an especially active phase they can be spectacular.

SOLAR JEWELS
Just before and after totality, a small part of the Sun's surface may peek out from behind the Moon. This glint of light is known as the Diamond Ring. Sometimes, an entire string of such gems—known as Baily's Beads—can appear.

INVISIBLE SUN
The Sun's outer corona becomes visible during an eclipse, but its appearance varies according to the amount of solar activity. If this is high, streamers emerge in all directions; at other times they are mostly confined to the solar equator.

SKY SPECTACLE
Although totality lasts only minutes, the entire show, including partial phases, can last up to 5 hours. Astronomers and other enthusiasts travel around the world in pursuit of solar eclipses and some tour firms make eclipses a specialty.

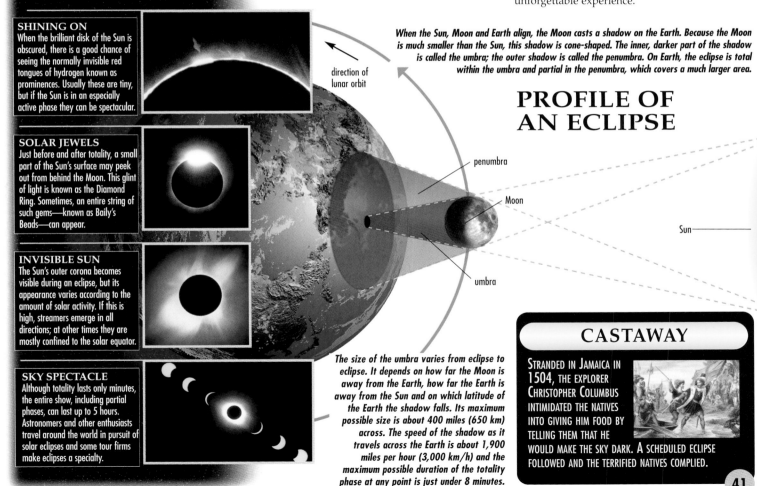

When the Sun, Moon and Earth align, the Moon casts a shadow on the Earth. Because the Moon is much smaller than the Sun, this shadow is cone-shaped. The inner, darker part of the shadow is called the umbra; the outer shadow is called the penumbra. On Earth, the eclipse is total within the umbra and partial in the penumbra, which covers a much larger area.

direction of lunar orbit

penumbra

Moon

umbra

Sun

The size of the umbra varies from eclipse to eclipse. It depends on how far the Moon is away from the Earth, how far the Earth is away from the Sun and on which latitude of the Earth the shadow falls. Its maximum possible size is about 400 miles (650 km) across. The speed of the shadow as it travels across the Earth is about 1,900 miles per hour (3,000 km/h) and the maximum possible duration of the totality phase at any point is just under 8 minutes.

NIGHT AND DAY

Despite the darkness, the path of an eclipse is narrow enough for light to remain visible on the horizon. Stars appear overhead and the air temperature becomes several degrees cooler.

As the bright disk of the Sun becomes obscured, the dimmer reddish prominences and outer layer, or chromosphere, become visible. Beyond them is the Sun's corona, which appears as a halo of bright, white light around the black disk of the Moon and is a truly awe-inspiring sight. For those lucky enough to watch, it is an unforgettable experience.

PROFILE OF AN ECLIPSE

CASTAWAY

STRANDED IN JAMAICA IN 1504, THE EXPLORER CHRISTOPHER COLUMBUS INTIMIDATED THE NATIVES INTO GIVING HIM FOOD BY TELLING THEM THAT HE WOULD MAKE THE SKY DARK. A SCHEDULED ECLIPSE FOLLOWED AND THE TERRIFIED NATIVES COMPLIED.

41

SUNSPOTS

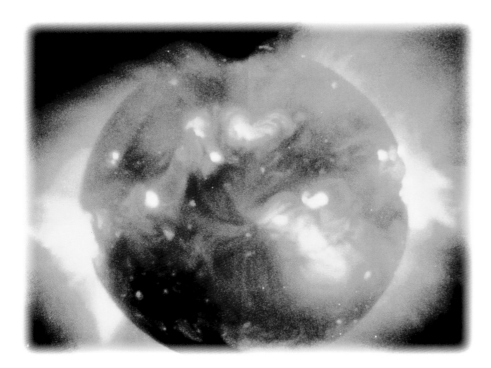

The Sun is a far from perfect star. Its surface is blemished by sunspots—gigantic blotches, often larger than the Earth, where the temperature is 3,500°F (2,000°C) cooler than the surrounding area. Sunspots usually occur in groups, up to 100 at a time, which can last from half a day to several weeks. But despite the vast range of their size and duration, and their dark appearance, all occur in active zones where the Sun's seething magnetic activity penetrates the outermost layer of hot gas, its photosphere. Through these sunspot windows we can see the soul of the Sun.

WHAT IF...

...THERE WERE SPOTS ON OTHER STARS?

Most stars are so far away that they are never much more than a speck of light, even under powerful magnification. Despite the distance, astronomers have worked out several ways to tell that some have "starspots" on them, just as the Sun does.

One way to find spots on stars is to look for a variation in the star's brightness as it spins. Spots are cooler and therefore darker than the rest of the star, so when a dark region rotates into the Earth's line of sight, the star appears to dim slightly. If the spot is particularly large, or if there are many spots in one location, the changes in brightness of the star as it spins can be quite pronounced. In the so-called RS Canum Venaticorum (RS CVn) stars, for example, astronomers have deduced the presence of starspots that often cover as much as 30% of the stars' surface.

Astronomers can also use a star's magnetism to tell if there are starspots. Some stars, like the Sun, have calcium in their atmospheres. Its precise signature—a certain combination of dark lines in the spectrum of sunlight—depends on the strength of the star's magnetic field. In the Sun, the calcium signature varies with the solar cycle over a period of about 11 years. Other stars display a similar pattern. The star 107 Piscium is a good example. Its calcium signature strengthens and weakens in a similar way, suggesting that the Pisces-constellation star has a magnetic cycle—and a related starspot cycle— lasting nine years. Often, spectroscopy can even specify how a star's spots are distributed across its surface.

Spots, then, are not unique to our Sun. They are a result of stellar magnetism and, as such, are likely to be present on many, if not most, Sun-like stars.

A bloated starspot dominates the face of a distant sun. The same magnetic field that caused the spot also generates a searing solar wind, and its high-energy particles have sterilized the star's planets.

SUNSPOT FACTS

PENUMBRA TEMPERATURE	9,400°F (5,200°C)
UMBRA TEMPERATURE	6,400°F (3,500°C)
TYPICAL SUNSPOT DIAMETER	8,000 MILES (13,000 KM), OR THE EARTH'S DIAMETER
MINIMUM DIAMETER OF SUNSPOT VISIBLE TO THE NAKED EYE	27,152 MILES (43,696 KM), OR SEVEN TIMES THE EARTH'S DIAMETER
TYPICAL SUNSPOT LIFETIME	A FEW DAYS
SHORTEST-EVER SUNSPOT LIFETIME	LESS THAN AN HOUR
LONGEST-EVER SUNSPOT GROUP LIFETIME	200 DAYS
EARLIEST RECORDED SUNSPOT SIGHTING	800 B.C., IN CHINA

IMPERFECT STAR

When Chinese astronomers saw dark spots on the Sun's disk one sunset 2,800 years ago, they had no real explanation. The great Italian astronomer Galileo Galilei (1564–1642) claimed that he first saw the strange markings in 1610. He wisely protected his eyesight by projecting the image of the Sun from his new telescope onto a card, and was the first to realize that the marks were on the "spotty and impure" solar globe itself. But it took a 20th-century science called spectroscopy to show that sunspots are a by-product of the Sun's magnetism.

Spectroscopy is the tool that reveals what stars are made of. Each element in the Sun's atmosphere absorbs light of certain wavelengths. So if sunlight is split into its spectrum of component colors, the missing wavelengths leave dark lines—showing which elements are in the Sun.

WILDLY ATTRACTED

Often, light can give more information. In 1908, the American astronomer George Hale (1868–1938) noticed that the "signature" lines in the spectrum split in the presence of sunspots. Where there should have been one spectral line, he saw several. This phenomenon was known to occur in the presence of a strong magnetic field. Hale concluded that sunspots are vast regions of magnetic turmoil where the Sun's own magnetic field is a thousand times stronger than average.

MAUNDER MINIMUM

BETWEEN 1645 AND 1715, ASTRONOMERS NOTED VERY FEW SUNSPOTS. THERE WAS A MATCHING DECLINE IN AURORAL DISPLAYS, WHICH ARE LINKED TO SOLAR ACTIVITY. SURPRISINGLY, THE UNUSUAL LULL PASSED ALMOST UNNOTICED UNTIL 1894, WHEN THE ENGLISH SOLAR ASTRONOMER EDWARD MAUNDER (1851–1928) DISCOVERED IT IN THE RECORDS. AS A REWARD FOR HIS DILIGENT HISTORICAL RESEARCH, THE EVENT IS NOW KNOWN AS THE MAUNDER MINIMUM.

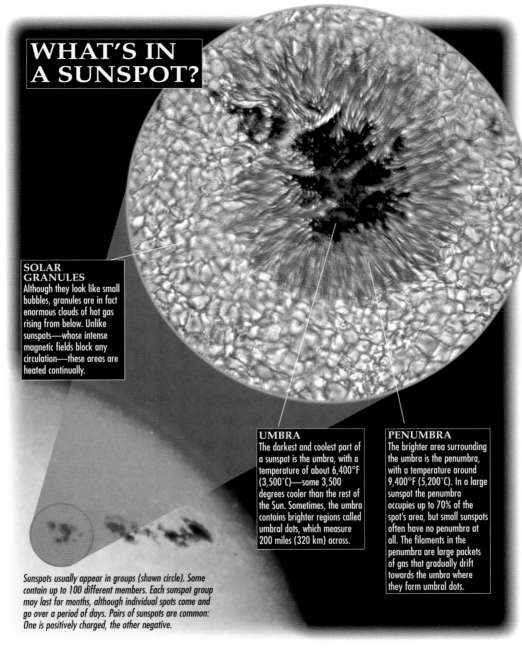

WHAT'S IN A SUNSPOT?

SOLAR GRANULES
Although they look like small bubbles, granules are in fact enormous clouds of hot gas rising from below. Unlike sunspots—whose intense magnetic fields block any circulation—these areas are heated continually.

UMBRA
The darkest and coolest part of a sunspot is the umbra, with a temperature of about 6,400°F (3,500°C)—some 3,500 degrees cooler than the rest of the Sun. Sometimes, the umbra contains brighter regions called umbral dots, which measure 200 miles (320 km) across.

PENUMBRA
The brighter area surrounding the umbra is the penumbra, with a temperature around 9,400°F (5,200°C). In a large sunspot the penumbra occupies up to 70% of the spot's area, but small sunspots often have no penumbra at all. The filaments in the penumbra are large packets of gas that gradually drift towards the umbra where they form umbral dots.

Sunspots usually appear in groups (shown circle). Some contain up to 100 different members. Each sunspot group may last for months, although individual spots come and go over a period of days. Pairs of sunspots are common: One is positively charged, the other negative.

Part of the disruption comes from the Sun's own rotation. It spins just as the Earth does, completing a full turn in a month. But the Sun is not solid; its equator turns faster than its poles. This so-called *differential rotation* plays havoc with the solar magnetic field.

SUNSPOT LIFE CYCLE

JUST LIKE MANY SPECIES OF ANIMALS, SUNSPOTS HAVE REGULAR CYCLES OF BIRTH AND MIGRATION. FOR SUNSPOTS, SPRING ARRIVES ABOUT ONCE A DECADE. AT THE BEGINNING OF A NEW CYCLE, MANY APPEAR AT LATITUDES OF AROUND 40°. EACH SPOT NORMALLY LASTS A FEW DAYS AND THEN DIES. AS THE CYCLE PROGRESSES, MORE AND MORE SPOTS ARE BORN AT LOWER AND LOWER LATITUDES—THOUGH THEY NEVER ACTUALLY FORM ON THE EQUATOR ITSELF. FOR OVER A CENTURY, SCIENTISTS HAVE KNOWN THAT THE CYCLE PEAKS EVERY 11 YEARS, BUT THEY ARE STILL NOT SURE WHY IT HAPPENS.

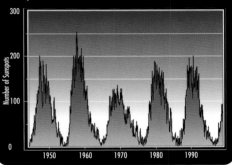

Magnetic field lines spanning the Sun resemble sections of an orange. The Sun pulls these vertical lines around with it as it spins. Since the Sun spins faster at its equator than near its poles, the lines begin to stretch and eventually become so twisted that they poke out of the Sun's outer layers, its *photosphere*, and form huge loops.

Sunspots occur at the bases of these magnetic loops—just where theory would predict them. Astronomers believe that the loops inhibit charged particles of hot gas and so prevent them from carrying heat to the surface. Cut off from this circulation, areas at the base of loops grow much cooler—and darker—than the rest of the photosphere. We see these areas as sunspots.

Theory also predicts that magnetic fields will tend to tangle most around the equator. Again, this is exactly what observations show: Sunspots do indeed appear to migrate from high to low latitudes on a timescale of about 11 years. The only question that remains is why, exactly, the process should take 11 years.

SOLAR FLARES

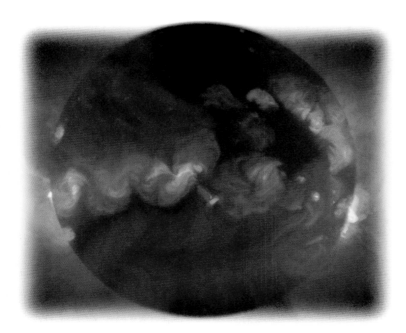

The surface of the Sun is in a state of perpetual turmoil, creating loops of magnetic field lines that arch high into the Sun's atmosphere, the corona. Sometimes, for reasons that astronomers still cannot fully explain, one of these loops becomes unstable, causing the field lines to snap together around a seething cauldron of energized plasma. The result is a solar flare: a gigantic outpouring of electromagnetic radiation across a broad range of wavelengths that takes less than 30 minutes to reach the Earth.

WHAT IF...

...AN UNUSUALLY LARGE SOLAR FLARE OCCURRED?

The date is July 23, 2040. William Doors, chairperson and CEO of Microcorp, the world's largest computer conglomerate, turns his attention to the business of the day, noting with satisfaction that Microcorp's royalty income from TOUCAN—the TOtally Unified Communications Assistance Network—is running at $20 billion a week. Without his company's vision, he muses, the current network of 360 orbiting comsats might never have existed and online communication would still be in the dark ages. As it is, 30 billion subscribers are each receiving data from space at the rate of 40 gigabytes every second of every day—and paying Microcorp 10 cents a week for the privilege. They get movies and music on demand, around 3,000 TV channels, online shopping and Microcorp's own "Live the Dream" virtual travel agency. It's a communications revolution.

Lost in thought, Doors almost misses the story on CNN about a "solar flare warning" issued by the National Solar Observatory's recently established lunar base. Rapid-fire descriptions of solar wind speeds over 1,000 miles per second (1,600 km/s) pass him by. But the words "unprecedented disruption to electronic communications" somehow catch his ear. Suddenly, the multipurpose screen on his desk goes blank. The phones stop ringing. He hears his personal assistant shriek.

Doors steps out of his office to see what's happening. No one knows. "The lines are down," mutters a systems technician, feverishly trying to reboot the office network's main servers. "But there aren't any lines!" exclaims Doors, his heart beating faster. "There must be some way we can find out what's going on out there."

"Take a taxi?" suggests the technician.

Data transmission by satellite is already a key part of the world's telecommunications networks. But few in the world's business community have spared a thought for the vulnerability of such networks to solar flares.

SOLAR FLARE STATISTICS

First Recorded Sighting	England, 1859
Average Energy Release	1032 ergs*
Maximum Energy Release	1037 ergs
Average Local Temperature	20–40 million°F (11–22 million°C)
Maximum Local Temperature	200 million°F (110 million°C)
Average Duration	15 minutes
Frequency	30–40 per day when the Sun is most active
Frequency of Major Flares	3–5 per year

1032 ergs = 26 million times the energy released in the bombing of Hiroshima

FLASH IN THE PAN

Like many of the other spectacular events that take place within the active regions of the Sun, solar flares are caused by irregularities in the Sun's magnetic field that disrupt the visible surface, or photosphere. Flares are linked with prominences, where the localized pull of the magnetic field counteracts the Sun's gravity and causes the photosphere to bulge. But in a flare, the effect is more sudden—and more violent: A big flare can liberate more energy than a billion thermonuclear explosions in just a few minutes, raising the temperature of the corona by tens of millions of degrees.

Despite their dramatic effects, solar flares remained a mystery until well into the 20th century because very little of their energy takes the form of visible light. The development of radio astronomy in the 1950s revealed that most flares radiate a combination of low-frequency radio waves and ultra-high-frequency X- and gamma- rays. But since most of the high-frequency waves are absorbed by the Earth's atmosphere, it took the latest generation of solar observation satellites to reveal solar flares in all their glory. In fact, at X-ray frequencies and above, a flare can briefly outshine the entire Sun.

Solar flares also create gusts in the solar wind—the stream of charged particles that constantly pours forth from the Sun. Traveling at up to 70% of the speed of light, some of these particles collide with the the Earth's magnetic field to produce colorful auroral displays in the

atmosphere and disrupt radio communications.

Astronomers have only a partial understanding of what triggers solar flares, but they seem to originate in the type of solar prominences known as coronal loops. These arise

PARTICLE ACCELERATOR

As the field lines reconnect, the plasma around them is turned into maelstrom of highly energized particles—mostly electrons and

protons. Some of these particles are fired straight out into the corona, later to become gusts in the solar wind; others are fired downward into the photosphere, where they collide with other particles to trigger a massive burst of high-frequency X- and gamma-ray radiation that takes only minutes to reach the Earth.

The entire event is over in a matter of minutes. Within half an hour, the field lines in the coronal loop have reformed, leaving the never-ending cycle of turbulence on the Sun's surface to start afresh.

when distorted lines of the Sun's magnetic field reach out high into the corona, drawing material with them and stirring up the photosphere below. But sometimes, the turbulence creates localized changes in the polarity of the field that in a split second cause the field lines to reconnect with each other. As they do so, there is an enormous release of pent-up energy—rather like an elastic band that is stretched until it snaps.

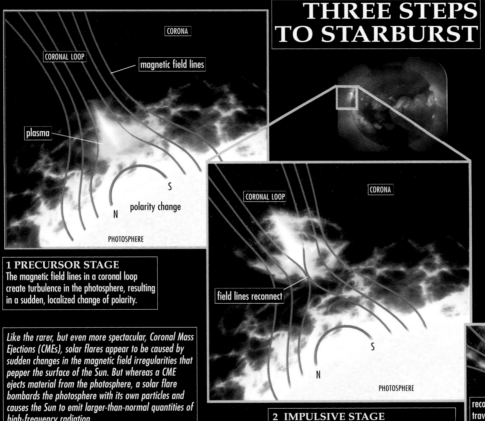

THREE STEPS TO STARBURST

1 PRECURSOR STAGE
The magnetic field lines in a coronal loop create turbulence in the photosphere, resulting in a sudden, localized change of polarity.

Like the rarer, but even more spectacular, Coronal Mass Ejections (CMEs), solar flares appear to be caused by sudden changes in the magnetic field irregularities that pepper the surface of the Sun. But whereas a CME ejects material from the photosphere, a solar flare bombards the photosphere with its own particles and causes the Sun to emit larger-than-normal quantities of high-frequency radiation.

2 IMPULSIVE STAGE
The field lines at the base of the loop reconnect, triggering a massive release of energy that turns the surrounding plasma into a soup of subatomic particles.

3 EJECTION STAGE
The field line reconnection shoots up the loop, catapulting particles outward into the corona and downward into the photosphere at near light-speed.

THE SOLAR CYCLE

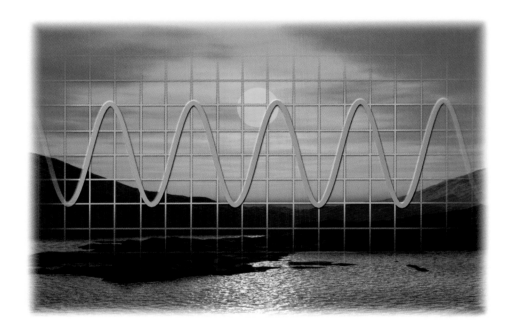

As the Sun beats down on a hot summer day, it is all too easy to believe—as our ancestors did—that our parent star is an inexhaustible and never-changing source of energy. But in fact, like most stars, the Sun goes through a series of regular changes in structure and output that astronomers refer to collectively as the solar cycle. Though we rarely notice them, these changes can have dramatic consequences on Earth. They also reveal much about the internal workings of the Sun's nuclear furnace.

WHAT IF...

...THE SOLAR CYCLE CHANGED?

Before astronomers can predict the future of the solar cycle, they have to understand how it works. At present, the most popular theory is called the solar dynamo model. This theory assumes that the solar cycle is driven by changes in the Sun's magnetic field, which extends between north and south poles that lie below the Sun's visible surface.

At the start of a solar cycle, the field resembles a series of lines that divide the Sun into segments like an orange. If the Sun were solid, the field lines would remain in the same position as it rotated. But because the Sun is a ball of gas, it rotates differentially—in other words, faster at the equator than at the poles. Over about four years' worth of rotations (each taking roughly 26 days at the equator, 33 days at 75° latitude), the magnetic field becomes twisted around the Sun. Eventually,

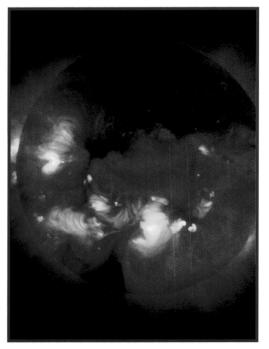

the tangled magnetic loops push their way through the photosphere—the visible surface of the Sun—to manifest themselves as dark pairs of sunspots.

At first, the loops burst through close to the Sun's magnetic poles. But as the field becomes more tangled, the loops increase in number and appear at lower latitudes. Solar prominences are hot gas streams that link one end of a loop with another. Solar flares occur when the magnetic field lines collide and join together, releasing colossal amounts of energy as they do so.

As the field lines become entangled, the Sun's activity increases. But at the same time, the magnetic field itself breaks up. Depending on their polarity, field lines drift toward a pole or toward the equator (obeying the principle that opposite poles attract and like poles repel). At the equator, they meet and cancel out; at the poles, they cause the overall field strength to diminish.

Eventually, after 11 years, the Sun loses its magnetic field altogether. But movements of electrically charged gas deep within it swiftly produce a new field, this time with the polarity reversed. So the time taken for the Sun to return to its original magnetic state is around 22 years.

If the theorists are right, the solar cycle will continue far into the future. But some of its secrets still elude us. For example, the length of the cycle varies by as much as five years, and so does its strength. More puzzlingly, it seems that the cycle can at times be suspended. During the so-called Maunder minimum, which lasted from 1645 to 1715, hardly any sunspots were observed on the Sun—a period that coincided with a mini-ice age on Earth.

It may be that such quiet periods are part of a much larger solar cycle. But only after many more years of careful solar observation will we know for certain.

This X-ray image, taken at solar maximum, shows the intense activity in the Sun's corona. It may be that solar cycles are blips in a much longer cycle of solar activity that spans anything from 70 to 500 years.

11-YEAR ITCH

The Sun is a more volatile place than most of us realize. The surface changes from day to day, as sunspots come and go and prominences send loops of hot gas arching into the Sun's atmosphere (corona). Then there are solar flares—gusts in the solar wind of charged gas particles that are ejected into space along the lines of the Sun's magnetic field. And, more rarely, there are coronal mass ejections, when the Sun explosively rids itself of billions of tons of matter.

When the appearance of these solar phenomena is logged over time, a pattern emerges. The Sun undergoes periods of intense activity and relative calm that alternate every 11 years or so in a pattern called the solar cycle.

The cycle is most obvious in the changes to sunspots—areas that appear darker because they are thousands of degrees cooler than the surrounding surface (even though they are still around 6,400°F/3,500°C). At the start of the 11-year cycle, the Sun's disc is relatively clear: There are just a few sunspots gathered at roughly 40° north and south of the solar equator. Most of these appear in pairs, one of which (the preceding spot) moves ahead of the other and is slightly closer to the equator. Individual sunspots left over from the old cycle may linger, while new ones appear and fade in anything from half a day to a few weeks. But as the cycle continues, the average number of sunspots gradually increases: They last longer, and groups of them appear closer to the equator.

MAXIMUMS AND MINIMUMS

Sunspots reach a peak at the so-called solar maximum, around four years after the start of a cycle, at which point they are concentrated around 15° north and south of the equator. Over the next seven years, their numbers decline and they draw closer to the equator. By the end of the cycle, at the solar minimum, just a few remain around the equator, while a new generation of spots begins to form at higher latitudes.

The fluctuations in sunspots reflect changes that take place in the Sun's magnetic field over

COMPONENTS OF THE SOLAR CYCLE

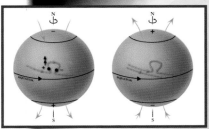

MAGNETIC FIELD
Parts of the Sun rotate at different rates. In time, this causes the magnetic field lines to kink and burst through to the surface, where they form sunspots. Over a single solar cycle, it also causes the Sun's magnetic field to reverse polarity (above). So a full cycle of magnetic events lasts for two cycles, or about 22 years.

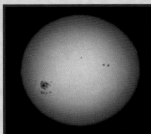

SUNSPOTS
These cooler areas of the surface reveal the effect of the loops of magnetic field lines at different parts of the cycle. The view at left is typical of the Sun's appearance near its solar maximum: The spots are concentrated at latitudes 15° north and south, in groups that run roughly parallel to the equator.

CORONA
The corona, or outer atmosphere, also changes during the solar cycle. The eclipse of July 11, 1991 (right), which occurred at near solar maximum, showed the corona extended for millions of miles in all directions. At solar minimum, it is more flattened along the equator, with prominent streamers that resemble wings.

TURNING BACK THE CLOCK

GEOLOGICAL EVIDENCE FOR THE REGULARITY OF THE SOLAR CYCLE EXTENDS MUCH FARTHER BACK IN TIME THAN ASTRONOMICAL RECORDS. ONE CLUE COMES FROM MEASURING THE LEVELS OF RADIOACTIVE CARBON-14 IN THE GROWTH RINGS OF VERY ANCIENT TREES, SUCH AS THOSE FOUND IN PARTS OF AUSTRALIA (RIGHT). TRACES OF CARBON-14 IN VEGETATION ARE NORMALLY THE RESULT OF COLLISIONS BETWEEN HIGH-ENERGY COSMIC RAYS AND NITROGEN ATOMS IN THE EARTH'S ATMOSPHERE. WHEN THE SUN'S MAGNETIC FIELD IS AT ITS STRONGEST, THE EARTH IS BETTER PROTECTED FROM COSMIC RAYS AND LESS CARBON-14 IS PRODUCED. THE EFFECTS OF THIS CAN BE MEASURED AND USED TO TRACE SOLAR CYCLES A FEW THOUSAND YEARS BACK. OVER LONGER PERIODS, TRACES OF

CARBON-14 IN THE FOSSIL RECORD OF SOIL SEDIMENTS LAID DOWN IN RIVER BEDS ALSO REVEAL AN 11-YEAR PATTERN THAT GEOLOGISTS THINK MAY BE LINKED TO THE SOLAR CYCLE.

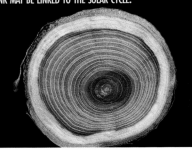

the length of the cycle. The Sun reverses polarity every 11 years, so the full magnetic cycle lasts 22 years. Other aspects of the Sun's behavior also change. Although sunspots signify cooler areas of the surface, the Sun's energy output actually increases slightly during a solar maximum. This variation amounts to only tenths of a percent in the visible part of the spectrum, but there is a tenfold increase in ultraviolet radiation and a one hundredfold increase in X-ray radiation. The Sun is also generally far more active at its solar maximum: Prominences and solar flares both increase in size and become more frequent.

The effect of these changes on the Earth is the subject of "heated" debate. In the past, surges in the Sun's output and in the solar wind have been smoothed out by the protective shroud of our atmosphere. Magnetic storms—disruptions to the Earth's magnetic field that occur during high levels of solar activity—have occasionally taken their toll on power lines and electronic equipment. But there is little evidence that the solar cycle affects global temperatures, the ozone layer, or any of the other "hot potatoes" that currently concern climatologists.

Things could be different though, as we venture farther into space and place increasing reliance on communications satellites. With more and more spacecraft exposed to the full fury of the Sun, we may have to follow the solar cycle more closely.

SOLAR WIND

Heat and light are not the only things that reach our planet from the Sun. There is also the solar wind—a fluctuating gale of subatomic particles. These fragments of electrically charged matter are hurled into space at colossal speeds by the Sun's magnetism. For the most part, we are shielded from them by the Earth's own magnetic field. But the solar wind still has the power to knock out communications satellites, and in polar regions it draws out shimmering bows of color: the aurorae.

WHAT IF...

...THE WIND WAS A GALE?

When scientists proved the existence of the solar wind in the 1950s, it was only seen as the benign cause of aurorae. But then came solid state circuitry and the space age. As electronic devices spread and grow more sensitive, they make it more likely that we will be affected by the Sun's stormy weather.

The Sun's "weather" seems to have a cycle of about 11 years. During the solar maximum around 1990, magnetic storms caused power failures in Canada and Sweden, and disrupted several satellites. The maximum around the year 2000 was milder in comparison, and satellite operators were better prepared, switching computers down on some satellites so that they performed only vital functions during major storms. However, it is still a wide precaution for astronauts to avoid spacewalks during major bursts of solar activity.

As was expected, the 2000–01 peak produced spectacular auroral activity at both north and south poles. But this time, the Sun had a surprise in store. As astronomers expected solar activity to die away during 2003, it suddenly returned with renewed vigor, as the Sun produced some of the biggest flares ever recorded. The magnetic storm of March 2003 was among the most spectacular in living memory, and was accompanied by widespread interference.

It now seems the Sun's activity is far more complex than we had imagined. If it can produce enormous flares with relatively little notice and out of sync with its usual cycle, future space missions may have to take this into account. For example, any planned human mission to Mars might well be timed to avoid solar maximum, protecting the astronauts on board from the potentially damaging solar wind particles. Without the reassurance of a reliable solar cycle, it may be necessary to equip future long-duration space missions with heavier shielding.

There is another risk, too. Since the early 1980s scientists have known that the Earth's upper atmosphere leaks oxygen, helium and hydrogen ions. In September 1998, scientists using NASA's Polar spacecraft also observed that air from the Earth's atmosphere was pushed out by violent gusts of solar wind. The upper atmosphere lost several hundred tons into space. Fortunately, the Earth can soon make up the loss.

STRENGTH OF THE WIND

FREQUENCY OF WINDS	27-DAY CYCLE, MATCHING THE SUN'S ROTATION
SOLAR WIND SEASON	11-YEAR CYCLE, MATCHING SUNSPOT AND OTHER SOLAR ACTIVITY
VELOCITY OF LOW-SPEED FLOW THROUGH CORONAL HOLES	900,000 MILES PER HOUR (1,500,000 KM/H)
VELOCITY OF HIGH-SPEED FLOW THROUGH CORONAL HOLES	1,900,000 MILES PER HOUR (3,100,000 KM/H)
SUN'S MASS LOST THROUGH SOLAR WIND, PER SECOND	1,000,000 TONS (910,000 TONNES)
TIME TAKEN FOR SUN TO LOSE 1% OF ITS MASS	1 BILLION YEARS
DENSITY OF SOLAR WIND AT THE EARTH'S ORBIT	140,000 PARTICLES PER CUBIC FOOT (0.03 CUBIC M)
WIND SPEED AT THE SUN	150,000 MILES PER HOUR (250,000 KM/H)
WIND SPEED AT THE EARTH	500,000 MILES PER HOUR (800,000 KM/H)
DISTANCE REACHED BY SOLAR WIND	MORE THAN TWICE PLUTO'S DISTANCE FROM THE SUN

OUR BLUSTERY STAR

Like sunlight, the solar wind comes from the Sun and blows over all the planets in the solar system. But unlike sunlight, it is invisible, composed of charged particles, and varies with the Sun's 11-year activity cycle.

The wind is an extension of the corona—the region of hot, ionized gas that surrounds the Sun. At a temperature of over a million degrees, coronal matter is moving fast enough to escape the Sun's gravity at a rate of about a million tons a second. Just why the corona becomes so hot—the temperature of the Sun's surface is only about 10,900°F (6,000°C)—is still a mystery. But Sun-observing satellites have identified the magnetic paths by which some of it makes its escape into interplanetary space.

There are two types of magnetic fields in the Sun's outer atmosphere, both fluctuating with general solar activity. One type is a "closed" field. Its lines of force emerge from the surface, loop up into the corona, and then return to the surface. Like bars on a cage, closed fields restrain the coronal gas. But the lines of force in the second kind of field are open. They stretch right through the corona and out into space. The solar wind passes through these magnetic gateways.

The solar wind is similar to Earthly winds in one respect: Both are prone to gusts created by atmospheric changes. Massive ejections of coronal gas usually coincide with eruptions of solar flares, sunspots and solar filaments.

BLOWING THROUGH SPACE

The hot gas of the solar wind continues to expand into space. Since the backward pull of the Sun's gravity decreases with distance, the solar wind actually travels faster the farther it goes. By the time it reaches Earth, the wind is traveling at half a million miles per hour.

By that stage, the coronal gas is very thin—only a few hundred protons and electrons to every cubic inch (16.4 cubic cm) of space. All the same, it is still powerful enough to produce some dramatic effects.

From most angles, the Earth's magnetic field protects us from the onslaught of the wind's charged particles. But at the North and South poles, the lines of magnetic force curve down toward the planetary surface. Just as the wind escaped the sun through open magnetic fields so it slides into the Earth's upper atmosphere. As the solar wind's ions strike molecules of air far above the ground, the energy that they discharge creates the spectacular light shows of the aurorae.

The solar wind whistles past the outer planets at nearly a million miles an hour. Spreading itself ever thinner, the wind will continue accelerating until far beyond the orbit of Pluto. At one hundred times the distance from Earth to the Sun—more than double the distance to Pluto—the solar wind at last merges with the almost imperceptibly thin gas of the interstellar medium. This point—the heliopause—marks the final frontier of the solar system.

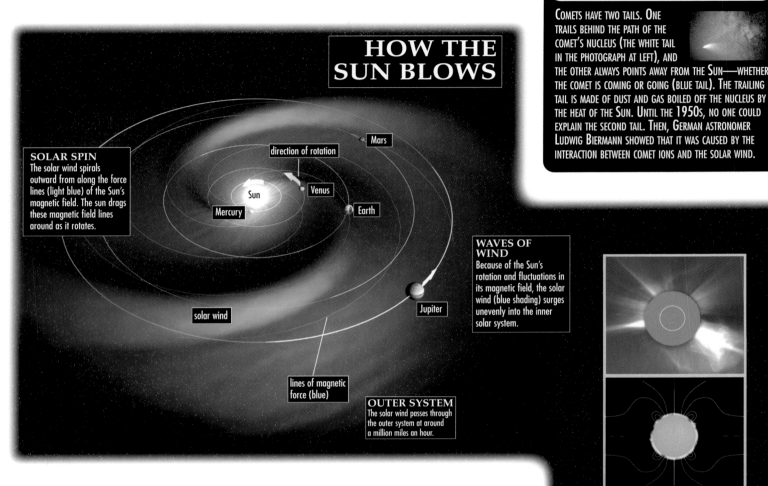

HOW THE SUN BLOWS

SOLAR SPIN
The solar wind spirals outward from along the force lines (light blue) of the Sun's magnetic field. The sun drags these magnetic field lines around as it rotates.

direction of rotation

Mars

Venus

Sun

Earth

Mercury

WAVES OF WIND
Because of the Sun's rotation and fluctuations in its magnetic field, the solar wind (blue shading) surges unevenly into the inner solar system.

solar wind

Jupiter

lines of magnetic force (blue)

OUTER SYSTEM
The solar wind passes through the outer system at around a million miles an hour.

TWO TAILS

COMETS HAVE TWO TAILS. ONE TRAILS BEHIND THE PATH OF THE COMET'S NUCLEUS (THE WHITE TAIL IN THE PHOTOGRAPH AT LEFT), AND THE OTHER ALWAYS POINTS AWAY FROM THE SUN—WHETHER THE COMET IS COMING OR GOING (BLUE TAIL). THE TRAILING TAIL IS MADE OF DUST AND GAS BOILED OFF THE NUCLEUS BY THE HEAT OF THE SUN. UNTIL THE 1950s, NO ONE COULD EXPLAIN THE SECOND TAIL. THEN, GERMAN ASTRONOMER LUDWIG BIERMANN SHOWED THAT IT WAS CAUSED BY THE INTERACTION BETWEEN COMET IONS AND THE SOLAR WIND.

ESCAPE ROUTE
The solar wind escapes from the regions of the Sun that appear darkest in the coronagraph image at top. These regions correspond to points where the Sun's magnetic fields do not loop back to the surface (bottom image).

THE HELIOSPHERE

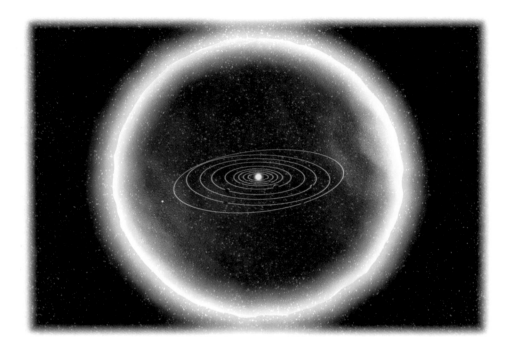

The Sun is cocooned at the center of a magnetic plasma bubble of its own making, known as the heliosphere. Extending far beyond the planets, the heliosphere is made up of the solar wind and the Sun's magnetic field. It fills the solar system with a thin dusting of material and touches all the planets and moons. Only when spacecraft reach the heliopause—the edge of the heliosphere and the boundary between the solar system and interstellar space—will we know exactly how enormous the heliosphere is.

WHAT IF...

...VOYAGER 1 LEAVES THE HELIOSPHERE?

In early 2005, NASA scientists announced that the Voyager 1 space probe, 8.7 billion miles (14 billion km) from the Sun, had begun its journey into the final reaches of the solar system, sending back data that indicated it had crossed the termination shock—the boundary region of the sun's heliosphere. When it passes the heliopause it will leave the last vestiges of the Sun's effects behind, and will begin its exploration of space between the stars.

The first hints that Voyager was leaving the region of the Sun's influence came in the early 2000s, as the probe's plasma science experiments, which measure the particle environment around it, began to send back data unlike scientists had ever seen before. This suggested that Voyager was approaching the termination shock, where the solar wind slows to subsonic speed and becomes turbulent before reaching the heliopause. As the wind of particles mixes with the interstellar wind of different speed, composition and temperature, carrying its own magnetic field, the remnants of the Sun's own magnetic field also become deformed.

Voyager is now in a region known as the heliosheath that could be hundreds of millions of miles thick. Here, the Sun's influence has almost disappeared. Only when Voyager crosses the heliopause will it leave behind the last remnants of the solar wind and magnetic field. The probe will enter a thin interstellar medium that may be filled with cosmic rays, produced by the remnants of tremendous supernova explosions.

The heliosphere itself could change over time. Astronomers do not yet know the current shape of the heliosphere, but they think that interstellar winds may blow it into a teardrop shape with a long tail. As the solar system orbits within the Milky Way Galaxy, the heliosphere could one day run into a denser region of interstellar dust and gas. This might make the Sun's magnetic bubble change shape under the pressure. It could grow an even longer tail, rather like the magnetotail stretching downwind from the Earth, or it could be compressed. At the same time, intense cosmic rays could pierce through the heliosphere to a much greater extent than they do now. A space probe of the future may have a much shorter trip before leaving the Sun's sphere of influence forever.

As Voyager heads to the heliopause, around 13 billion miles (21 billion km) away, tracking stations on Earth will find it difficult to stay in contact. The probe has only 20 watts of power—less than a light bulb—to broadcast its radio signals.

SOLAR WIND STATISTICS

COMPOSITION	GASEOUS PLASMA OF SUBATOMIC PARTICLES
ORIGIN	SUN'S ATMOSPHERE (THE CORONA)
SPEED	SLOW SOLAR WIND FROM SUN'S EQUATORIAL REGION: 900,000 MPH (1.5 MILLION KM/H)
	FAST SOLAR WIND NEAR POLES: 1,728,000 MPH (2,780,946 KM/H)
DENSITY	5 PARTICLES PER CUBIC CM (0.06 CUBIC INCH) WHEN THE SUN IS QUIET
TEMPERATURE	ABOUT 180,000°F (100,000°C)
TERMINATION SHOCK	AROUND 10 BILLION MILES (16 BILLION KM) FROM THE SUN
HELIOPAUSE	AROUND 13.5 BILLION MILES (22 BILLION KM) FROM THE SUN

RADIO WAVES

In 1992, the two Voyager probes relayed a strange burst of radio waves to the Goldstone tracking station in California. About 400 days previously, the Sun had experienced violent outbursts. Scientists realized that the radio waves were caused by blasts of solar wind smashing into the heliopause. They estimated that the heliopause was around 145 AU or 13.5 billion miles (21.5 billion km) away.

WINDS OF TIME

Until a probe ventured into interplanetary space, we had no proof that the heliosphere existed. A few scientists thought the Sun sent out particles and not just light and heat, but most believed its gravity was too great for anything to escape from it. So when Mariner 2 detected a blanket of particles (the solar wind) streaming from the Sun and filling space in all directions, understanding of the solar system deepened. The solar wind is a plasma, a super-hot, electrically charged gas, that erupts from the Sun in high-velocity gusts, depending on activity at the Sun's surface.

No one knows exactly how far the wind blows, but estimates have suggested over 100 astronomical units (AU), or 100 times the distance from the Earth to the Sun. This means the heliosphere goes far beyond the remotest planet, Pluto, which is at a mere 40 AU.

The heliosphere also marks out the extent of the Sun's magnetic field. Near the surface of the Sun, the magnetic field is so strong that plasma blasts erupt from the boiling surface only to be dragged back. But there are gaps in the magnetic field through which the wind escapes. When it does so, the electrical charges of the plasma particles carry the Sun's magnetic field with them. The magnetic field can then extend as far as the wind itself. It is weaker the farther it goes, but at 60 AU the heliospheric current sheet—the boundary that separates the northern and southern halves of the Sun's magnetic field—is still detectable by spacecraft.

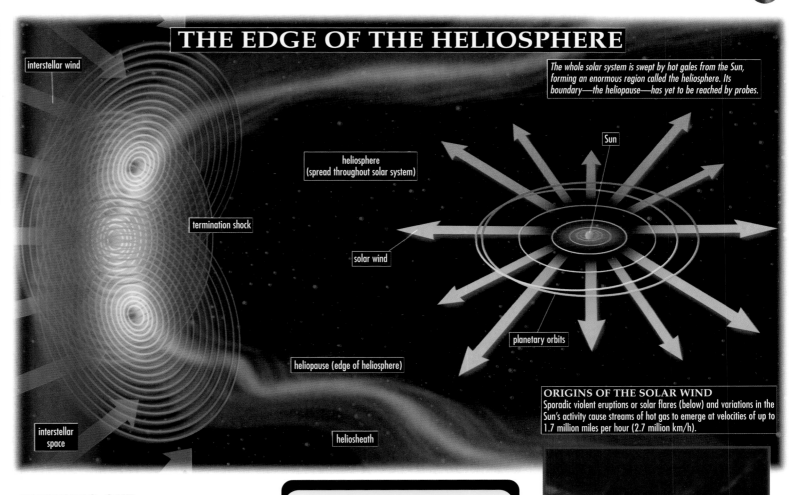

THE EDGE OF THE HELIOSPHERE

interstellar wind

The whole solar system is swept by hot gales from the Sun, forming an enormous region called the heliosphere. Its boundary—the heliopause—has yet to be reached by probes.

Sun

heliosphere (spread throughout solar system)

termination shock

solar wind

planetary orbits

heliopause (edge of heliosphere)

interstellar space

heliosheath

ORIGINS OF THE SOLAR WIND
Sporadic violent eruptions or solar flares (below) and variations in the Sun's activity cause streams of hot gas to emerge at velocities of up to 1.7 million miles per hour (2.7 million km/h).

THINNING OUT

Data from the Voyager and Pioneer probes indicates that there are other changes occurring in the farthest reaches of the heliosphere. Freed from the immense solar gravity, the solar wind races even faster, and the speed variations between streams are reduced as the gas combines with faster or slower moving particles. The solar wind also thins, becoming as thin as the best vacuum on Earth. The two Voyagers may tell us where the wind becomes as thin as the interstellar matter, marking the end of the heliosphere.

THEORIES PROVED

Comet tails always point away from the Sun. U.S. scientist Eugene Parker linked this fact to the ideas of Sidney Chapman, who showed that gas from the Sun's corona should extend beyond Earth. Parker related the extended corona to the solar particles thought to cause aurorae and comet tails, calling this phenomenon the solar wind. Four years later, in 1962, Mariner 2 (above) confirmed the solar wind existed.

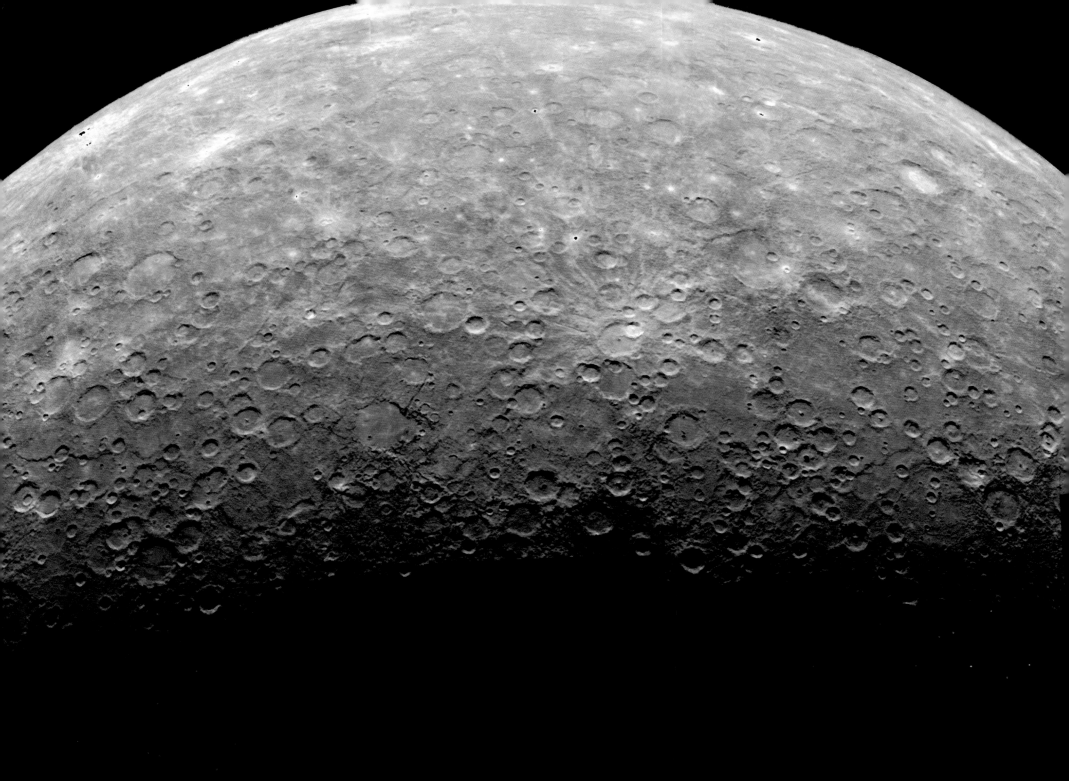

MERCURY

Mercury is the closest planet to the Sun, completing an orbit once in every 88 days. Mercury is named after the winged Roman messenger of the gods, due to its rapid movement in Earth's skies, where it is only ever briefly seen before sunrise or after sunset. Because it is always appears so close to the Sun, telescope images of the planet are frustratingly blurry. The details of Mercury's surface remained a mystery until the arrival of the Space Age. In 1974, NASA's Mariner 10 space probe made a series of flybys in which it photographed half of Mercury's surface, revealing an airless world with a strong initial resemblance to our Moon. Mercury has had a very different history from our satellite, however, as revealed by the great geological faults that crisscross its surface. One side of the planet is also dominated by a huge impact scar called the Caloris Basin.

Mariner 10's first image of Mercury, acquired on March 24, 1974, when the spacecraft was 3,340,000 miles (5,380,000 km) from the surface of the planet. Temperatures on the sunward side of Mercury (top) can reach 800°F (420°C), while in the perpetual shadow of its craters the temperature can be as low as −300°F (−180°C)

MERCURY

Barren Mercury, the first planet out from the Sun, is a small, rocky world that is scarred by impact craters and scorched by solar radiation almost seven times fiercer than that on Earth. With no substantial atmosphere to counteract the Sun's rays, the surface can reach a searing 800°F (420°C) during the day and plunge to –290°F (–140°C) at night. But the near-vacuum on the surface of Mercury has also helped to preserve the planet's contours, and the imprints left there by countless meteoroid scars have provided valuable information about the early days of the solar system.

WHAT IF...

...YOU WORKED ON MERCURY?

Not many people volunteer for a third tour on Mercury. But robotic mining is your specialty, so it makes sense to go back now that the mines are up and running at last, three years after your last visit in 2124. The mining company promised that the crew quarters are less squalid these days—but they said that last time. Right now, though, your main worry is surviving your first trip on a solar lightsail spaceship.

"The crazy thing about visiting Mercury," someone once told you, "is that you never see the Sun—except on a telescopic viewing screen. If you look at it directly, it'll probably be the last thing you ever see." You gaze around you: As on your last trip, the ship's observation windows have been shuttered since you passed Venus.

As the ship makes its final approach, even the crew seem edgy as they turn the Mercury ferry's lightsail to face the Sun. The intense pressure of the sunlight near Mercury pushes on the sail to brake the ship and counteract the Sun's enormous gravitational pull—though no one seems quite sure if it will work. Earlier ferries were rocket-powered and burned most of their fuel fighting the Sun's gravity. Somehow,

though, they were more reassuring.

A shuttle delivers you to the mining station that hides in the permanent shade of the crater Chao Meng-Fu, near the south pole. On this most sunburned of planets, Mercurians rely on simulated sunlight to keep their skin healthy. But the dependable cold of the crater's shadow is actually safer than a space station or Moon base, and heat collectors on the scorched terrain beyond the crater provide as much power as the station can handle.

Only the retractable instruments of the Chao Meng-Fu observatory dare to poke their heads over the crater's walls. It was the massive magnetic storms, caused by the 2076 solar activity peak, that prompted world governments to pour money into putting early-warning observatories as close to the Sun as possible. Today, Chao Meng-Fu Station is one of three observatories around Mercury's south pole that ensure constant monitoring of solar activity.

The mining consortium moved in to exploit the new infrastructure. The deal is that the company maintains the observatories while its robot prospectors search for mineral extraction sites. The MercuriTours Travel Corporation has been less successful, though: Sightseeing in a space suit never really caught on.

MERCURY FACTS

MERCURY		EARTH
3,031 MILES (4,878 KM)	DIAMETER	7,973 MILES (12,831 KM)
0°	AXIS TILT	23°27'
87.97 EARTH DAYS	LENGTH OF YEAR	365 DAYS
58.64 EARTH DAYS	ROTATION PERIOD	23.93 HR
36 MILLION MILES (58 MILLION KM)	AVERAGE DISTANCE FROM SUN	93.5 MILLION MILES (150 MILLION KM)
+806°F TO −292°F (+430°C TO −145°C)	SURFACE TEMPERATURE	AVERAGE 59°F (15°C)
0.38 G	SURFACE GRAVITY	1G
POTASSIUM, SODIUM, OXYGEN, ARGON, HELIUM	ATMOSPHERE	NITROGEN (78%), OXYGEN (21%), ARGON (1%)
NEGLIGIBLE	ATMOSPHERIC PRESSURE	1,000 MILLIBARS
LARGE IRON CORE, MANTLE OF SILICATE MINERALS	PROBABLE COMPOSITION	MAINLY IRON, NICKEL AND SILICATES

SUNBURN AND SCARS

At first sight, Mercury has more in common with the Moon than with any of the major planets. Both share jagged, impact-scarred landscapes, thanks to the absence of wind and water that once softened the contours of the Earth and Mars. Mercury is also relatively close to the Moon in size, a little under 1½ times larger. This is significantly smaller than both Jupiter's moon Ganymede and Saturn's moon Titan.

Yet Mercury is much denser than the Moon, and apart from the Earth itself, it is the densest body in the solar system. Scientists believe this is explained by a massive body of iron at the planet's core. In fact, the Earth may only be denser than Mercury because of its superior mass, which increases the strength of its gravitational field and pulls it together more tightly. It is possible that Mercury's iron core accounts for most of the planet's interior.

Pressures and temperatures at Mercury's core are likely to be so high that at least some of the iron remains in a permanently liquid state. Further evidence for this comes from Mercury's magnetic field, which is much stronger than those of fellow rocky worlds, Venus and Mars, but only about 1% as strong as the Earth's. Scientists believe that the magnetic fields of all the rocky inner planets are generated by ripples in the molten metal at their cores.

SUN DANCE

A "DAY" ON MERCURY IS UNLIKE ANY OTHER IN THE SOLAR SYSTEM: IT LASTS TWO YEARS, AND THE SUN REGULARLY APPEARS TO MOVE BACKWARD IN THE SKY. THE REASON IS MERCURY'S UNIQUE COMBINATION OF SLOW ROTATION ABOUT ITS AXIS AND RAPID PERIOD OF ORBIT AROUND THE SUN. TOGETHER, THESE EXTEND THE MERCURIAN "SOLAR DAY"—FROM NOON TO NOON—TO 176 EARTH DAYS, OR TWO MERCURIAN YEARS. FROM CERTAIN SPOTS ON THE PLANET, AN OBSERVER WOULD SEE THE SUN RISE, THEN DIP BACK BELOW THE HORIZON BEFORE FINALLY RISING FOR GOOD. SIMILARLY, OTHER POINTS ON THE PLANET WOULD WITNESS A "DOUBLE SUNSET."

MYSTERIES OF THE PLAINS

Mercury's surface is not totally peppered with craters: There are also smoother plains, like the so-called seas of the Moon. No one is sure how these plains were formed. One theory is that after cataclysmic impacts like the one that created the mighty Caloris Basin, lava from the youthful planet's molten interior gushed out over the surface. Other plains may consist of matter that was sprayed out after lesser meteoroid impacts.

With no erosion by wind and water and no tectonic plates to shift and crumple, Mercury has never experienced the natural forces that constantly reshape the surface of the Earth. As a result, the pattern of impacts across the surface provides important clues to the evolution of the solar system. When Mercury's plains were formed, they provided a fresh surface for impacts. But the largest craters on Mercury are found only in the older, more rugged regions that pre-date the plains. This suggests that massive impacts by enormous chunks of interplanetary debris were a regular occurrence during the early days of the solar system, but that by the time Mercury's plains were formed, they had petered out.

Mercury's most unusual features are the huge ridges or scarps that snake their way across the planet's surface—sometimes for over 100 miles—and rise to nearly 10,000 feet (3,000 m) above the surrounding landscape. They are probably wrinkles, a sign that Mercury once shrank by over 3,000 feet (900 m)—0.1% of the planet's surface area—as the core cooled.

Because it is so close to the Sun, Mercury is both difficult and dangerous to observe from Earth. Little was known about the planet until the Mariner 10 space probe paid a visit in the mid-1970s; and even then, Mercury's slow rotation allowed Mariner 10 to map only 40% of the surface as it flew by. The Hubble Space Telescope, too, is dazzled by the light of the Sun, so we shall have to wait until the Mercury-probing missions planned for the first decade of the 21st century to learn more about the solar system's first planet.

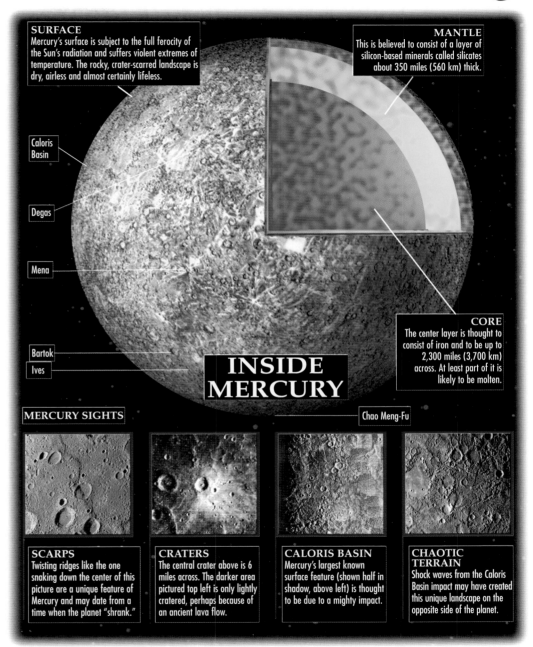

SURFACE
Mercury's surface is subject to the full ferocity of the Sun's radiation and suffers violent extremes of temperature. The rocky, crater-scarred landscape is dry, airless and almost certainly lifeless.

MANTLE
This is believed to consist of a layer of silicon-based minerals called silicates about 350 miles (560 km) thick.

Caloris Basin

Degas

Mena

Bartok

Ives

CORE
The center layer is thought to consist of iron and to be up to 2,300 miles (3,700 km) across. At least part of it is likely to be molten.

INSIDE MERCURY

MERCURY SIGHTS

Chao Meng-Fu

SCARPS
Twisting ridges like the one snaking down the center of this picture are a unique feature of Mercury and may date from a time when the planet "shrank."

CRATERS
The central crater above is 6 miles across. The darker area pictured top left is only lightly cratered, perhaps because of an ancient lava flow.

CALORIS BASIN
Mercury's largest known surface feature (shown half in shadow, above left) is thought to be due to a mighty impact.

CHAOTIC TERRAIN
Shock waves from the Caloris Basin impact may have created this unique landscape on the opposite side of the planet.

CHANGING VIEWS OF MERCURY

Although Mercury lies tantalizingly close to the Earth, the planet is hard to see through the blinding light of its nearest neighbor, the Sun. Details of Mercury's surface can only be resolved in narrow orbital windows—a restriction that has led to several misconceptions in the past. But as improving technology slices through the Sun's glare, our perceptions of Mercury have sharpened. Modern radio telescopes have even discerned possible polar ice caps—but the planet's interior remains as impenetrable as ever.

WHAT IF...

...OUR VIEWS OF MERCURY CHANGED AGAIN?

The three flybys of the Mariner 10 spacecraft in 1974–5 provided our best look yet at Mercury, and returned one of the planet's biggest surprises. Mariner discovered a weak magnetic field around Mercury, indicating that the planet's core may be at least partially molten like the Earth's. As Mercury spins slowly on its axis, the motion of this fluid core probably creates a dynamo effect, generating electrical currents deep inside the planet—and a magnetic field in space.

The results were unexpected because Mercury is so tiny—less than one-and-a-half times the size of our own Moon. Most theories of planetary formation hold that such a small world should lose heat fast, so that its core solidifies soon after the planet forms.

To resolve this apparent dilemma, future missions to Mercury will carry more sensitive instruments than those aboard Mariner 10, allowing them to produce detailed maps of the planet's magnetic field and its gravity. This data should yield a more detailed view of Mercury's interior.

But even an orbiting spacecraft may not fully resolve the mystery. The best way to study Mercury's interior would be to drop seismographs on the planet's surface to listen for "Mercury-quakes." Shock waves behave differently as they travel through different types of material below the surface, so a network of listening stations around Mercury could help scientists to peer into the planet's interior.

The sensitivity of such a network might even reveal whether volcanoes—extinct or active—dot the Mercurian surface. Mariner 10 found no evidence of recent volcanic activity, but it photographed less than half of the planet.

A recent theory suggests that Mercury formed a thick, solid crust early in its history. The crust trapped most of young planet's heat, allowing at least part of the core to remain molten. Calculations show that the layer between the crust and the core, called the mantle, might be partially molten, too. If so, then some of this thick, hot rock might have bubbled to the surface in the fairly recent past. And if the crust has not grown too thick, molten rock might still occasionally force its way to the surface as volcanoes today.

Mission controllers of the future find out too late that their suspicions were right—Mercury does have the occasional active volcano, and the last thing that this sophisticated Mercury lander reports is a stream of hot lava.

MAPPING OUT MERCURY

1882–4	GIOVANNI SCHIAPARELLI MAKES OBSERVATIONS THAT SUGGEST THE SAME HEMISPHERE OF MERCURY ALWAYS FACES THE SUN
1934	E.M. ANTONIADI DRAWS A NEW MAP OF MERCURY
1949	RALPH B. BALDWIN SUGGESTS THAT THE SURFACE OF MERCURY IS HEAVILY CRATERED, LIKE THE MOON'S
1962	RADIO TELESCOPES MEASURE SURFACE TEMPERATURE ON MERCURY'S NIGHT SIDE
1965	RADAR TECHNIQUES ARE USED TO DISCOVER MERCURY'S TRUE ROTATION PERIOD
1974–5	MARINER 10 DISCOVERS MAGNETIC FIELD, PHOTOGRAPHS SURFACE AND MEASURES DENSITY
1991	RADIO TELESCOPE DISCOVERS POSSIBLE EVIDENCE OF THIN ICE IN CRATERS AT MERCURY'S POLES

SMALL WONDER

Mercury makes its first appearance in professional stargazing as a fleet-footed messenger in ancient Egyptian, Greek and Mayan myth. The planet's proximity to the Sun whizzes it around a complete orbit in just 88 days, and to ancient astronomers, the fast-track "star" seemed to have important news to deliver.

With the benefit of telescopes, 19th-century scientists started to debate the realities of the innermost planet. One source of dispute was whether tiny Mercury had the gravity to hang onto an atmosphere. From the blurring of light to the corners of its disk, German astronomer J.H. Schroter concluded that the planet did indeed possess a fairly dense atmosphere. More evidence came in the form of a dark "halo" that Mercury projected on the Sun on its transit of November 9, 1802. But on the same day British astronomer William Herschel saw the planet's limb "cut the solar clouds with perfect sharpness"—an observation consistent with an airless world. The planet was also found to reflect only 13% of the sunlight it received: The clouds typical of a dense atmosphere reflect much more.

We know now that Mercury's atmosphere is all but nonexistent. By the late 19th century, it was clear that better telescopes made for less conspicuous haloes, and Mercury's supposed atmospheric distortion was chalked up to optical

illusion. But astronomers still did not accept a totally airless Mercury. The occasional disappearance of spots seen on the planet by Italian scientist Giovanni Schiaparelli was attributed to atmospheric effects.

FALSE EVIDENCE

Schiaparelli noticed that Mercury's faint markings always appeared in the same place—meaning that the planet must take as long

to turn on its axis as it did to orbit the Sun. This was a revelation. It meant that one of Mercury's hemispheres was in permanent darkness, with the other bathed in sunlight. Astronomers did not see either side as a suitable place for life. On the hot side, they thought, any liquid water would have boiled off long ago, while the dark side would be covered by ice.

The belief that Mercury's day was equal to its year was accepted until 1962, when microwave studies showed that the "dark" side was actually quite hot. It turned out that astronomers had been misled by a quirk of Mercury's rotation. The planet does not keep the same face pointed to the Sun. Instead, it spins three times on its axis for every two revolutions of the Sun. Mercury's orbital motion is also coupled to that of the Earth, and by coincidence, the same hemisphere always faces our way when the planet is easiest to observe.

Many of our views of Mercury have yet to form, let alone change. The Mariner 10 mission of 1974 may have mapped out Mercury's surface detail, but our views of its interior are still unclear.

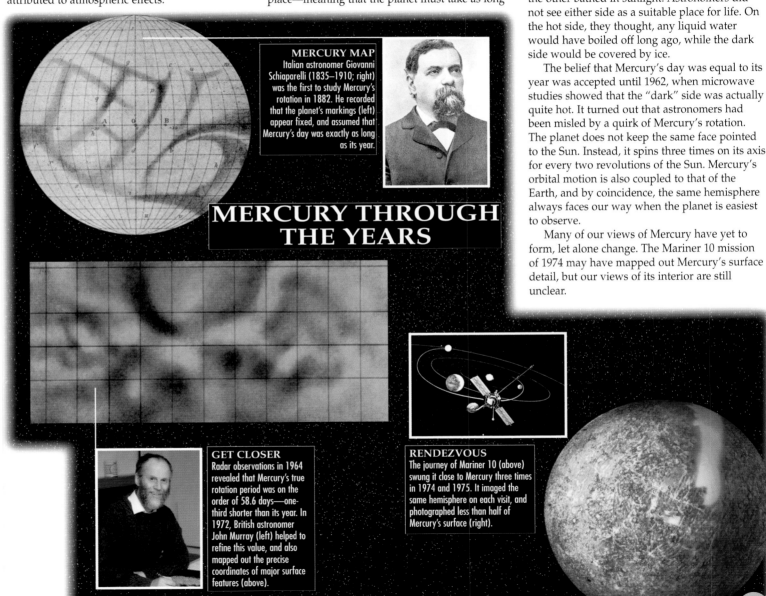

ICY REALITY

S<small>CIENTISTS USED TO THINK THAT NO ICE WOULD EVER BE FOUND ON THE PLANET CLOSEST TO THE</small> S<small>UN.</small> B<small>UT IN</small> 1991, <small>RADAR</small> <small>SIGNALS BOUNCED OFF THE PLANET'S SURFACE REVEALED AREAS OF HIGH REFLECTIVITY THAT MAY BE DEPOSITS OF ICE AT</small> M<small>ERCURY'S POLES (CIRCLED, ABOVE).</small> M<small>ERCURY'S ROTATION AXIS IS NEARLY PERPENDICULAR TO ITS ORBITAL PLANE, WHICH MEANS THAT THE FLOORS OF SOME POLAR CRATERS ARE NEVER EXPOSED TO SUNLIGHT.</small> T<small>HE ICES MIGHT ARRIVE IN THE COMETS THAT SLAM INTO</small> M<small>ERCURY.</small>

MERCURY THROUGH THE YEARS

MERCURY MAP
Italian astronomer Giovanni Schiaparelli (1835–1910; right) was the first to study Mercury's rotation in 1882. He recorded that the planet's markings (left) appear fixed, and assumed that Mercury's day was exactly as long as its year.

GET CLOSER
Radar observations in 1964 revealed that Mercury's true rotation period was on the order of 58.6 days—one-third shorter than its year. In 1972, British astronomer John Murray (left) helped to refine this value, and also mapped out the precise coordinates of major surface features (above).

RENDEZVOUS
The journey of Mariner 10 (above) swung it close to Mercury three times in 1974 and 1975. It imaged the same hemisphere on each visit, and photographed less than half of Mercury's surface (right).

SURFACE OF MERCURY

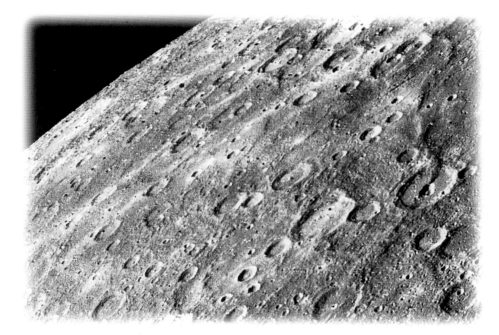

L ittle changes on the surface of Mercury: Some of its craters are over 4 billion years old. Yet comparisons with the Moon— Mercury's cosmic lookalike—show that the planet has been resurfaced more than once. Our first good look at Mercury came in 1974, when Mariner 10 imaged 45% of its surface. This data provided evidence of large scale impacts and volcanism. We are getting to know the rest of Mercury through radar. And one day, the innermost planet could attract a crewed mission.

WHAT IF...

...WE HAD A MERCURY BASE?

B y the early part of the 22nd century, crewed missions have compiled detailed profiles of our closest planetary neighbors, and there is little left to discover about Mars, Jupiter and Venus. Now it is time to scrutinize the more inhospitable corners of the solar system, and the cratered wastes of Mercury are as good a place as any to start.

This rocky little world makes the best solar observatory of all. With their instruments positioned at an average of just 36 million miles (58 million km) from the Sun, the newly landed crew of the HEAT (Helio Extremes Analysis Team) module can study the workings of our star in close-up. But such a good view can get hot. The power of the full Sun on Mercury would melt lead, and the team have chosen their living quarters wisely.

The base is set up in a crater near the south pole, where the sun rises and sets above the crater rim with Mercury's slow spin. Every day and every night lasts for three Earth months—but the position of the HEAT crater means that this part of Mercury is always in the dark. There is no shortage of solar power, and the team heats their frozen crater with solar panels, which they move to keep ahead of the long Mercurian nights.

HEAT researchers stick to the night side to make studies of Mercury's surface, but they must venture into the heat to look at the Sun. Although the Sun scorches the rock, there is no atmosphere to hold heat. As long as they reflect the solar rays, the scientists are protected. They keep their cool with umbrella-like sunshades and insulated boots, made of high-shine aluminum foil.

One day, while the crew is out on the sunny side, catastrophe strikes: A solar flare develops. Within minutes, it blows outward in a 2 billion-megaton blast. The scorching flare races toward Mercury at a temperature of 18 million°F (10 million°C). When a similar flare occurred on March 6, 1989, the radiation surge put out the lights for 6 million people in Quebec, Canada. But the HEAT crew is 57 million miles (92 million km) closer to the Sun. The radiation arrives at the speed of light, followed by a pulse of deadly cosmic rays made up of protons and electrons. It would prove fatal to anyone shielded by a sunshade alone.

But the emergency is not unexpected. Large flares occur every year, and the team knows what to do. Even as the pulse arrives, the staff are in their Mercury Excursion Modules. They speed beyond the pole's extended shadows and park in the safety of darkness—protected by the planet itself from the incoming blast of radiation. Then they make the return journey, restore power and get back to work.

MERCURY'S MAIN FEATURES

MAXIMUM SURFACE TEMPERATURE	806°F (430°C)
MINIMUM SURFACE TEMPERATURE	−292°F (−145°C)
LENGTH OF YEAR	87.97 EARTH DAYS
AGE OF INTERCRATER PLAINS	4.2–5.0 BILLION YEARS
AGE OF LOWLAND PLAINS	3.8 BILLION YEARS
DIAMETER OF LARGEST IMPACT BASIN	830 MILES (1,300 KM)
SIZE OF LARGEST IMPACTOR	100 MILES (160 KM)
LENGTH OF LARGEST SCARP	300 MILES (480 KM)
HEIGHT OF LARGEST SCARP	1 MILE (1.6 KM)

THE SUN'S GLARE MEANS THAT ALMOST NO SURFACE DETAIL ON MERCURY CAN BE SEEN FROM EARTH. BUT THAT DID NOT STOP ITALIAN ASTRONOMER GIOVANNI SCHIAPARELLI (RIGHT) FROM TRYING. BETWEEN 1881 AND 1889, HE MADE SKETCHES THAT SUGGESTED THAT MERCURY ALWAYS KEEPS THE SAME FACE TOWARD THE SUN. THIS WAS DISPROVED IN 1962, WHEN MICROWAVE STUDIES SHOWED THAT THE "DARK SIDE" OF MERCURY WAS HOT, AND SO MUST SOMETIMES SPIN INTO THE SUN.

LUNAR LOOKS

At first glance, the surface of Mercury looks very similar to that of the Moon. Meteors and asteroids have left a lasting impression on both bodies: Neither one possesses the weather or the geological activity that erode impact craters on the Earth. Many of Mercury's scars date back more than 4 billion years, when rocks left over from planetary formation—known as planetesimals—crashed into the newly formed planets.

These ancient collisions threw up mountains on Mercury that are slowly pulled down. A steady rain of meteorites and dust grains wears the peaks away, while they are assaulted by a 1,170°F (630°C) variation in temperature as the planet's surface spins in and out of the Sun. At the same time, gravity pulls the dust stirred up by meteorite strikes into lowlands and craters.

But Mercury's extra gravity—over twice as strong as the Moon's—makes for different crater shapes. Impacts on Mercury and the Moon both eject blankets of molten rock. The ejecta does not travel as far on Mercury— gravity slams it down closer to the crater rim. This causes more damage to the new crater than occurs on the Moon. Scientists can work

out the age of Mercury's craters from the rate of their decay.

The most interesting craters on Mercury are the missing craters. Although the planet must

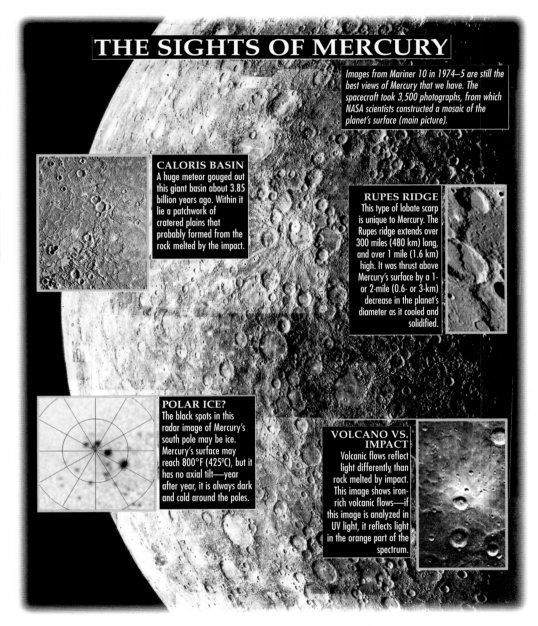

THE SIGHTS OF MERCURY

Images from Mariner 10 in 1974–5 are still the best views of Mercury that we have. The spacecraft took 3,500 photographs, from which NASA scientists constructed a mosaic of the planet's surface (main picture).

CALORIS BASIN
A huge meteor gouged out this giant basin about 3.85 billion years ago. Within it lie a patchwork of cratered plains that probably formed from the rock melted by the impact.

RUPES RIDGE
This type of lobate scarp is unique to Mercury. The Rupes ridge extends over 300 miles (480 km) long, and over 1 mile (1.6 km) high. It was thrust above Mercury's surface by a 1- or 2-mile (0.6- or 3-km) decrease in the planet's diameter as it cooled and solidified.

POLAR ICE?
The black spots in this radar image of Mercury's south pole may be ice. Mercury's surface may reach 800°F (425°C), but it has no axial tilt—year after year, it is always dark and cold around the poles.

VOLCANO VS. IMPACT
Volcanic flows reflect light differently than rock melted by impact. This image shows iron-rich volcanic flows—if this image is analyzed in UV light, it reflects light in the orange part of the spectrum.

have taken a battering much like the Moon's, Mercury bears far fewer impact scars, and is especially short of craters of less than 30 miles (50 km) across.

BLOTTED OUT

Many of Mercury's missing craters are buried beneath its rolling plains. Intercrater plains are mixed into the planet's highlands. When this terrain formed 4 billion years ago it blotted out many craters: Small impact sites were covered more fully due to their shallow depths.

Lowland plains stretch across the floors of huge impact basins. These plains carry fewer craters than the intercrater ones, so they probably formed more recently.

The plains solve the missing crater mystery, but where did these landscapes come from? Some scientists believe that they were formed by collisions. A large impact can melt huge volumes of rock. Giant sheets of impact melt may have oozed across Mercury to solidify as the lowland plains. Others suggest that the plains came from within: They could be the result of enormous volcanic eruptions. But so far, no volcanic features have been found on Mercury—so the plains question remains.

That we know even this much is largely due to images from Mariner 10: The spacecraft delivered our first good look at Mercury in 1974. Now, reanalysis of Mariner data may at last prove the origin of the planet's plains. Minerals in Mercury's surface reflect light at different wavelengths. In some areas researchers have found the signature of iron—a sure sign of a volcanic flow—while other plains can now be defined as impact melt.

Mariner has taught us a great deal, but it only scanned 45% of Mercury's surface. In 1991, radar images of the unseen areas made an amazing discovery. They picked up two bright spots at Mercury's north and south pole. These spots appear similar to the south polar ice cap on Mars. No one expected to find ice on the closest planet to the Sun. But Mercury's orbit means that some polar craters may never see the light—and the minus 290°F (145°C) low experienced by the planet's night side is cold enough to keep ice permanently frozen. It seems we still have a few secrets left to map out on the surface of Mercury.

GEOLOGY OF MERCURY

Among the four inner planets, Mercury is the least explored. A visit from just one flyby spacecraft has revealed only half of Mercury's surface and the other hemisphere still remains a total mystery. At first glance, the geology of Mercury appears distinctly Moon-like. Yet the resemblance is only skin deep: Mercury's interior is unlike that of any other planet. To discover Mercury's geological secrets will take a hardened spacecraft able to orbit above its baked and barren surface—a difficult and perilous task.

WHAT IF...

...WE RETURN TO MERCURY?

At present, Mercury is a mysterious little world. Many of its features are poorly understood, and only half of it has been photographed. But all this is likely to change in 2011, when NASA's Messenger probe (MErcury Surface, Space ENvironment, GEochemistry, and Ranging mission) becomes the first manmade object to enter orbit around Mercury, and begins the first comprehensive survey of the planet. From its polar orbit, circling the planet every 12 hours for at least one Mercury year, the probe will send back pictures to revolutionize our view of its host.

Such a mission is essential for understanding the geologic processes which have created the closest planet to the Sun. The 55% of Mercury's surface not imaged by Mariner 10 remains the only planetary landscape not yet seen by humans. Planetary scientists know more about moons that orbit far-flung worlds in the depths of interplanetary space than they do about Mercury's geology. A successful Mercury orbiter would produce a wealth of new data, including the first complete visual image map of the surface, the surface rock composition, and a topographic map showing the roughness of Mercury's landscape.

Scientists could also measure Mercury's magnetic field and search for ice at its poles.

Yet the mission will be a tough one. Mercury offers one of the harshest environments in the solar system. The sensitive instruments onboard the orbiter must be able to withstand extreme temperatures—scorching on the sunlit side of the planet and then plummeting into temperatures hundreds of degrees below zero in the planet's shadow. Such immense thermal stresses are enough to fracture the hardest rocks on the surface of Mercury and would soon destroy an unprotected spacecraft. To combat the fluctuating temperatures, a series of sun-shields and radiators will have to maintain a constant temperature onboard, whether the craft is in sunshine or shadow. A further danger comes from lethal doses of radiation emitted by the Sun. Mercury's close proximity to the Sun means that the radiation would soon fry sensitive electronic circuitry unless it was protected by insulating coatings.

Launched in 2005, Messenger is already on its way, taking a circuitous route that swings it past Earth and Venus to pick up speed and allow it to catch up with fast-moving Mercury. If all goes well in 2011, Mercury should finally reveal some more of its secrets.

GEOLOGICAL FACTS

	MERCURY	EARTH
CORE COMPOSITION	Mostly iron (some nickel)	Iron, nickel
CRUST COMPOSITION	Believed to be silicate rocks	Silicate rocks, mainly basalts
AVERAGE DENSITY	5.44 G/CC (3.144 OZ/CU IN)	5.5 G/CC (3.146 OZ/CU IN)
AVERAGE SURFACE AGE	Very old, ±4 billion years	Young, most <200 million years
ACTIVE VOLCANOES	No, ceased ±4 billion years ago	Yes
EARTHQUAKE ACTIVITY	Unknown	Yes, relatively high
PLATE TECTONICS	No	Yes, eight major plates
SURFACE WEATHERING	Rock fracture by extremes of temperature	Wind, water and ice

THE IRON WORLD

Until the Mariner 10 probe swung by in 1974–5, our geological understanding of Mercury was sketchy. Like Earth, Mars and Venus, Mercury was known to be a terrestrial-type planet, made of silica-rich rocks. But other geological questions—such as its internal structure, or whether it had ever been volcanically active— were still unanswered.

Images returned from Mariner 10 revealed about 45% of the Mercurian surface and answered some perplexing questions. And Mariner provided more than images. On any interplanetary mission, the precise gravitational pull on a spacecraft as it passes a planet or moon can tell a great deal about that body's interior structure. From the motion of Mariner, the average density of Mercury was calculated to be similar to that of Earth and Venus and greater than that of Mars. Such a small, dense world implies a large core made of heavy elements such as iron and nickel. In fact, Mercury's iron core may make up about 70% of the planet—the largest core relative to its size of any planet in the solar system.

Like Earth, Mercury probably underwent a process of "differentiation" early in its history when it was still molten. Heavy minerals sank toward the interior of the planet and lighter materials floated upward. It is similar to what happens when a jar of oil and water is shaken and left to stand—the lighter oil rises and floats on the surface of the denser water. From Mercury's dense iron core, lighter silicate materials have migrated upward to form a thick rocky silicate crust.

SIGNS OF INNER ACTIVITY

The other terrestrial planets show abundant evidence of volcanic activity, but we need to look a little harder on Mercury. Some of its areas, such as the intercrater plains and smooth plains, are linked to volcanic activity. The craters and ridges found there resemble those in the maria of the Moon, which are believed to have a volcanic origin. Yet few volcanic domes, calderas

(collapsed volcano basins) or fresh-looking lava flows are visible on Mercury. The planet may have been volcanically active soon after its formation, but the activity soon died down. Any sign of it is now hidden beneath the myriad impact craters that scar its surface.

Unlike volcanism, evidence for tectonic activity is everywhere. Steep, curving fractures— called arcuate or lobate scarps, depending on their shape—crisscross the surface for hundreds of miles. They are thought to have formed either as the planet cooled and contracted, or by the effects of the Sun's gravity as it slowed Mercury's spin from a few hours to its present 58.6 days. The tidal forces involved could have cracked the planet's ridged crust. But any answer to this or other Mercurian mysteries will probably have to await the arrival of a 21st-century space probe.

DISCOVERY SCARP

UNIQUE TO THE PLANET MERCURY ARE THE "LOBATE SCARPS"—A WEB OF SURFACE FRACTURES DISCOVERED BY THE MARINER 10 SPACECRAFT. ONE OF THE LARGEST, DISCOVERY SCARP, IS A HUGE FAULT WHERE LAND HAS BEEN THRUST UPWARD AND FORWARD. THE SNAKING WALL IS 310 MILES (500 KM) LONG AND STANDS UP TO 1.5 MILES (2.5 KM) ABOVE THE MERCURIAN SURFACE. BUT WHAT CAUSED THE SCARPS? AFTER MERCURY WAS FORMED, ITS LARGE IRON CORE AND SURROUNDING MANTLE COOLED AND CONTRACTED, SHRINKING THE PLANET BY A SMALL PERCENTAGE OF ITS ORIGINAL SIZE. AS THE PLANET SHRANK, THE SURFACE CRUST WAS COMPRESSED AND WRINKLED AROUND IT INTO AN EXTRAORDINARY NETWORK OF SCARPS.

Discovery Scarp

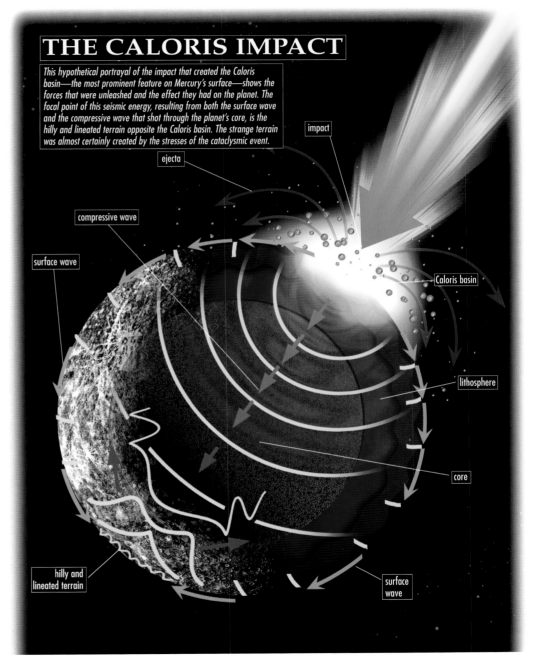

THE CALORIS IMPACT

This hypothetical portrayal of the impact that created the Caloris basin—the most prominent feature on Mercury's surface—shows the forces that were unleashed and the effect they had on the planet. The focal point of this seismic energy, resulting from both the surface wave and the compressive wave that shot through the planet's core, is the hilly and lineated terrain opposite the Caloris basin. The strange terrain was almost certainly created by the stresses of the cataclysmic event.

impact

ejecta

compressive wave

surface wave

Caloris basin

lithosphere

core

hilly and lineated terrain

surface wave

MYSTERIOUS MERCURY

Mercury—closest planet to the Sun and slightly larger than the Moon—is as hard to study as a speck of dust on a searchlight. In the mid-1970s, the Mariner 10 probe removed some misconceptions. For instance, it was thought that a spacecraft landing on the planet would have a stark choice between the extreme heat of eternal day or the extreme cold of eternal night. Mariner revealed that such a lander would experience both, in turn. But other mysteries—such as Mercury's highly elliptical orbit—still remain.

WHAT IF...

...WE WERE TO FIND OUT MORE ABOUT MERCURY?

Another mission to Mercury would surely provide new insights into Mercury's greatest mysteries—the reasons for its eccentric orbit and rotation speed. The answers could come from new clues to the planet's origins. Astronomers accept that the solar system evolved from a cloud of gas flattened into a disc by centrifugal forces. This neatly explains why the planets lie roughly on the same plane, and why they all orbit in the same direction. But Mercury, like Pluto, is an odd-ball. Among the many Mercurian quirks that have mystified astronomers are: Why is Mercury's orbit much more elliptical than the orbits of other major planets? Why is its orbit set at a sharper angle to the plane of the solar system? Why does it spin three times on its axis while circling the Sun only twice? And why does it have a magnetic field?

If a new probe were to provide further data, it may lead to a more defined theory of Mercury's origins. The scenario could run something like this:

In the early days of the solar system, the planet nearest to the Sun condenses out of the primordial cloud of gas and debris left over from the Sun's formation. It is a small planet, comparable to Mars. Although the Sun scorches away any atmosphere, it may have first evolved similarly to the Red Planet. Like all the inner planets, Mercury suffers a terrible battering from stray space debris. But it falls victim to something much larger—a planetoid only slightly smaller than itself, which slices it open to the core. Half the mantle is ripped away and scatters to join the asteroids that are slowly being blotted up by other nascent planets. What remains is an irregular blob of gel-like consistency, blasted into a new and eccentric orbit, with a violent spin. As the planet reforms under the influence of its own gravity, the core settles beneath the softer mantle. But the Sun's powerful gravitational field forces the mantle to bulge with every turn, as the oceans do on earth. Tidal drag slows the planet's spin while its gel-like body sets, becoming gradually more solid. A continuing, but decreasing bombardment from asteroids keeps the planet plastic for eons. But eventually the conflicting forces—spin, drag and plasticity—reach a balance.

However it happened, Mercury found stability some 4,000 million years ago. Since then, the little planet has remained virtually unchanged since the time of its birth—a fossil awaiting the scientific attention that will unlock its remote past.

MERCURIAN RECORDS

c.1900 B.C.	BABYLONIANS RECORD SYNODIC PERIOD OF MERCURY
1845	FRENCH ASTRONOMER URBAIN LEVERRIER MEASURES ADVANCE OF MERCURY'S ORBITAL AXIS
1882	ITALIAN ASTRONOMER GIOVANNI SCHIAPARELLI SUGGESTS MERCURY'S DAY EQUALS ITS YEAR
1965	RADAR ECHOES SHOW MERCURY'S DAY IS 58.6 EARTH-DAYS
1973	MARINER 10 LAUNCHES TO MERCURY VIA VENUS ON NOVEMBER 3
1974	MARINER 10 PASSES AT 468 MILES (753 KM) ON MARCH 29
	MARINER 10 PASSES AT 29,800 MILES (48,000 KM) ON SEPTEMBER 21
1975	MARINER 10 PASSES AT 198 MILES (319 KM) ON MARCH 24

QUICK & QUIRKY

Swamped by the Sun's glare, Mercury was the object of myth and misunderstanding until three decades ago. In ancient times, almost all that was known about Mercury was that it was the Sun's closest companion and that it moved fast. Astrologers picked up on these traits. The Roman god who gave Mercury his name was a messenger of the gods and was often shown with winged heels. For alchemists, "mercury" was an alternative name for the strange liquid metal, quicksilver.

When telescopes revolutionized astronomy in the 17th century, astronomers saw that Mercury's orbit was the most eccentric of all the known planets. It is 28.5 million miles (45.9 million km) from the Sun at its closest—its perihelion—and 43.3 million (69.7 million km) at its farthest—its aphelion. That 50% difference in distance means that around 230% as much light and heat reaches Mercury at perihelion as arrives at aphelion.

In 1845, the French astronomer Urbain Leverrier discovered another Mercurian oddity. Mercury's orbit traces out a path like an egg spinning slowly in space, making one rotation every 3 million years. Part of the displacement is caused by the gravitational tug of the other planets. But Leverrier found that Mercury's orbit was rotating slightly faster than could be accounted for in this way. He suggested that Mercury was being influenced gravitationally by an unknown planet even closer to the Sun, which he named Vulcan. But Vulcan's existence has never been confirmed.

HOT POLES

MERCURY'S UNIQUE ORBIT AND ROTATION CAUSE IT TO HAVE FOUR EXTRA POLES, CALLED "HOT POLES," AT 90-DEGREE INTERVALS ALONG ITS EQUATOR. ONE SEARINGLY HOT POLE (RED AREA) ALWAYS FACES THE SUN WHEN MERCURY IS CLOSEST—AT PERIHELION.

MAGNET IN A TEAR-DROP

THE DETECTION OF A WEAK MAGNETIC FIELD ON MERCURY BAFFLES SCIENTISTS. CURRENT THEORY SUGGESTS THAT MAGNETIC FIELDS FOR ROCKY PLANETS WITH MOLTEN CENTERS ARE CREATED WHEN A SPINNING IRON CORE ACTS AS A DYNAMO. BUT MERCURY ROTATES VERY SLOWLY. HOW CAN ITS CORE MOVE FAST ENOUGH TO GENERATE MAGNETISM? WHATEVER THE CAUSE, ITS WEAK FIELD—1/100 AS STRONG AS THE EARTH'S—IS ENOUGH TO CREATE A MAGNETOSPHERE OF CHARGED PARTICLES FROM THE SUN, MOLDED INTO THE SAME TEAR-DROP SHAPE AS THE EARTH'S BY THE PRESSURE OF SOLAR RADIATION.

DECIPHERING THE CLUES

In the late 19th century, the Italian astronomer Giovanni Schiaparelli thought he saw vague markings on Mercury that led him to believe Mercury rotated in a time equal to its period of revolution around the Sun—88 days. In other words, it kept one face permanently turned Sun-ward, as the Moon keeps one side always facing Earth. This idea held sway until the 1960s, and inspired many science-fiction stories based on the supposed existence of two hellish half-worlds—one of fire, the other of ice.

But truth can be stranger than fiction. In 1965, radar beams bounced off Mercury established the real rotation period: 58.6 days. This is exactly two-thirds of 88, the orbital period. Mercury rotates on its axis three times while going around the Sun twice. The planet's day is 176 Earth-days—twice the length of its 88-Earth-day year.

As yet, no one knows why. But the puzzle of Mercury's advancing orbit has been solved. In 1915, Albert Einstein's General Theory of Relativity explained gravity as the result of bodies like planets and stars distorting space

THEORIES ON MERCURY'S FORMATION

CONDENSATION
Like all the other planets in our solar system, Mercury coalesced from material left over from the formation of the Sun. The heavier materials sank to the new planet's center to form a dense core, and the lighter materials cooled and hardened to form the crust. But Mercury seems to be out of balance: Its core is far too large for a planet of its size.

COLLISION
Mercury may once have been a larger planet, with a better balance between its core and crustal materials. But an early collision with another body could have knocked away much of the lighter crust. A similar cataclysm may have helped to shape the Earth. In our case, debris from the impact eventually formed the Moon.

RICOCHET
Another version of the collision theory suggests that Mercury could have been torn from the body of another planet—either the Earth, Mars or Venus—in the very early years of the solar system. The impact sent the "stolen" material into Mercury's current orbit, where it slowly formed into the planet we can see today.

and time. He predicted that the axis of a planet's elliptical orbit would slowly revolve around its star. The closer to the star, the greater the distortion and the more extreme the effect.

In the case of Mercury, Einstein's calculations agreed almost exactly with the observed displacement. Mercury's orbit remains one of the key proofs of Relativity.

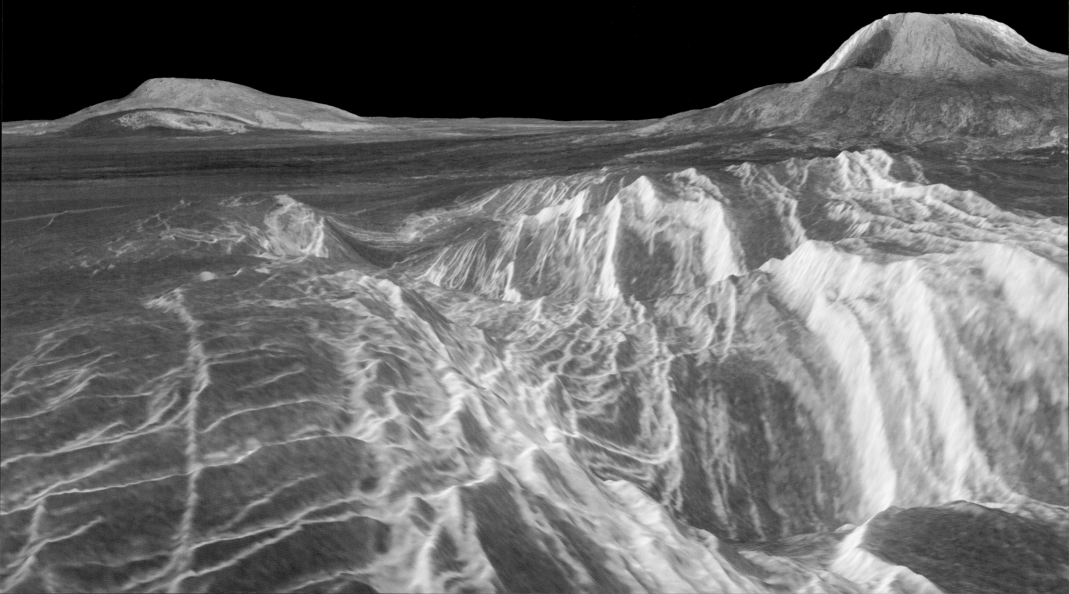

VENUS

Venus has been described as Earth's "deadly twin." With its orbit a little closer to the Sun than Earth, and its diameter and mass slightly smaller, astronomers long suspected that the brightest planet in Earth's skies would be a hospitable place, perhaps even a home to alien life. The arrival of the first space probes in the 1960s shattered this illusion, revealing instead a world permanently veiled in brilliant yellow-white clouds of sulfur dioxide, and with a surface temperature far hotter than even Mercury. Robot landers broke down during descent or after just a few minutes on the ground, but revealed that the atmosphere was dominated by carbon dioxide, with sulfuric acid raining from its clouds. Now that orbiting probes have used radar to look through the clouds and map the surface, we know that Venus is a very different world from Earth—one periodically racked by huge volcanic eruptions that apparently resurface the entire planet every few hundred million years.

This three-dimensional, computer-generated view of the surface of Venus shows two spectacular volcanoes. Gula Mons, the volcano on the right horizon, reaches 1.8 miles (3 kilometers) high, while Sif Mons, the volcano on the left horizon, has a diameter of 186 miles (300 kilometers) and a height of 1.2 miles (2 kilometers).

VENUS

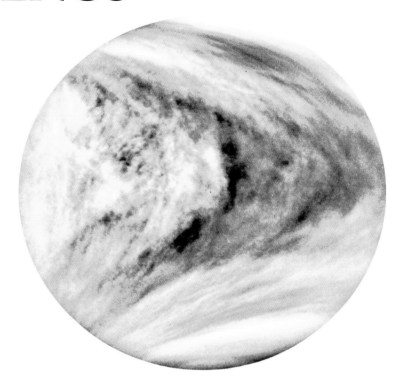

Venus, the second planet from the Sun, is similar to the Earth in size and mass. But there the resemblance ends. Both planets formed at about the same time from similar materials, and once had similar atmospheres. Now, Venus is dry and lifeless with not a trace of water on its surface, but the Earth teems with life and over two-thirds of it is covered in oceans. Searing temperatures, crushing pressures and a suffocating atmosphere have made Venus a very different world from our own.

WHAT IF...

...WE COULD GO TO VENUS?

Try to imagine standing in an enormous oven filled with choking, noxious fumes—nasty, but loitering on Venus would be far worse. The planet's surface is at least twice as hot as a typical domestic oven, and the atmospheric pressure is tremendous: 1,300 pounds (590 kg) per square inch (2.54 square cm), which is 90 times what it is on Earth and about the same pressure that you would feel if you were at the bottom of a lake over 3,000 feet (915 m) deep. On Venus, without protection you would be fried and crushed to death in an instant.

If you did land on the surface of Venus, the thick clouds of sulfuric acid that cover the planet would prevent you from seeing the Sun, the stars and other planets. But from orbit around Venus, the Sun would appear roughly 1.4 times larger than it does from Earth. Mercury would be the brightest object in the sky aside from the Sun, and the Earth would appear as a bright blue speck.

Not surprisingly, none of the world's space agencies are even dreaming of sending astronauts to Venus. The surface conditions are simply too hostile. Yet not so long ago, Venus was thought of as the best hope for finding life elsewhere in the solar system.

One common view was that Venus was covered in swamps much like the Earth was during the Carboniferous period of the Paleozoic era, some 350 million years ago. And as recently as the 1960s, many astronomers speculated that Venus might have extensive oceans of water, perhaps covering the whole planet. Where there were oceans, or swamps, could there also be life?

Sadly, no. The data received by space probes sent to Venus in the 1960s and 1970s finally put an end to such speculation. Not only is Venus extremely unlikely to have life of any kind, but there is little prospect of people from Earth going there for a very long time; and building a base there would be impossible until we are able, if ever, to transform the planet's hostile environment to suit our purposes.

If Venus is to be explored further in the foreseeable future, it will be done by automated probes and robots. One proposal, the Venus Multi-Probe Mission, envisions a spacecraft dropping 16 tiny probes into the atmosphere. Ground stations back on Earth will monitor meteorological data transmitted from the probes as they sink to the surface. This project is, however, still only a suggestion, and it may be many years before it becomes a reality.

VENUS PROFILE

VENUS		EARTH
7,523 MILES (12,108 KM)	DIAMETER	7,973 MILES (12,831 KM)
177°	AXIS TILT	23° 27'
225 EARTH DAYS	LENGTH OF YEAR	365 DAYS
243 EARTH DAYS	LENGTH OF DAY	24 HR
67.2 MILLION MILES (108.1 MILLION KM)	DISTANCE FROM SUN	93.5 MILLION MILES (150.4 MILLION KM)
855°F (457°C)	SURFACE TEMPERATURE	59°F (15°C)
0.9 G	SURFACE GRAVITY	1 G
CARBON DIOXIDE (96.5%), NITROGEN (3.5%)	ATMOSPHERE	NITROGEN (78%), OXYGEN (21%), ARGON (1%)
90,000 MILLIBARS	ATMOSPHERIC PRESSURE	1,000 MILLIBARS
SILICON, ALUMINUM, IRON, NICKEL	COMPOSITION	SILICON (60%), ALUMINUM (15%)

A BARREN WASTELAND

Our knowledge of the cloud-shrouded surface of Venus comes from radar images produced by Earth-based radio telescopes and by orbiting spacecraft. These images have revealed a landscape of massive volcanoes, surrounded by extensive lava plains crossed by lava flow channels thousands of miles long. The few impact craters are large, because only the most massive meteorites have been able to penetrate the atmosphere. Volcanic activity may still be occurring here and there, and the curious blisterlike features called coronae are believed to be bulges caused by heat within the planet melting and blistering the crust.

Venus' huge number of volcanoes seems puzzling at first, given its resemblance to the Earth. From measurements of the gravitational field of Venus, scientists conclude that the planet has an iron core, about the same size as the Earth's, overlaid by a rocky mantle, again just like that of the Earth. Both planets should produce about the same amount of internal heat, largely from the decay of radioactive elements. So why, then, is the surface of Venus dominated by volcanoes while the Earth's is not?

The explanation may lie in a crucial difference revealed by radar mapping from the Magellan spacecraft. Where the Earth's crust is fractured into constantly moving "plates," with earthquakes and volcanoes occurring along their margins, the crust of Venus seems to be intact. Instead of internal heat being lost through volcanoes at plate margins, as it is on Earth, it is thought to escape through the numerous "hot spot" volcanoes that dot the entire surface of Venus.

Scientists think that the original atmospheres of both Venus and the Earth were created from gases released by volcanoes, when both planets were very young and volcanic activity was much more intense. But the closeness of Venus to the Sun meant that the "greenhouse effect," in which heat is trapped within the atmosphere, resulted in the temperature rising so high that all the surface water evaporated. With all the water now in the atmosphere, the intense ultraviolet radiation from the Sun split the water molecules into hydrogen and oxygen. The hydrogen escaped into space and the oxygen combined with other chemicals in the atmosphere. So eventually, Venus lost virtually all its water.

In contrast, the Earth cooled down, oceans formed, and life began to develop. The Earth became a living planet while Venus remained barren.

HOT PROBE

THE TEMPERATURES AND PRESSURES ON VENUS ARE SO EXTREME THAT THE FIRST THREE PROBES THAT WERE SENT INTO THE ATMOSPHERE WERE DESTROYED ON THE WAY DOWN. VENERA 7 WAS THE FIRST TO LAND SAFELY, IN DECEMBER 1970, BUT ITS SIGNALS WERE LOST AFTER 23 MINUTES.

TRANSIT

IN 1769, OBSERVERS WATCHING VENUS PASS IN FRONT OF THE SUN SAW THAT IT APPEARED ELONGATED AS IT CROSSED THE EDGE OF THE SUN'S DISK. IT WAS LATER REALIZED THAT THIS COULD ONLY HAVE HAPPENED IF VENUS HAD AN ATMOSPHERE.

SURFACE FEATURES

Labels: Aphrodite Terra, Atalanta Planitia, North Pole, Tethys Regio, Beta Regio, Ishtar Terra, Lakshmi Planum, Cleopatra Patera, Maxwell Montes, Gula Mons, Eistla Regio

RIFT VALLEYS
The large rift valley in the west of the Eistla Regio area is an indication of past movements in the crust. It was formed when two parts of the crust moved apart and the ground between them sank.

RADAR MAP
Although the surface of Venus is shrouded with clouds, its features can be mapped with radar. This map was based on data gathered by the Magellan radar-mapping spacecraft that went into orbit around Venus in 1990.

MOUNTAIN HIGH
Gula Mons, an extinct volcano in the western Eistla Regio area, rises about 9,800 ft (3,000 m) above the surrounding plains.

CHANGING VIEWS OF VENUS

Except for Mars, no planet in the solar system captured the imaginations of early astronomers more than Venus. Two hundred years ago, Venus was a mystery. No telescope could penetrate the clouds that permanently blanket the entire surface, and imaginations ran wild. The planet was visualized as a balmy swamp, replete with primitive alien mammals or reptiles. But when radar penetrated the clouds for the first time in the 1960s, Venus was unveiled—and the dream was shattered.

WHAT IF...

...VENUS "CHANGED" AGAIN?

Our understanding of Venus was completely transformed when probes from the 1960s onward gradually revealed the surface beneath the clouds. Out went the 18th-century tropical paradise idea; in came the desolate vision of Hell we now know Venus to be. Or do we? The progress of science never stops. Could it be that the model of Venus we currently have so much faith in will undergo another dramatic change?

Our present knowledge of the planet seems concrete—we now know that Venus rotates once in 243.01 days, has thick clouds of sulfuric acid and no magnetic field, and possesses a crushing, blistering atmosphere. We have sent probes there and have detailed maps of the surface—and these basic findings can never be refuted.

But we may be misled by the dramatic depictions of Venus's surface, derived not from photography but from radar data. The only true pictures of the surface come from the four Venera probes, and they only lasted a total of a few hours in the hostile conditions. If the only views of Earth sent to some alien craft were radar maps and a handful of images of the surface, they could gain a very skewed impression of our planet—so perhaps we are similarly mistaken about Venus.

The Venera images are mostly of flattened rocks, with tantalizing glimpses of a more distant view at the frame edges. Based on these, the atmosphere at the surface is thought to be fairly clear, but the light filtering through the clouds is strongly orange-tinted. The true color of the rocks is thought to be gray, like Earth basalts.

The spectacular views derived from radar data are definitely wrong. When they were reconstructed, the scientists exaggerated the vertical scale by a factor of up to 20 to make the scenes look more interesting, and took the orange color from the limited Venera data. So our present views of Venus are deliberately falsified in the interests of good publicity for the space program. And radar images miss a lot of information—the presence of lava pools, clouds of vapor or atmospheric effects, for example. We will not know about any weird Venusian rainbows until we get there. And although no scientist expects to find life on Venus, radar maps of Earth might not even reveal its presence here, so we could be in for some surprises.

ZOOMING IN ON VENUS

DATE	INSTRUMENT OR PERSON	RESULT OR CONCLUSION
1761	MIKHAIL LOMONOSOV	DISCOVERY OF ATMOSPHERE SUGGESTS A HOT AND STICKY SWAMP PLANET
1841	FRANCESCO DE VICO	VENUS SPINS IN 23 HOURS, 21 MINUTES
1890	GIOVANNI SCHIAPARELLI	VENUS ROTATES ONCE IN AN ORBITAL PERIOD (EVERY 224.7 EARTH DAYS)
1962	EARTHBOUND RADIO TELESCOPES	VENUS' ROTATION PERIOD FOUND TO BE CLOSE TO 240 EARTH DAYS
1962	MARINER 2	HARSH TEMPERATURES AND NO SIGNIFICANT MAGNETIC FIELD
1967	VENERA 4	ATMOSPHERIC CHEMICAL COMPOSITION, PRESSURE AND TEMPERATURE
1972	VENERA 8	LIGHT LEVELS ON VENUS AS ON EARTH ON A RAINY DAY
1975	VENERAS 9 AND 10	PICTURES FROM SURFACE REVEAL A ROCKY, ARID LANDSCAPE
1979	PIONEER-VENUS	FIRST TOPOGRAPHIC RADAR MAPPING OF PLANET
1982	VENERAS 13 AND 14	FIRST ANALYSIS OF CRUSTAL COMPOSITION
1990–4	MAGELLAN	HIGH-RESOLUTION RADAR MAPPING SUGGESTS ACTIVE OR RECENT VOLCANISM

VENUS SKY TRAP

Venus is perhaps the place in the solar system most like hell. Yet according to 18th-century scholars—armed only with their primitive telescopes—it was a very different place. Even with modern instruments, Venus is a featureless white globe. And so astronomers of old, unable to see surface detail, used their imagination.

Since the planet was so bland through a telescope, the 18th-century astronomers rightly believed that Venus was permanently covered in clouds. In 1761, when Russian astronomer Mikhail Lomonosov (1711–65) watched Venus as it passed in front of the Sun, he saw a fuzzy edge that told him the planet was enshrouded in a thick atmosphere, and the cloud hypothesis gained even more standing. However, scientists in those days were only familiar with water vapor clouds. They also knew that Venus was 30% closer to the Sun than Earth is, so they naturally concluded it must be slightly warmer. Thus the image that developed was one of a humid world lush with vegetation. Some even speculated that there might be life in a state of evolution like that during Earth's Carboniferous Period, 300 million years ago.

To strengthen this "Earth twin" view of Venus, observations by the Italian Francesco de Vico (1805–48) falsely hinted that the planet rotated once every 24 hours. Indeed, the picture of a swamp planet was even entertained by 20th-century scientists as recently as the early 1960s. But the 1960s also saw the beginning of the space age, and our views of Venus were about to change.

DISGUISED

THE ITALIAN ASTRONOMER GIOVANNI DOMENICO CASSINI (1625–1712) MADE THESE TELESCOPIC SKETCHES OF VENUS IN 1666–7. THEY REPRESENT ONE OF THE EARLIEST ATTEMPTS TO MEASURE THE PLANET'S PERIOD OF ROTATION. EARLY ASTRONOMERS OFTEN THOUGHT THEY COULD DETECT SURFACE DETAILS, BUT THEY WERE IN FACT LOOKING AT TRANSITORY SHAPES IN VENUS' CLOUDS. THESE ROTATE MUCH FASTER THAN THE PLANET ITSELF, LEADING TO THE MANY INACCURATE ESTIMATES OF ITS ROTATION.

END OF THE DREAM

The change in our conception of Venus is one of the most dramatic turnarounds in the history of astronomy. It is a fine example of how early astronomers were dependent on—and limited by—their imagination.

MARINER AND VENERA
In the 1960s and 1970s, Mariner and Venera probes (Venera 1, below) found Venus to be a very hostile place. It is completely dry and excessively hot, has an immense atmospheric pressure and is totally sterile. It also lacks a magnetic field and has the slowest planetary rotation period.

FOREST PLANET
Believing that Venus' proximity to the Sun would make it warmer than Earth, and encouraged by the discovery of a thick atmosphere by Mikhail Lomonosov (above) in the 18th century, astronomers once imagined a lush carboniferous forest beneath the Venusian clouds.

MAGELLAN
The Magellan probe (right), launched in 1989, revealed more surprises when it mapped 98% of Venus' surface in microwave radar. Venus has few impact craters. The planet was once extremely volcanically active, and most of the impact craters seem to have been obliterated by lava. The planet may still have volcanic activity.

WELCOME TO HELL

In 1962, the U.S. probe Mariner 2 flew past Venus and recorded a temperature way in excess of that expected. Though it is farther from the Sun than Mercury, Venus is hotter. Temperatures close to 900°F (480°C)—enough to melt lead and tin and to turn steel red-hot—are common even at the poles. That same year, radio telescopes on Earth penetrated the Venusian cloud cover for the first time and recorded the planet's rotation period as close to 240 days—greater than its year and far longer than that of any other planet. A little later, in 1970, the Soviets' Venera 7 confirmed Mariner 2's temperature reading and measured an atmospheric pressure of 90 bars—equivalent to the pressure at the bottom of a sea 3,000 feet (1,650 m) deep. If all this were not enough to rid us of the misguided belief in a vegetative Venus, the planet is bone dry. Far from the humid jungle of the 18th century, Venus has no water at all on its surface and only traces in its carbon-dioxide atmosphere—the clouds are made of sulphuric acid, not water vapor.

The last probe to visit Venus was NASA's Magellan, arriving in 1990 to perform the first high-resolution radar-mapping of the planet's surface. Magellan discovered impact craters, lakes of solidified lava, and other evidence of intense volcanism in the recent past—the planet may be active even now, but this remains unconfirmed.

THE SURFACE OF VENUS

Beneath Venus' mask of sulfuric acid cloud lies a planetary surface like none other in the solar system. Due to high temperatures and rampant volcanism, the Venusian landscape is a boiling brew of volcanoes, craters, rifts, ridges and chasms. Untouched by the erosive forces that smooth out the Earth's landscapes, these geological contortions appear in raw and rugged detail. U.S. and Soviet probes have sent back images of these extraordinary surface features, but there are mysteries still unsolved.

WHAT IF...

...WE DECIDED TO MINE VENUS?

The day may come in the not-too-distant future when Earth's natural resources will be exhausted—the final drop of oil burned, the last nugget of gold dug from the ground. By then, we could be searching the solar system for new materials to plunder. We may turn to our nearest planetary neighbors first—Mars and Venus. What would we find if we were to mine Venus? Would the first scouting parties discover natural resources on Venus that would be of use to us on Earth?

Venus is often seen as the Earth's sibling. It formed at the same time, in a similar area of the solar system, and is also of a similar size and density as the Earth. Given these circumstances, one might think it would be a good place to prospect for minerals. But Venus is the Earth's ugly sibling—a hot, lifeless place, with a searing temperature that has prevented liquid water from existing on its surface for hundreds of millions of years.

In the 1950s, before spacecraft had discovered the real nature of the Venusian environment, there were many theories suggesting that Venus might be covered with oceans of oil. But without life, Venus could never have formed the reservoirs of hydrocarbons—oil, coal and gas—that existed on Earth, ripe for human exploitation. Perhaps we should look for minerals deposited by volcanic activity instead.

Although Venus has undergone vast amounts of volcanic activity, its lack of water may dash any hopes of finding the rich terrestrial mineral deposits we now take for granted on Earth. Hydrothermal activity—the work of heat and water—are important in leaching, concentrating and depositing minerals. On Venus, mineral abundances are likely to be far less concentrated than on Earth. Mining is likely to be both difficult and expensive.

Even if a valuable mineral were found, Venus would be a tough place to work. Mineral extraction and refinement under temperatures hot enough to melt lead and pressures 90 times that at sea level would make conditions horrendous. Finally, a vast amount of energy would be needed to haul the ore back to Earth.

Venus is probably safe from human exploitation. To mine the planet successfully, we would need some amazing technology. But if we had that kind of technology, we would never need to go to such an enormous effort.

VENUS FACTS

FEATURE	NAME	LOCATION	SIZE
LARGEST CRATER	MEAD	12° N, 57° E	175 MILES (280 KM) IN DIAMETER
HIGHEST MOUNTAINS	MAXWELL MONTES	65° N, 3° E	7.5 MILES (12 KM) HIGH
HIGHEST VOLCANO	MAAT MONS	1° N, 194° E	5 MILES (8 KM) HIGH
LARGEST CORONA	ARTEMIS	35° S, 135° E	1,625 MILES (2,615 KM) IN DIAMETER
LONGEST CANALE	BALTIS VALLIS	49–51° N, 165–8° E	ABOUT 4,300 (6,920 KM) MILES LONG
LARGEST PLAINS AREA	GUINEVERE PLANITIA	NORTHERN HEMISPHERE	ABOUT 4,700 (7565 KM) MILES ACROSS
DEEPEST VALLEY	DALI CHASMA	18° S, 167° E	ABOUT 4.4 MILES (7 KM) DEEP

TRUE COLOR

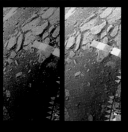

THE FIRST COLOR IMAGE FROM THE VENUSIAN SURFACE WAS MADE BY VENERA 13 IN 1982. IF WE COULD STAND ON VENUS, WE WOULD SEE A LANDSCAPE BATHED IN AN EERIE ORANGE GLOW, THE ONLY LIGHT THAT FILTERS THROUGH THE THICK CLOUDS. THE PHOTO TO THE FAR RIGHT IS ADJUSTED TO SHOW THE SAME AREA IN NORMAL SUNLIGHT.

VENUS UNVEILED

Until the space age, the surface of Venus—our nearest planetary neighbor—remained hidden. Now, planetary probes—the U.S. Magellan mission and the Russian Venera orbiters—have used powerful radars to pierce the planet's obscuring clouds. Russian spacecraft have even landed on the surface. Between them, these probes have given us a global picture of Venus' tortured landscapes.

The radar images show a world where volcanoes run rampant. More than 80% of the landscape was formed by volcanic activity, itself the result of heat rising from the planet's interior. Most of the surface consists of volcanic plains called planitiae with other volcanic features scattered across them. The largest are huge shield volcanoes (volcanoes with gently sloping sides) surrounded by lava flows hundreds of miles long. Formations such as the spidery arachnoids and coronae (Latin for crown), created by upwelling molten rock, have warped the crust. So-called "pancake domes" have formed where a sticky type of lava oozed onto the surface, like toothpaste from a tube. And enigmatic snake-like channels called canali meander for hundreds of miles—streambeds cut by extremely hot, fluid lava.

The first detailed pictures of the surface of Venus were taken during the 1970s by four

A FIRE-TORMENTED LANDSCAPE

GULA MONS VOLCANO
Located in the Eistla Regio area of Venus, Gula Mons is a large shield volcano, 1.8 miles (2.9 km) high. It was photographed during the Magellan mission, which was named after Portuguese navigator Ferdinand Magellan (1480–1521).

LAVA FLOW
Lava flows from the base of the Gula Mons volcano for hundreds of miles across a cracked plain of arachnoids.

SIF MONS
This shield volcano is 1.2 miles (1.9 km) high with a diameter of 180 miles (290 km). Photos from the Magellan probe indicate that Sif Mons is still active in a few hot spots, although most of it has been dormant for the past few hundred million years.

ARACHNOID FORMATIONS
The arachnoids in this radar image from the Magellan orbiter are from 31 to 143 miles (50 to 230 km) in diameter. Arachnoids were given their name because of their resemblance to spiders and their webs.

GIANT DOME VOLCANOES
These seven volcanoes, with an average diameter of 15 miles (24 km) and a maximum height of 2,500 feet (760 m), are the culmination of a series of thick lava flows that built up into a dome-like formation.

LAVA FLOWS
A caldera named Ammavaru sends out lava flows in the Lada region. The lava crosses a ridge belt (running north to south in the photograph) and spreads over an area of 38,564 square miles (99,877 square km).

METEORITE CRATERS
In the Lavinia region of Venus lie three large craters, ranging from 23 to 31 miles (37 to 50 km) in diameter. The bright ejecta around each crater rim was thrown out by the impact of a meteorite.

Soviet Venera landers. After a perilous descent to the hostile surface, with its extreme temperatures and pressures, they survived just long enough to transmit images back to Earth. The surface had a slab-like appearance, covered in dark rock and surrounded by soil which, at some landing sites, contained gravel-sized rock fragments. In the short time before the Venusian environment destroyed them, two of the landers were able to analyze the surface rocks. The results suggested a similarity with basaltic rocks found beneath the Earth's oceans.

Later, in the early 1990s, orbital radar on the U.S. probe Magellan returned images of the surface that showed a maze of fractures, faults and rifts. Crisscrossing the planitiae is a fine network of ridges known as wrinkle ridges. Mountain belts in the northern hemisphere mark where powerful compression forces have buckled the crust. Elsewhere, stretching forces have torn the surface apart, producing rift valleys.

Approximately 1,000 craters have been counted on Venus, far more than on Earth (where crater impact sites are soon eroded) but many

fewer than on the Moon, Mercury or Mars. Since craters accumulate over time, the more craters, the older the surface must be. The same simple rule can also be applied to a specific area of a planet. An area with more craters is probably an older surface than one with fewer craters. But impact craters appear to be randomly distributed all over Venus, which makes it difficult to tell the age of different parts of the surface. Overall, the number of craters implies an age of up to 500 million years—so Venus has a surface much younger than most of the Earth's.

GEOLOGY OF VENUS

For good reason, Venus has sometimes been called the "volcano planet." Unlike the Earth, where eruptions are mainly a byproduct of plate movements, volcanoes seem to be the principal geologic driving force on Venus. The ceaseless motion of subterranean lava cracks Venus' surface and scars it with volcanic domes. But volcanism in its past was more violent still. Outpourings of lava may have resurfaced the entire planet 500 million years ago, and Venus could be headed for another volcanic makeover today.

WHAT IF...

...VOLCANOES RESURFACED VENUS AGAIN?

For the majority of the last 500 million years, Venus has been fairly quiet. A new volcano erupts now and then, but for the most part, an even spread of impact craters suggests that the surface remains relatively unchanged.

Images of Venus' now-settled landscape do point to a violent past, though, when a series of eruptions may have completely resurfaced the planet. In this scenario, Venus gained a new surface when the planet literally blew its lid.

The radioactive decay of elements within a planet generates heat, and on Earth the motions of crustal plates are an important means of letting off steam. Heat from the Earth's interior bubbles up along ridges in the middle of the oceans, where two plates separate to make room for new crust. Elsewhere, one plate dives beneath another, which serves to plunge cool rock into the hot magma below. In these ways, movements of the Earth's plates provide a conveyor belt for heat loss.

But Venus' surface is one continuous whole that does not lend itself very well to the disposal of internal heat. In the past, surface heat loss caused the planet's upper crust, or lithosphere, to stiffen into a kind of "lid" that keeps most of the heat inside. Some heat can still escape into space by conduction through the crust. But this cooling process is only effective if the crust is thin, and without the plate movements that recycle it on the Earth, Venus' lithosphere has thickened with age.

If internal energy is not dissipated, radioactive decay ensures that the heat continues to rise. This could be the situation on Venus today, and if so, the lithosphere may not be able to keep a lid on it for much longer. Eventually, molten rock below the surface will become hot enough to blast its way through thin spots in the crust and begin to resurface the planet all over again. Some simulations of Venus's interior suggest that these episodes repeat every 350 million to 700 million years.

If Venus' geology made a fresh start today, we might not notice the change with our eyes. The planet would only brighten slightly as its atmosphere expanded. But the extra heat would make Venus glow more brightly at infrared wavelengths, which telescopes on the Earth and in space could easily detect.

Venus' crust cracks apart as magma surges out of the planet's interior. Some scientists think volcanism gives Venus a new surface as often as every 350 million years.

GEOLOGICAL FACTS

Core composition	PROBABLY IRON AND NICKEL
Crust composition	VOLCANIC ROCKS
Average surface age	MORE THAN 500 MILLION YEARS
Active volcanoes	PROBABLY, ALTHOUGH NO ERUPTIONS HAVE ACTUALLY BEEN OBSERVED
Seismic activity	UNKNOWN
Planet tectonics	NO EVIDENCE SO FAR—VENUS IS PROBABLY A 1-PLATE PLANET
Surface weathering	BY CHEMICAL WEATHERING, VOLCANIC ACTIVITY, MINOR WIND EROSION AND ROCK FRACTURE BY HIGH TEMPERATURES

HOT ROCKS

Around half a billion years ago, Venus turned itself inside out. Molten rock from the interior bubbled to the surface—and covered the entire planet to a depth of up to six miles (10 km). At the same time, rock at the bottom of Venus' crust plunged downward to cool the hot depths and slow the resurfacing process. When it was over, Venus was left with a new skin, unblemished by the ridges and impact craters that mark old age.

Of all the theories that describe Venus' current face, catastrophic resurfacing on a planetary scale is the most dramatic. Other models hold that the planet's surface was recycled into its hot interior over a more prolonged period, which only stopped as the crust stiffened.

Both of these scenarios are consistent with the widely held view of Venus as a "1-plate planet." Before space probes took a closer look, scientists thought that Venus—so Earth-like in terms of mass and size—might have familiar geology, too. The main geologic process on the Earth is called plate tectonics. Miles-thick "plates" of lightweight rock float slowly across the denser rock below. Mountains and volcanoes form where the plates collide, while new crust wells up where they pull apart.

But Venus' geology does not bear any family resemblance to Earth's. High temperatures keep the planet's crust plastic, or pliable, and its surface features seem to have been created by periods of contraction and extension. Volcanoes rule—and on Venus, they are not restricted to the surrounds of few plate margins.

VOLCANIC VENUS

Radar images taken through Venus' dense clouds by NASA's Magellan spacecraft show that over 80% of the planet's land forms have volcanic origins—from vast plains to giant shield volcanoes that cover areas of up to 25,000 square miles (65,000 square km). Magellan data suggested that some volcanoes are still active, although the probe did not capture any actual eruptions.

The planet within is even more mysterious. Scientists assume that the early Venus separated into layers, as Earth did, with iron and nickel sinking to form the core, while lighter minerals settled in the mantle and crust.

But once again, Venus may have grown apart from its planetary twin. Local variations in Venus' field of gravity were mapped out from the changes of speed that they cause in orbiting spacecraft. Variations in the Earth's field of gravity are paired to currents in the mantle. But on Venus, field anomalies are coupled to areas of altered crust density. This difference suggests that Venus lacks an asthenosphere, the partially molten layer that separates convection patterns in the mantle from plate motions on Earth. Venus may not qualify for this layer because it is too dry. Water lowers the melting point of rocks on Earth to make an asthenosphere possible. But Venus' water probably escaped into space.

The lack of water may give Venus rocks extra structural strength. Some geophysicists think that Venus' dry crust can support larger loads, including mountains steeper than any on Earth, for longer than on Earth.

For now, this and virtually every other theory of Venus' geology is open to debate. And while we cannot explain it, we do know that the resurfacing of Venus obliterated important clues to the planet's past. In this way, at least, Venus is like the Earth: Whether it is remodeled by volcanoes, plate tectonics or weather, neither planet's surface stays the same for long.

MAGELLAN MAP

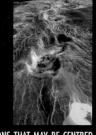

OUR MOST DETAILED VIEWS OF VENUS' GEOLOGY WERE TAKEN FROM NASA'S MAGELLAN SPACECRAFT, WHICH MAPPED THE PLANET'S SURFACE IN THE EARLY 1990S. MAGELLAN DATA WAS FED INTO COMPUTERS TO GENERATE PERSPECTIVE VIEWS OF VENUS' VOLCANIC PLAINS (RIGHT). THESE SHOW *CORONAE*—CIRCULAR DEPRESSIONS THAT MAY BE CENTRED OVER LOCALIZED "HOT SPOTS" IN VENUS' MANTLE.

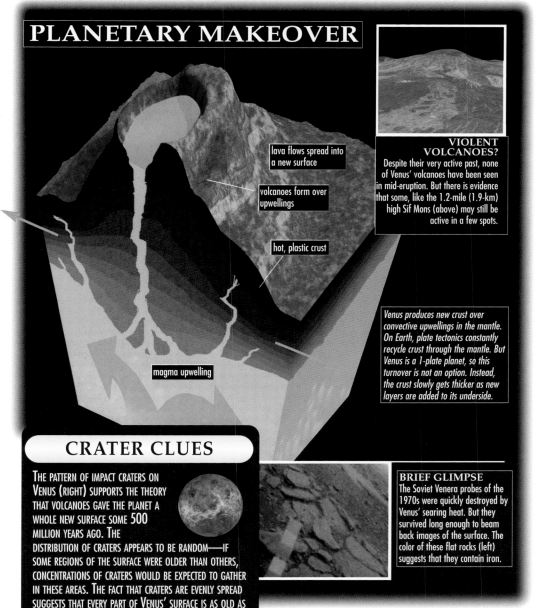

PLANETARY MAKEOVER

lava flows spread into a new surface

volcanoes form over upwellings

hot, plastic crust

magma upwelling

VIOLENT VOLCANOES?
Despite their very active past, none of Venus' volcanoes have been seen in mid-eruption. But there is evidence that some, like the 1.2-mile (1.9-km) high Sif Mons (above) may still be active in a few spots.

Venus produces new crust over convective upwellings in the mantle. On Earth, plate tectonics constantly recycle crust through the mantle. But Venus is a 1-plate planet, so this turnover is not an option. Instead, the crust slowly gets thicker as new layers are added to its underside.

CRATER CLUES

THE PATTERN OF IMPACT CRATERS ON VENUS (RIGHT) SUPPORTS THE THEORY THAT VOLCANOES GAVE THE PLANET A WHOLE NEW SURFACE SOME 500 MILLION YEARS AGO. THE DISTRIBUTION OF CRATERS APPEARS TO BE RANDOM—IF SOME REGIONS OF THE SURFACE WERE OLDER THAN OTHERS, CONCENTRATIONS OF CRATERS WOULD BE EXPECTED TO GATHER IN THESE AREAS. THE FACT THAT CRATERS ARE EVENLY SPREAD SUGGESTS THAT EVERY PART OF VENUS' SURFACE IS AS OLD AS THE NEXT. THE CRATERS ALSO TELL US THAT THE RESURFACING EPISODE WAS BRIEF COMPARED TO THE IMPACT RATE: ONLY BY COVERING THE ENTIRE PLANET QUICKLY COULD VOLCANOES MAINTAIN THE RANDOMNESS OF THE IMPACT RECORD.

BRIEF GLIMPSE
The Soviet Venera probes of the 1970s were quickly destroyed by Venus' searing heat. But they survived long enough to beam back images of the surface. The color of these flat rocks (left) suggests that they contain iron.

ATMOSPHERE OF VENUS

After the Sun and the Moon, Venus is the brightest object in our sky. The 17th-century Dutch astronomer Christiaan Huygens thought this brightness could be explained by an atmosphere surrounding Venus. He was right. Since then, Venus has been observed by the Mariner, Pioneer, Venus and Venera probes of the 1960s and 1970s and by Magellan in the 1990s. They have revealed Venus' atmosphere to be unbreathable and hot enough to melt lead, with a surface pressure 90 times greater than Earth's.

WHAT IF...

...ASTRONAUTS LANDED ON VENUS?

Asphyxiated. Incinerated. Crushed. These could be the multiple-choice answers to the question "What would happen to astronauts landing on Venus?" The correct response, though, would have to be "All of the above."

After the Moon, Venus is the most visited object in the solar system. Fifteen Soviet and six U.S. missions have made the trip, but those that have braved the descent into the atmosphere have had limited success. Several Soviet probes managed to land on the surface and send back photographs. But none of them was able to withstand the heat or pressure for much more than an hour. An astronaut would be extremely lucky to fare anywhere near that well.

But if it were possible for people to survive the journey through the atmosphere, what would it look like? The sky would be perpetually overcast. The unbreathable air would have a red tint, created by the scattering of sunlight through the clouds and a dull glow from the hot ground. There would be no wind, the temperature could be as high as 900°F (480°C), and the pressure would be crushing.

Obviously, this is not an atmosphere humans could withstand—unless it could be made more like that of the Earth. The process of transforming a hostile planet into one more hospitable to human life is called terraforming. In theory, the process could work on Venus. First we would have to learn how to steer comets to impact the planet: They would bring much-needed water, and perhaps blow away some carbon dioxide (CO_2). Or some as-yet unknown biological agent—algae or bacteria—could be introduced on Venus to convert the CO_2 into oxygen. Even if we were able to achieve all this, the weather patterns resulting from these changes might be unendurable to humans. And then there is the challenge of Venus' slow rotation—117 Earth days pass between noon and noon on Venus.

Terraforming Venus is certainly not a realistic option for the near future—it would take centuries and cost more than we can begin to estimate. The next Venus missions will probably be carried out by a series of orbital probes, atmospheric balloons or gliders and surface robots. For human astronauts to travel to Venus, a number of things would be absolutely vital: long-range spaceships, landing craft capable of withstanding enormous heat and pressure, and space suits beyond the dreams of present-day engineers. Until then, human observation of Venus will have to be done from a safe distance.

ATMOSPHERIC FACTS

VENUS		EARTH
96%	Carbon Dioxide	0.03%
3%	Nitrogen	77%
0.4%	Oxygen	21%
0.0007%	Argon	0.93%
180ppm*	Sulfur Dioxide	0.0002ppm*
0.001%	Water	71%
No	Magnetic Field	Yes
100%	Cloud Cover	50%
Retrograde	Rotation	Normal

* PARTS PER MILLION

LAYERS OF HEAT

Like all our solar system's planets, Venus shines with light reflected from the Sun. Venus bounces 75% of the sunlight that strikes it back into space, while the Earth reflects only 33%. This is because a thick layer of clouds circling high above Venus wraps the planet in an all-over reflective cloak. These clouds also kept Venus' surface screened from astronomers' gaze for centuries—until radar was used to penetrate the barrier.

If we could descend with the ease of a radar beam from the top of Venus' atmosphere, 250 miles (400 km) above the planet's surface, we would pass through five layers: the photochemical zone, the cloud deck, the evaporation zone, the thermochemical zone and the mineral buffering zone. Our first encounter would be with the photochemical zone. Composed chiefly of sulfur dioxide (SO_2) and water vapor, this layer is bombarded by the Sun's ultraviolet rays. This causes a photochemical reaction between the SO_2 and water vapor. The result—sulfuric acid—sinks to form the cloud deck.

Venus' clouds are yellow—the color of the sulfuric acid that created them. They are also thin and wispy, like a fine mist on Earth. Yet at the altitude of the cloud deck, we still can't see through the clouds to the surface. Towering 14 miles high from 44 miles (71 km) above the

AEROBRAKING

NASA WANTED ITS MAGELLAN PROBE (RIGHT) TO MEASURE THE GRAVITATIONAL PULL OF SURFACE FEATURES ON VENUS. TO DO THIS, THE PROBE HAD TO MOVE FROM AN ELLIPTICAL TO A CIRCULAR ORBIT—BUT IT DID NOT HAVE ENOUGH FUEL FOR THE MANEUVER. SO, IN A FIRST-EVER MANEUVER CALLED AEROBRAKING, MISSION CONTROLLERS FIRED MAGELLAN'S THRUSTERS TO SEND THE CRAFT SKIPPING THROUGH THE UPPER ATMOSPHERE LIKE A STONE ACROSS WATER, SLOWING IT DOWN AND THUS RESHAPING ITS ORBIT.

surface, what the clouds lack in density they make up for in depth. These are the clouds that reflect back most of the Sun's light, making Venus a beacon visible from the Earth's surface even in daylight.

FAST CLOUD, SLOW PLANET

The clouds circle Venus in just four days, reaching speeds of over 200 mph (300 km/h), while Venus itself takes 243 Earth days to rotate on its axis. This phenomenon—in which the clouds orbit much faster than the planet rotates—is called superrotation. Scientists think it may be caused by solar energy or photochemistry in the upper atmosphere, but they have no clear and convincing explanation for the clouds' behavior. They do know that superrotation is limited to the cloud deck. On the surface, winds barely exist, circulating about as much as water at the bottom of the Earth's oceans.

Like clouds on Earth, Venusian clouds also make rain. But on Venus, they drizzle sulfuric acid, not water. This acid rain is captured by the evaporation zone below. The extreme heat (200°F/93°C) here causes the acid rain to decompose. The rain molecules split into sulfur dioxide and water vapor; some rise back up to the clouds, starting the process all over again.

Below this boiling rain lies the thermochemical zone. At this level, the air is so hot (approaching 600°F/315°C) that it causes further chemical reactions among the gases. Carbon dioxide (CO_2)— the major gas here— creates a runaway greenhouse effect. The sunlight that manages to get past the clouds heats the planet's surface. This energy is then re-emitted as infrared radiation, which rises to the thermochemical zone. There, it is absorbed by the CO_2. Because the heat is unable to escape into space, temperatures on Venus' surface are maintained at a hellish 900°F (480°C).

Finally, we reach that searing surface. We are in the atmospheric region called the mineral buffering zone, made up of minerals—sulfur is the most abundant—that seep out of surface rocks. As the sulfur molecules become airborne, they interact with the CO_2 above. This strong

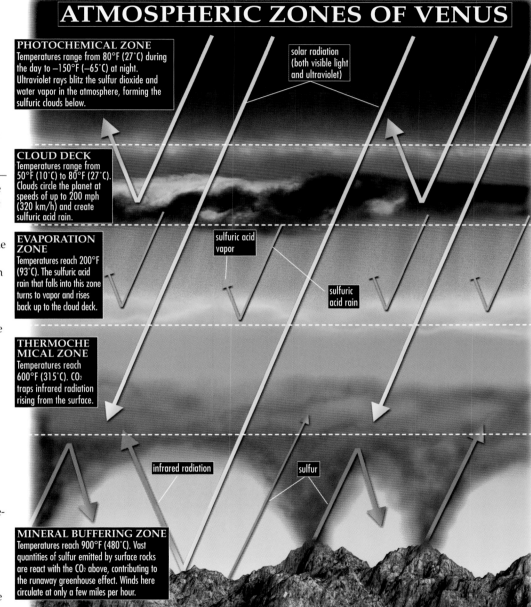

ATMOSPHERIC ZONES OF VENUS

PHOTOCHEMICAL ZONE
Temperatures range from 80°F (27°C) during the day to −150°F (−65°C) at night. Ultraviolet rays blitz the sulfur dioxide and water vapor in the atmosphere, forming the sulfuric clouds below.

solar radiation (both visible light and ultraviolet)

CLOUD DECK
Temperatures range from 50°F (10°C) to 80°F (27°C). Clouds circle the planet at speeds of up to 200 mph (320 km/h) and create sulfuric acid rain.

sulfuric acid vapor

EVAPORATION ZONE
Temperatures reach 200°F (93°C). The sulfuric acid rain that falls into this zone turns to vapor and rises back up to the cloud deck.

sulfuric acid rain

THERMOCHE MICAL ZONE
Temperatures reach 600°F (315°C). CO_2 traps infrared radiation rising from the surface.

infrared radiation

sulfur

MINERAL BUFFERING ZONE
Temperatures reach 900°F (480°C). Vast quantities of sulfur emitted by surface rocks are react with the CO_2 above, contributing to the runaway greenhouse effect. Winds here circulate at only a few miles per hour.

alliance between the hot sulfur and the CO_2 drives the runaway greenhouse effect that produces Venus' extreme surface temperatures. It is also responsible for Venus' lack of liquid water—the surface heat would have boiled away any Venusian lakes or seas long ago.

MYSTERIOUS VENUS

Astronomers once had high hopes that Venus might be the Earth's twin. Since it was almost the same size as our home planet, and only a little closer to the Sun, the chances seemed good that it contained water—perhaps even life. During the 1960s and 1970s, a succession of space probes delved into the thick, cloudy Venusian atmosphere to reveal one of the most hostile environments in the entire solar system. But the probes could not account for all of Venus' mysteries, some of which baffle scientists to this day.

WHAT IF...

...WE COULD CHANGE THE ATMOSPHERE OF VENUS?

The year is 2159, and engineers on the hopelessly overcrowded Earth and Mars are about to embark on the most ambitious space project ever attempted. Inspired by the knowledge that neighboring Venus acquired its unusually slow rotation due to a massive asteroid strike shortly after its formation, they plan to reverse cosmic history by bombarding the planet again—this time to make it hospitable. They call the plan, affectionately, "playing cosmic pool."

Phase 1 involves installing antimatter particle drives on each of a group of five rocky bodies that currently circle the Sun within the asteroid belt. The idea is to activate the drives simultaneously to kick the asteroids out of their present orbits and on to a collision course with Venus. Strangely, this is easier than it sounds. The asteroids themselves are peppered with tunnels left over from abandoned 21st century mining operations, so housing the recently developed drives is no problem. And the drives synthesize their antimatter from silicates, which the asteroids still possess in abundance. So far, so good.

Phase 2 is the nail-biter: It has taken three years to calculate the paths the asteroids will take when the drives are activated. And in theory, using the latest neural-net hypercomputers for the job should have removed any possibility for human error. But one inescapable fact remains: Between the asteroids and their target lies planet Earth. If something goes wrong, it will be the costliest mistake in the history of humanity.

Space agency chiefs comfort themselves by thinking ahead to Phase 3. Impacting Venus at around 45 miles (70 km) per second, predictions state that the asteroids will vaporize instantly and send vast columns of the planet's thick, noxious atmosphere gushing into space. If the computers are right, the fallout from the impacts will then mingle with the remaining atmospheric gases to form solid, sulfur-rich deposits that will rain down on Venus like hailstones. About a century later, the Venusian atmosphere should consist largely of carbon dioxide that can be "seeded" to release life-supporting oxygen using already proven technology. Nobody involved in the operation today will be alive to see it. But they look upon it as an investment in the future.

An antimatter particle drive propels an asteroid toward Venus. If the planet's atmosphere could be blasted away by hundreds of asteroid impacts, we might eventually be able to replace it with breathable air.

VENUS: THE MYSTERY TOP 5

PHENOMENON	DESCRIPTION	DISCOVERER/YEAR
ASHEN LIGHT	A DIM, GREY GLOW ON VENUS' NIGHT SIDE	GIOVANNI RICCIOLI (ITALY), 1643
PHASE DISCREPANCY	HALF DISK VARIES IN APPEARANCE FROM WHAT PREDICTIONS SUGGEST	JOHANN SCHRÖTER (GERMANY), 1793
DAY LONGER THAN YEAR	NOT TRUE FOR ANY OTHER PLANET	RADAR MEASUREMENTS, 1961
BACKWARD ROTATION	THE REVERSE OF THE OTHER PLANETS	RADAR MEASUREMENTS, 1961
RACING CLOUDS	CLOUD TOPS SPIN 60 TIMES FASTER THAN SURFACE	CHARLES BOYER (FRANCE), 1957

LIGHT SHOW

SCIENTISTS HAVE PUT FORWARD MANY THEORIES OVER THE YEARS TO EXPLAIN THE MYSTERIOUS ASHEN LIGHT (RIGHT), BUT FEW WERE AS INVENTIVE AS THAT OF A 19TH-CENTURY AUSTRIAN ASTRONOMER NAMED FRANZ VON PAULA GRUITHUISEN. HE PROPOSED THAT THE LIGHT WAS CREATED BY VENUSIANS CELEBRATING THE ACCESSION OF A NEW EMPEROR.

VEILED IN SECRECY

Although Venus is the brightest planet visible to the naked eye, gaze at it through a telescope and all you will see is a mysterious blank globe. Early astronomers, frustrated by Venus' apparent refusal to reveal itself, simply imagined that they saw features there—and some even went so far as to draw maps of the "surface." Not all these early observations were imaginary, though: Some were accurate—and baffling.

The first Venusian mystery to be observed was recorded in 1643 by a Jesuit astronomer named Giovanni Riccioli. Studying Venus through one of the recently invented telescopes, he noticed what later became known as the Ashen Light—a strange, gray glow that becomes visible on the dark side when the planet is a crescent. Not every astronomer who has gazed at Venus since then has seen the Ashen Light, but many have. Yet even now we are as much in the dark about what causes this effect as Riccioli was in the 17th century.

One theory is that the Ashen Light is some kind of aurora—a natural light show in the night sky that occurs on Earth when the solar wind hits particles trapped by the planet's magnetic field. But as far as we know, Venus does not have a magnetic field, which casts doubt on the aurora theory. Among other explanations for the Ashen Light is that it is a form of lightning—though this would require continuous planetwide thunderstorms—or an unknown effect of solar

radiation on the Venusian atmosphere that causes it to glow intermittently.

FIRST OF MANY

The Ashen Light was the first, but certainly not the last, of Venus' mysteries. Like the Moon, Venus goes through phases from new to full, and in the late 18th century the astronomer Johann Schröter carefully noted the times at which it reached dichotomy—the point at which half its disk is lit. To his surprise, Schröter found that his predictions didn't match reality: When Venus was visible in the evening, dichotomy arrived early; when it was seen in the morning, dichotomy arrived delayed.

Later astronomers confirmed Schröter's findings, establishing that dichotomy could arrive up to 10 days early in the evening, and six days late in the morning. There was no disputing the accuracy of Schröter's math, so the discrepancy had to be classed as an observational phenomenon. The best explanation is that it is due to the way light is refracted by the thick Venusian atmosphere.

Other strange features of Venus were only discovered much later, when radar astronomy and spacecraft finally probed beneath its thick atmospheric shroud. When the planet's rotation period was discovered, it was found to be 243 Earth days—an unusually slow rate that means that a Venusian day lasts longer than a Venusian year. Venus was also found to rotate "backward," causing the Sun to rise in the west and set in the east. The reasons remain unknown. One theory is that at some point in its past, Venus collided with a massive asteroid that hit it hard enough to reverse its rotation.

Stranger still were photographs taken in 1957 by a French astronomer, Charles Boyer, that appeared to show clouds racing around Venus in just four days. Boyer's findings were later confirmed when Mariner 10 flew by the planet in 1974. But exactly why the cloud tops on Venus spin an astonishing 60 times faster than the planet's surface is yet another Venusian mystery. It may be many years before we know the answer.

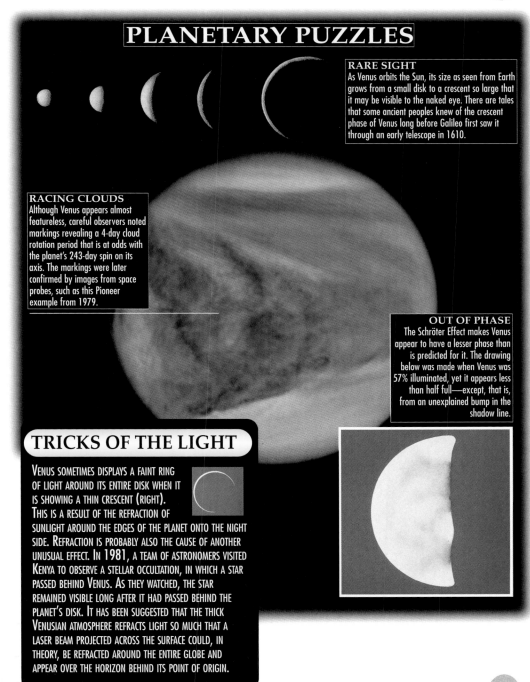

PLANETARY PUZZLES

RARE SIGHT
As Venus orbits the Sun, its size as seen from Earth grows from a small disk to a crescent so large that it may be visible to the naked eye. There are tales that some ancient peoples knew of the crescent phase of Venus long before Galileo first saw it through an early telescope in 1610.

RACING CLOUDS
Although Venus appears almost featureless, careful observers noted markings revealing a 4-day cloud rotation period that is at odds with the planet's 243-day spin on its axis. The markings were later confirmed by images from space probes, such as this Pioneer example from 1979.

OUT OF PHASE
The Schröter Effect makes Venus appear to have a lesser phase than is predicted for it. The drawing below was made when Venus was 57% illuminated, yet it appears less than half full—except, that is, from an unexplained bump in the shadow line.

TRICKS OF THE LIGHT

VENUS SOMETIMES DISPLAYS A FAINT RING OF LIGHT AROUND ITS ENTIRE DISK WHEN IT IS SHOWING A THIN CRESCENT (RIGHT). THIS IS A RESULT OF THE REFRACTION OF SUNLIGHT AROUND THE EDGES OF THE PLANET ONTO THE NIGHT SIDE. REFRACTION IS PROBABLY ALSO THE CAUSE OF ANOTHER UNUSUAL EFFECT. IN 1981, A TEAM OF ASTRONOMERS VISITED KENYA TO OBSERVE A STELLAR OCCULTATION, IN WHICH A STAR PASSED BEHIND VENUS. AS THEY WATCHED, THE STAR REMAINED VISIBLE LONG AFTER IT HAD PASSED BEHIND THE PLANET'S DISK. IT HAS BEEN SUGGESTED THAT THE THICK VENUSIAN ATMOSPHERE REFRACTS LIGHT SO MUCH THAT A LASER BEAM PROJECTED ACROSS THE SURFACE COULD, IN THEORY, BE REFRACTED AROUND THE ENTIRE GLOBE AND APPEAR OVER THE HORIZON BEHIND ITS POINT OF ORIGIN.

EARTH

Earth has been called the Solar System's "Goldilocks planet." At just the right distance from the Sun, and with just the right mass and gravity for liquid water to exist in large amounts, it is also the only rocky planet large enough to retain a molten interior, and this drives the plate tectonics that steadily redraw its surface. These two influences have shaped the planet we know more than any others, and laid the way for another great force—the evolution of life itself. The first living organisms seem to have arisen almost as soon as the Earth became habitable, and through billions of years the action of single-celled organisms (and more recently multicellular plants and animals) have transformed Earth's atmosphere, seas, and landscape. Earth is our key to understanding the other rocky planets and moons of the Solar System—despite gaps in our knowledge, it is still the planet we understand best, and all serious models for the histories of other planets seek to take the processes seen on Earth and apply them elsewhere.

This stunning image of sunlight striking a cloud-covered Earth was taken with a simple 35-mm camera by the crew of the space shuttle Discovery. The view of Earth from space is a common photographic image today, but less than 50 years ago no human being had ever witnessed this sight.

EARTH

The blue-green Earth is the only place in the solar system known to have large quantities of water in liquid form. Water was almost certainly a prerequisite for Earth's unique characteristic: its living things. Life on Earth is responsible for its unusual atmosphere, rich in the highly reactive gas oxygen. Earth is the birthplace and current sole residence of a race of intelligent bipeds. Recently, spacecraft built by these humans have allowed them to see their planet from space—and recognize its place in the solar system.

WHAT IF...

...THE EARTH WAS INHABITED BY INTELLIGENT LIFE?

Back in 1990, in the course of its 6-year journey to Jupiter, the Galileo spacecraft swung around the inner planets in a gravity slingshot maneuver to gain speed for the long haul to Jupiter. When Galileo swooped from Venus back to our planet, NASA noted that it was "the first confirmed interplanetary visitor to Earth." Since the Galileo team used the opportunity to try out the spacecraft's systems, the flyby meant that a state-of-the-art planetary probe had a chance to investigate the Earth.

Suppose that Galileo really had come from another world. At its 1990 approach, its cameras swept by just 620 miles (1,000 km) above the atmosphere. What could the probe have told its makers about the blue world beneath it?

Galileo's cameras found no sign of intelligence. And they did not even detect conclusive evidence for the existence of life. Of course, Galileo was built to investigate conditions around Jupiter, and the instruments it carried were not what designers would have put aboard for a life-sniffing mission. The spacecraft sent back masses of data on the Earth's magnetic field and its interaction with the solar wind, for example: interesting and valuable material, but not much help in the quest for life.

The one clue that an alien Galileo could give its controllers was its accurate analysis of the Earth's atmosphere. The ingredients are a long way from chemical equilibrium, with an astonishing 21% of highly reactive oxygen, an odd preponderance of nitrogen and only traces of carbon dioxide. That improbable mixture would certainly have started a debate going. Life-on-Earth proponents would argue that only living organisms could create such an atmosphere and keep it stable. Conservative scientists would look for a straightforward geological explanation—perhaps even a link with Earth's unusually large moon.

The arguments would heat up when the aliens looked at the radio noise detected by other Galileo instruments. The probe was designed to pick up radio crackle from thunderstorms. But its recorders were swamped by radio output from all over the Earth. The planet was emitting more radio frequency energy than the Sun. And as alien computers would discover, most of that radio was encoded signal, not random static—life for sure.

Then the aliens, pouring over Galileo's data, might find another curiosity. At the cloud-tops, a layer of ozone acts as a shield against incoming ultraviolet radiation. But near the planet's south pole, Galileo had detected a peculiar hole in the ozone layer. Obscure chemicals were eating it away. So the aliens could reach an agreement. Yes, there was life on Earth—industrial life, at that. But intelligence? Impossible. Any creature that wrecks its own environment is, by definition, not intelligent.

EARTH STATISTICS

DIAMETER . . . . .7,926 MILES (12,756 KM)	SURFACE TEMPERATURE (MEAN). . 59°F (15°C)
AXIS TILT. . . . . .23°27'	SURFACE GRAVITY (MEAN). . . . . 1 G (32.2 FEET/SEC²)
LENGTH OF YEAR . . . . .365.24 DAYS	ATMOSPHERE . . . . .NITROGEN (78%), OXYGEN (21%), ARGON (1%)
LENGTH OF DAY (MEAN SOLAR). . 24 HOURS	ATMOSPHERIC PRESSURE . . . . . .1,000 MILLIBARS (AT SEA LEVEL)
DISTANCE FROM SUN (MEAN) . . . .93 MILLION MILES (150 MILLION KM)	COMPOSITION . . . . .IRON, NICKEL, SILICON, ALUMINUM

LIVING WORLD

Of the four rocky worlds that make up the inner solar system, Earth is probably the only one that is still geologically active—4.5 billion years after its formation. Vast, continent-sized plates of crustal rock float on top of molten magma, continually replenished by new material that pushes upward along mid-ocean ridges from the planet's interior. These plates push powerfully against each other, forcing up new mountain ranges and constantly rebuilding the planet's surface.

Earth's living geology is one reason why the planet bears so few of the craters that mark the faces of the other inner planets and Earth's own Moon. Back in the early years of the solar system, Earth must have received its fair share of the asteroid bombardments that scarred its neighbors. But crustal movement has long since healed the damage and replaced it with new landscapes.

In any case, on Earth both meteoric craters and new-made mountains are under attack by erosion as soon as they form. The planet's surface is dominated by a vast blanket of liquid water, on average several miles deep. The Earth's water is a powerful scouring force, especially coupled with the winds that are a constant feature of its atmosphere. The atmosphere itself is far less dense than neighboring Venus. But

coupled with the Earth's pronounced axial tilt and its speedy, 24-hour daily rotation, it is more than enough to give the planet powerful weather systems that are visible as swirling cloud patterns from far off in space.

It is the composition of Earth's atmosphere, rather than its density, that distinguishes it from Mars and Venus. Earth's air is 21% oxygen, a reactive gas discernible only in minute quantities elsewhere in the solar system. There are also traces of other gases—notably methane—that should not be able to coexist for long with oxygen. And carbon dioxide, the major component of the atmospheres of Mars and Venus, exists only as a few hundred parts per million—just enough to provide a modest greenhouse effect. Oxygen provides energy for the Earth's extraordinary array of life—and is in turn renewed by the plants that depend on it.

The unusual atmosphere is a clear indication of the Earth's most remarkable feature: life. The first microorganisms appeared more than 3.5 billion years ago, and ever since then, they have been at work on the Earth's atmosphere—and a lot of other things, too. Living organisms are

almost certainly responsible for the high oxygen content of the air, which without constant replenishment would soon be locked up in chemical oxides, just as on Mars and Venus. Over time, life has diversified into millions of genetically diverse species, some of which can exist in even the most inhospitable terrestrial environments.

One of Earth's life-forms has even found ways of sending itself or its artifacts outside the atmosphere: A few have escaped Earth's gravity altogether. But it is too early to say whether Earth life will prove an entirely local phenomenon, or whether it will spread throughout the solar system and even across the gulf of interstellar space to other stars.

GOLDILOCKS AND GAIA

THE SIMPLEST EXPLANATION FOR THE EARTH'S COMFORTABLE TEMPERATURE AND ATMOSPHERE IS THE SO-CALLED GOLDILOCKS THEORY. VENUS, A LITTLE CLOSER TO THE SUN, WAS JUST TOO HOT TO ALLOW THE APPROPRIATE CHEMISTRY. MARS, AN EXTRA 50 MILLION MILES (80 KM) OUT, WAS JUST TOO COLD. BUT EARTH, LIKE THE PORRIDGE GOLDILOCKS ATE AND THE BED IN WHICH SHE SLEPT IN THE FAIRY TALE (ABOVE), WAS JUST RIGHT. MOST SCIENTISTS NOW THINK THAT THE INFLUENCE OF LIFE ON ITS OWN ENVIRONMENT WAS ALSO VERY IMPORTANT. THE EXTREME VERSION OF THIS VIEW IS THE GAIA THEORY, IN WHICH THE EARTH IS CONSIDERED TO BE A SELF-REGULATING SUPERORGANISM.

COLD POLES
Because the Earth's axial tilt restricts the sunlight falling on the poles, both are covered with ice. Antarctica's ice sheet (right) is up to 3 miles (5 km) deep. The poles are also cold because snow and ice reflects most of the little sunlight that they receive.

TWIN PLANET

THE EARTH'S OVERSIZED SATELLITE IS A SOLAR SYSTEM ODDITY. WHEREAS MOST PLANETARY MOONS ARE A TINY FRACTION OF THEIR PRIMARY'S DIMENSIONS, THE MOON IS MORE THAN A QUARTER OF THE EARTH'S DIAMETER—THOUGH LESS THAN 2% OF ITS MASS. SEEN FROM DEEP SPACE, THE EARTH-MOON SYSTEM APPEARS ALMOST AS A DOUBLE PLANET—AS IS STRIKINGLY APPARENT IN THIS PICTURE TAKEN BY THE GALILEO SPACECRAFT IN 1992, 4 MILLION MILES (6.5 KM) OUTBOUND ON ITS WAY TO JUPITER.

INSIDE OUT
Magma from the Hawaiian volcano of Kilauea flows into the sea in an explosion of sparks (above). Such outbursts of molten rock—which can reach temperatures of 3,500°F (1,920°C)—can transform landscapes very quickly.

IN THE WIND
Water evaporates off seas and forms clouds, which, pushed by wind, travel inland and distribute water to the surface. Water vapor also transports vast quantities of energy around the Earth's atmosphere.

CRUMPLED CRUST
The Earth's mountain chains (the Alaska Range is shown above), are driven upward by collisions between adjacent crustal plates. Around the Pacific rim, some 10 plates rub against each other. The area is so volcanically active that it is called the ring of fire.

In this satellite image of the east coast of Oman, the sea dominates the view. If humans had arrived on their planet as colonists from space, they would never have called the place Earth: "Ocean" would be a far better name. More than 70% of the surface is covered with liquid water.

EARTH'S INTERIOR

The Earth beneath our feet may seem steady enough, but anyone who has witnessed an erupting volcano or experienced an earthquake will know that our planet's restless interior harbors violent and destructive forces. Beneath the thin shell that we call the crust, the Earth is a seething mass of molten rock. And scientists now believe that around the Earth's core there may be continents and oceans like those on the surface.

WHAT IF...

...WE VISIT THE EARTH'S CENTER?

We have a long way to go if we want to visit the Earth's core. The deepest borehole yet drilled is on the Kola Peninsula in Arctic Russia. But at 7.5 miles (12 km)—just a quarter of the way through the crust—it barely pricks the surface. The Kola borehole would have to be over 500 times deeper to reach the center of the Earth and would encounter increasingly tough conditions along the way. Aside from the hardness of the rocks, the pressure and temperature increase rapidly with depth, which makes drilling very difficult.

Another borehole now being drilled in Oberpfalz, Germany, will go no deeper than 6 miles (10 km). Engineers have calculated that at this depth the rocks are so hot and soft that the hole will close up as fast as it is drilled. It seems highly unlikely, then, that we will ever get through the 30-odd miles (50-odd km) of the crust, let alone reach the Earth's core. The pressure there is thousands of times higher than on the surface and the temperature can reach more than 12,000°F (6,650°C).

...CRUST SINKS TO THE CORE?

The rocks of the ocean crust—the seabed—are gradually being pushed towards the continents, as new material is added to the crust by volcanoes in mid-ocean. Where the oceanic crust meets a continent, it is pushed under the continent and down into the Earth's interior—a process called subduction.

For years, scientists have debated what happens next. They were astonished to discover in 1991 that a slab of subducted oceanic crust in the western Pacific had plunged all the way to the core beneath North America. Not every piece of subducted crust makes it this far—some pieces dissolve in the mantle about 400 miles (650 km) down. It is also a slow process: Any part of the sea bed that finishes the 2,000-mile (3,200-km) journey to the core will take almost 200 million years to get there.

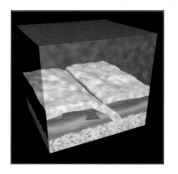

THE EARTH, LAYER BY LAYER

LAYER	DEPTH (MILES/KM)	PROPORTION OF EARTH'S MASS
OCEANIC CRUST	0–6/0–10	0.099%
CONTINENTAL CRUST	0–30/0–50	0.374%
UPPER MANTLE	6–410/10–660	15.3%
LOWER MANTLE	410–1,790/660–2,880	49.2%
"D" LAYER	1,670–1,790/2,690–2,880	3.0%
OUTER CORE	1,790–3,190/2,880–5,130	30.8%
INNER CORE	3,190–3,950/5,130–6,360	1.7%

GOING DEEP

Most people think of the Earth as a solid ball of rock, as if what lies below the soil continued all the way to the center. In fact, our planet is more like a soft-boiled egg. The outside is a thin, hard shell of rock called the crust. Immediately below it, no more than 30 miles (50 km) down, is the mantle, where the rock is hot and soft. Far beneath the mantle, some 1,860 miles (2,990 km) down, is a core of metal. In the outer part of this core the metal is molten, since it is heated by natural radioactivity to temperatures approaching those of the Sun's surface. The enormous pressure at the very center of the Earth keeps the inner core solid.

Much of what lies deep within the Earth remains a mystery, for the simple reason that it is impossible to see it. Mining and drilling barely penetrate even a quarter of the way through the crust, and we may never be able to cope with the enormous pressures and temperatures in the regions beyond.

Instead, we have to rely on indirect evidence. Much of the interplanetary debris in our part of the solar system was formed from the same material as the Earth. When this debris falls to Earth in the form of meteorites, it yields important clues about the interior. Geoscientists also delve deep into the ocean to analyze the molten rock that is forced up from the mantle at mid-ocean ridges. Mantle minerals such as olivine often come to the surface in this way.

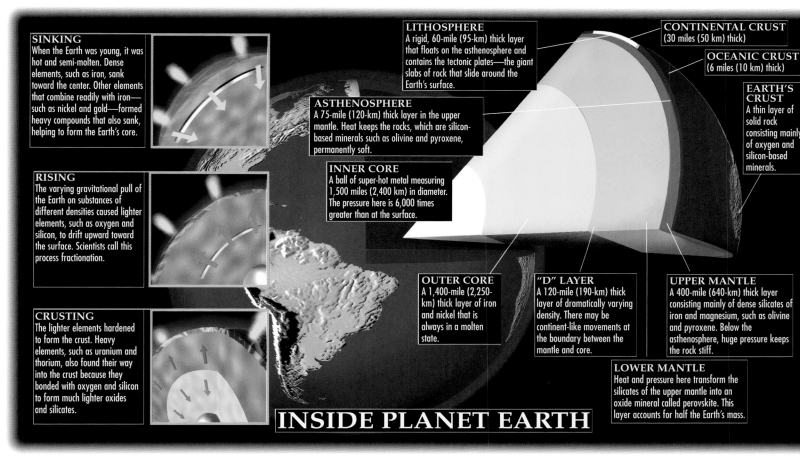

SINKING
When the Earth was young, it was hot and semi-molten. Dense elements, such as iron, sank toward the center. Other elements that combine readily with iron—such as nickel and gold—formed heavy compounds that also sank, helping to form the Earth's core.

RISING
The varying gravitational pull of the Earth on substances of different densities caused lighter elements, such as oxygen and silicon, to drift upward toward the surface. Scientists call this process fractionation.

CRUSTING
The lighter elements hardened to form the crust. Heavy elements, such as uranium and thorium, also found their way into the crust because they bonded with oxygen and silicon to form much lighter oxides and silicates.

LITHOSPHERE
A rigid, 60-mile (95-km) thick layer that floats on the asthenosphere and contains the tectonic plates—the giant slabs of rock that slide around the Earth's surface.

ASTHENOSPHERE
A 75-mile (120-km) thick layer in the upper mantle. Heat keeps the rocks, which are silicon-based minerals such as olivine and pyroxene, permanently soft.

INNER CORE
A ball of super-hot metal measuring 1,500 miles (2,400 km) in diameter. The pressure here is 6,000 times greater than at the surface.

OUTER CORE
A 1,400-mile (2,250-km) thick layer of iron and nickel that is always in a molten state.

"D" LAYER
A 120-mile (190-km) thick layer of dramatically varying density. There may be continent-like movements at the boundary between the mantle and core.

CONTINENTAL CRUST
(30 miles (50 km) thick)

OCEANIC CRUST
(6 miles (10 km) thick)

EARTH'S CRUST
A thin layer of solid rock consisting mainly of oxygen and silicon-based minerals.

UPPER MANTLE
A 400-mile (640-km) thick layer consisting mainly of dense silicates of iron and magnesium, such as olivine and pyroxene. Below the asthenosphere, huge pressure keeps the rock stiff.

LOWER MANTLE
Heat and pressure here transform the silicates of the upper mantle into an oxide mineral called perovskite. This layer accounts for half the Earth's mass.

INSIDE PLANET EARTH

THE FLOOD

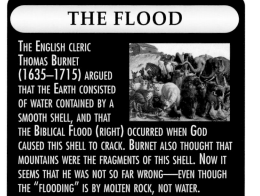

THE ENGLISH CLERIC THOMAS BURNET (1635–1715) ARGUED THAT THE EARTH CONSISTED OF WATER CONTAINED BY A SMOOTH SHELL, AND THAT THE BIBLICAL FLOOD (RIGHT) OCCURRED WHEN GOD CAUSED THIS SHELL TO CRACK. BURNET ALSO THOUGHT THAT MOUNTAINS WERE THE FRAGMENTS OF THIS SHELL. NOW IT SEEMS THAT HE WAS NOT SO FAR WRONG—EVEN THOUGH THE "FLOODING" IS BY MOLTEN ROCK, NOT WATER.

SHOCK TACTICS

Ironically, most of our knowledge about the Earth's interior has come from studying the seismic shock waves that accompany earthquakes. Seismic waves travel at different speeds through different rocks—for example, they travel much faster through the cold, hard rocks of the continental crust than they do through the softer, warmer rocks of the oceanic crust. So by measuring the speed of these shock waves, scientists can build up directional patterns of the rock formations below.

The most extraordinary discoveries of recent years have come from probing even

farther down, to the mysterious zone of transition between the mantle and the core known as the "D" layer—short for "D double prime." As the solid minerals of the mantle give way to the molten iron and nickel of the outer core, there is an astonishing leap in density—even greater than that between air and rock.

There are even more surprises at the bottom of the "D" layer, on the very boundary between mantle and core. Just as there are continents and ocean basins on the surface of the Earth, so there appear to be "anticontinents" on the core-mantle boundary that shift and change in just the same way.

SOUNDINGS

THE 25-TON (23-TONNE) "VIBROSEIS" TRUCK (RIGHT) PRODUCES CONTROLLED SHOCK WAVES THAT TRAVEL DEEP INTO THE EARTH. SCIENTISTS MEASURE THE FREQUENCY AND DIRECTION OF THOSE WAVES THAT RETURN TO THE SURFACE TO BUILD UP A PICTURE OF WHAT THE WAVES ENCOUNTERED DURING THEIR SUBTERRANEAN JOURNEY.

PLATE TECTONICS

How do mountain ranges rise? How do great valleys sink? What makes the land shake—and why do Africa and South America look like matching pieces of an enormous jigsaw puzzle? Some of the biggest questions in Earth science were answered by the 1960s discovery that our planet's thin crust is broken into gigantic slabs called plates. Floating on a global ocean of molten rock, these continent-size fragments drift and collide in a process called plate tectonics—sometimes with catastrophic results.

WHAT IF...

...CONTINENTAL DRIFT REVERSED?

Continental drift continues today. The Pacific is shrinking and the Himalayan mountains rise higher every year. In fact, Mount Everest has grown by 10.5 inches (26.7 cm) since 1900. But we know very little about the forces at work within the Earth. We do not fully understand the cause of continental drift, nor could we foresee a change in the process. If the patterns of movement changed it would take millions of years for the effects to be visible—but the eventual consequences for life on Earth would be enormous.

In the 1920s, German meteorologist Alfred Wegener proposed that around 200 million years ago, there was just one continent. He named this continent "Pangaea" ("all land"), and the ocean that surrounded it "Tethys." Wegener's theory is now widely accepted, although the mechanics behind the movement of plates remains poorly understood. Convection currents in the Earth's mantle, driven by heat from the core, are one possible explanation. If they exist, these currents may have been flowing in the same direction for the last 200 million years. But there is no guarantee they will continue to do so. If

they switch direction, continental drift may suddenly be put into reverse.

The Atlantic Ocean would cease to spread. Instead, the American Plates would begin to collide with the African and Eurasian plates, perhaps creating a new chain of islands in the mid-Atlantic. The Americas would start to close on Africa and Europe. Gradually, the continents would reunite until, after 200 million years, the giant continent of Pangaea might reappear.

But history cannot reverse itself so simply. One of the clues to continental drift is the fossil record, which suggested that branches of the same animal families evolved differently on separated continents. By the time Pangaea reformed, the old divisions between habitats would have disappeared again. Kangaroos would be able to meet polar bears, if either of them still existed. Plants and animals would have to adapt to the changing environment or face extinction. It is impossible to predict which species would benefit, which would dominate, and which would evolve. Human beings may be able to adapt. But we have had only had about 5 million years of practice so far, and measured against that, 200 million years is a very long time.

PRODUCTS OF TECTONICS

FEATURE	CAUSE	PLATE ACTION
HIMALAYAS	CONVERGENT PLATES	INDIAN PLATE COLLIDING WITH EURASIAN PLATE
MID-ATLANTIC RIDGE	DIVERGENT PLATES	AMERICAN PLATES PULLING AWAY FROM AFRICAN AND EURASIAN PLATES
MARIANAS TRENCH	SUBDUCTION ZONE	PACIFIC PLATE DRIVEN UNDER BY EURASIAN PLATE/PHILIPPINE PLATE
ANDES	SUBDUCTION ZONE	NAZCA PLATE DRIVEN UNDER BY SOUTH AMERICAN PLATE
SAN ANDREAS FAULT	TRANSFORM FAULT	PACIFIC PLATE RUBS AGAINST NORTH AMERICAN PLATE
GREAT RIFT VALLEY	DIVERGENT PLATES	SOMALI PLATE PULLING AWAY FROM AFRICAN PLATE
ICELAND	DIVERGENT PLATES	VOLCANIC OUTCROP OF MID-ATLANTIC RIDGE
MT. ST. HELENS	SUBDUCTION ZONE	PACIFIC PLATE DRIVEN UNDER BY NORTH AMERICAN PLATE
MT. PINATUBO	SUBDUCTION ZONE	PHILIPPINE PLATE DRIVEN UNDER BY EURASIAN PLATE

THE EARTH'S SHIFTING SURFACE

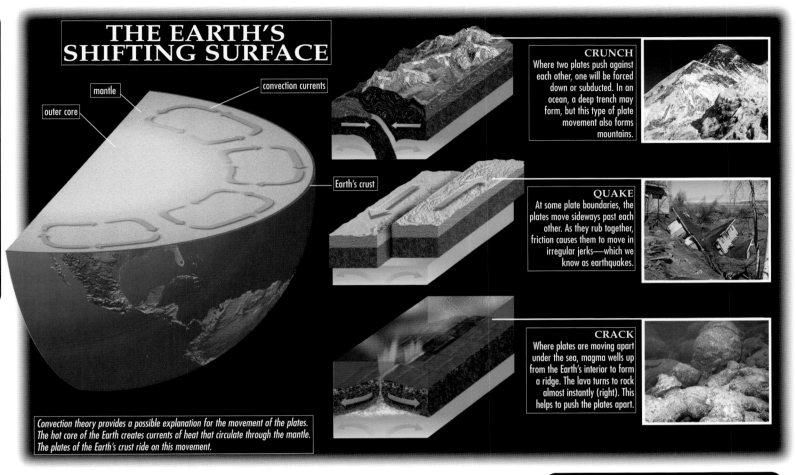

mantle

outer core

convection currents

Earth's crust

CRUNCH
Where two plates push against each other, one will be forced down or subducted. In an ocean, a deep trench may form, but this type of plate movement also forms mountains.

QUAKE
At some plate boundaries, the plates move sideways past each other. As they rub together, friction causes them to move in irregular jerks—which we know as earthquakes.

CRACK
Where plates are moving apart under the sea, magma wells up from the Earth's interior to form a ridge. The lava turns to rock almost instantly (right). This helps to push the plates apart.

Convection theory provides a possible explanation for the movement of the plates. The hot core of the Earth creates currents of heat that circulate through the mantle. The plates of the Earth's crust ride on this movement.

WHERE THE ACTION IS

THE EARTH'S SURFACE CAN BE DIVIDED INTO SIX OR SEVEN MAJOR PLATES: PACIFIC, ANTARCTIC, AFRICAN, EURASIAN, INDO-AUSTRALIAN AND AMERICAN (OFTEN DIVIDED INTO NORTH AMERICAN AND SOUTH AMERICAN). EXCEPT FOR THE PACIFIC PLATE, ALL ARE CONTINENTAL. THERE ARE A DOZEN OR SO SMALLER PLATES, AND EVEN MORE MICROPLATES. THE BOUNDARIES BETWEEN THEM ARE SITES OF MAJOR GEOLOGICAL ACTIVITY, SUCH AS RIFTS, RIDGES, TRENCHES, MOUNTAINS, FAULTS AND CHAINS OF VOLCANOES.

BROKEN PLATES

The concept of continental drift, put forward by the German meteorologist Alfred Wegener in 1912, suggests that the Earth's continents began as a single enormous landmass that broke up and drifted apart over millions of years. For decades, geological matches and fossil records scattered across the continents supported the theory, but the mechanics remained unclear.

The breakthrough came in the 1960s, when underwater geology showed that ocean floors are not fixed in place. They are slowly creeping away from central ridges where they are formed, toward the ocean's edge. Scientists realized that the Earth's crust is fragmented into plates that are forever on the move. This phenomenon, called plate tectonics, explained continental drift—and much else.

There are seven major plates—although there are many more "microplates"—floating on the Earth's molten mantle. Neighboring plates jostle each other as they drift. When plates pull apart on land, a rift valley appears—Africa's Great Rift Valley is the best-known example. But when the thinner ocean plates move apart, the cracks that form become portals to the underworld. Hot magma rises up from the mantle, then cools to form undersea mountain chains—the oceans' central ridges—and fresh ocean-floor rock. The fresh rock can push the plates apart by as much as 6 inches (15 cm) a year. This may not sound like very much, but over 10 million years—an instant in the Earth's lifespan—it will add up to a movement of 950 miles (1,530 km).

CHECKS AND BALANCES

Expansion of Earth's crust is balanced by destruction elsewhere. When plates push against each other, one is driven down, or subducted, and the other is pushed up. The upthrust may buckle the plates to build a mountain range—which is the how Himalayas have grown. The subducted plate is forced back into the Earth's mantle, a process which may create an ocean trench.

Plates make their presence felt most strongly when they slip and scrape against each other. Then, as at California's San Andreas Fault, devastating earthquakes prove the power of moving plates.

STRIPY FLOOR

IN 1963, TWO GRADUATE STUDENTS, DRUMMOND MATTHEWS AND FRED VINE (RIGHT), FROM CAMBRIDGE UNIVERSITY IN ENGLAND, CAME UP WITH THE EVIDENCE NEEDED TO BACK THE THEORY OF PLATE TECTONICS. THEY FOUND "STRIPES" OF REVERSED MAGNETIC FIELDS IN THE OCEAN FLOOR THAT MATCHED REVERSALS IN EARTH'S MAGNETIC FIELD OVER TIME. THE STRIPES SHOWED THAT THE FLOOR HAD NOT BEEN FORMED AT ONE TIME, BUT HAD MOVED OUTWARD FROM THE CENTRAL RIDGE.

EARTH'S IMPACT CRATERS

Every 200,000 years or so, a massive meteorite, comet or asteroid weighing at least several hundred tons slams into Earth's surface and gouges out a huge impact crater. Scientists have identified about 150 of these "extraterrestrial impressions," ranging in size from a few hundred yards to more than a hundred miles (160 km) in diameter. Many older impact craters, dating back to dinosaur times and beyond, will never be found. They have been erased from the landscape, etched away by erosion and geological activity.

WHAT IF...

...A METEORITE HIT THE ATLANTIC OCEAN?

In March 1989, the 1,500-foot (460-m) wide asteroid 1989FC passed within 435,000 miles (700,000 km) of the Earth—a near miss by solar system standards. It will be back, and next time it might not miss. If this asteroid hit a land surface, the impact energy—5,000 times more powerful than the nuclear bomb dropped on Hiroshima during World War II—would devastate 75,000 square miles (195,000 square km), an area roughly the size of Kansas. But if it misses the land and plunges into an ocean—the Atlantic, say—the consequences would be even more catastrophic. Although the initial blast might not claim so many victims, it will generate a tsunami, or wave surge, that would threaten the lives and properties of hundreds of millions of people in coastal regions.

The tsunami will radiate out from the impact site at up to 400 miles per hour (650 km/h), giving the inhabitants of the densely populated seaboards only hours to escape. Out in the open ocean the wave might be only about 35 feet (10 m) high but, as it reaches shallower coastal waters, it will slow down and rear up into a 300-foot (90-m) high wall loaded with debris churned up from the continental shelf.

The deluge would penetrate at least 10 miles (16 km) inland—much farther where channels and valleys concentrated its energy—battering every major city on the eastern seaboard of the U.S. On the other side of the Atlantic, the broad European continental shelf would partially sap the energy of the tsunami, reducing the scale of damage done to Ireland, France and Britain. But Portugal and northwest Spain, which are fringed by deep water, would be hit hard, while low-lying Holland and Denmark would be almost completely swamped. Even if people could be evacuated in time, the destruction wreaked on industry and transport systems would cripple the global economy for decades.

The last of New York's City's Chrysler Building slumps into a wall of water. The huge wave is the result of a mid-Atlantic asteroid impact: It will devastate cities on both sides of the ocean.

TOP TEN IMPACT CRATERS

LOCATION	DIAMETER	AGE
VREDEFORT, ORANGE FREE STATE, SOUTH AFRICA	187 MILES (301 KM)	2 BILLION YEARS
SUDBURY, ONTARIO, CANADA	156 MILES (251 KM)	1.85 BILLION YEARS
CHICXULUB, YUCATAN PENINSULA, MEXICO	125 MILES (201 KM)	65 MILLION YEARS
MANICOUAGAN, QUEBEC, CANADA	62 MILES (100 KM)	212 MILLION YEARS
POPIGAI, EASTERN SIBERIA, RUSSIA	62 MILES (100 KM)	35 MILLION YEARS
ACRAMAN, SOUTH AUSTRALIA, AUSTRALIA	56 MILES (90 KM)	570 MILLION YEARS
PUCHEZH-KATUNKI, WESTERN RUSSIA	50 MILES (80 KM)	220 MILLION YEARS
KARA, NORTHERN RUSSIA	40 MILES (64 KM)	73 MILLION YEARS
BEAVERHEAD, MONTANA, UNITED STATES	37 MILES (60 KM)	600 MILLION YEARS

SCARS FROM SPACE

Until the 1960s, most geologists believed that the craters dotted across the Earth were ancient volcanoes. Then analysis of lunar rock samples collected by Apollo astronauts proved that most of the Moon's craters had been gouged out by impacts of massive debris from space. Since our atmosphere provides no defense against objects larger than about 500 feet across, geologists were forced to conclude that the Earth must have suffered an equally severe pounding in the past.

The bombardment was most intense between about 4.6 and 3.8 billion years ago, when the solar system was forming and countless rocky lumps, some the size of planets, were orbiting in a disk of swirling dust and rubble around the Sun. Since then, the storm has subsided to a "drizzle," but the threat is still there. While hundreds of tons of harmless meteorite dust fall to the Earth's surface each day, estimates suggest that a meteorite half a mile wide hits the Earth every 200,000 years on average, and an object 6 miles (10 km) in diameter collides every 50–100 million years. The last major impact event on Earth was in 1908, in sparsely populated Siberia, Russia. Luckily, there was no loss of life that time.

So where are the other impact craters? Unlike the inert Moon, the Earth is very efficient at healing impact scars. Erosion wipes out the smaller depressions within a few hundred thousand years, and even craters spanning hundreds of miles are eventually obliterated by recycling of the Earth's crust. No impact craters older than two billion years have been found. Most surviving craters are either "young"—a few million years old—or are located in the geologically stable continental shields of Canada, Australia and Russia.

CRATER CREATION

While it may take millions of years to wipe away an impact crater, it only takes seconds to create one. A 60-mile (100-km) wide depression is formed in about 100 seconds. It all starts when the projectile—an asteroid or comet—hits the Earth's surface at a speed of hundreds of miles per second. At such a high velocity, crater formation is driven by shockwaves generated at the point of impact. The crater is circular even if the projectile hits the surface at an oblique angle, because the shockwaves automatically excavate a round hole and fling out a "curtain" of fragmented rock called ejecta that is dispersed over the surrounding terrain. The crater ends up about 20 times the diameter of the projectile. Fragments of the projectile survive only in small craters. Bodies big enough to produce a hole more than half a mile wide are vaporized or melted by impact pressures up to 9 million times atmospheric pressure and temperatures that may reach 8,000°F (4,430°C). But these forces inflict telltale shock damage on the rocks. Markers include deformed minerals and rock shock patterns that indicate the craters were created by forces from above. These signs can be read long after the obvious crater landmarks have gone. In Canada, the pattern of shock effects has helped geologists map the vanished rim of Manicouagan, a 62-mile (100-km) wide crater created 212 million years ago. Clearly, the Earth's impact craters have made their mark.

DOUBLE WHAMMY
The Clearwater Lakes in Quebec, Canada, conceal a rare phenomenon—twin impact craters. It is believed they were formed together by two separate but related impacts. This Shuttle view shows their close proximity.

CRATER LAKE
An infrared satellite image of Elgygytgyn Lake in Siberia, Russia, gives a clear view of what was once a fiery hole punched in the landscape by a giant meteorite. Many impact craters subsequently fill with water, which can speed the process of erosion.

Puchezh-Katunki, western Russia

Kara, northern Russia

Beaverhead, Montana, U.S.

Manicouagan, Quebec, Canada

Sudbury, Ontario, Canada

Popigai, eastern Siberia, Russia

The sheer size of most of the Earth's impact craters makes it hard to see them clearly from the ground. Orbital imagery has revealed the extent to which the Earth has been bombarded from space.

Chicxulub, Yucatan Peninsula, Mexico

Vredefort, Orange Free State, South Africa

EARTH ATTACKED

Acraman, South Australia, Australia

SANDBLAST
Located in the Namib Desert in southern Africa, the Roter Kamm crater—the bright circle at the upper center of this space radar image—is hard to see from the ground because its floor is covered by shifting sand dunes.

STANDING PROUD
The Wolfe Creek crater in Australia is one of only a few of Earth's impact craters that is well-preserved and prominently visible from the ground. The arid local environment slows down the weathering of rocks.

OCEANS

Spinning in the endless night of space, our planet looks utterly different from any other member of the Sun's family. Earth's appearance and nature are defined by oceans. These great bodies of liquid water were responsible for the beginning of life, and continue to nurture Earth's creatures, providing food and the climate necessary for life. But pollution and other changes brought about by humans affect the planet's water reservoir, and if the damage is too great, all of Earth's living things will suffer.

WHAT IF...

...THE SEA LEVEL FELL?

Throughout Earth's geological history, the level of the oceans has risen and fallen due to slow natural cycles of heating and cooling. When the temperature falls, water becomes locked up in glaciers and ice caps, and the sea level falls. But when the temperature rises, the glaciers melt and the ice caps recede. Water is freed up and the oceans rise. The natural causes of these cycles are not well understood, but may be associated with changes in solar activity.

Scientists have determined that we have been in a warming cycle for the past 20,000 years. Until the 19th century, this development was an entirely natural cycle. But, since then, there has been a noticeable increase in global warming. It is not certain that increased global warming is caused by air pollution. But it is a fact that the average temperature has increased 1°F, and the ocean level has risen 10 inches (25 cm), over the last century.

One expected result of global warming is that ice will melt and the sea level rise. But is a continuous increase in temperatures and ocean level inevitable? Some scientists have suggested that global warming could cause large-scale changes in ocean currents. Any major alteration of the finely balanced system of ocean currents could have incalculable consequences—and a new ice age is as likely as a hot "greenhouse" world.

A rise in sea level with global warming is widely forecast. But some scientists do not expect the period of global warming to last, although its aftereffects could continue for several thousand years. Without global warming, the natural cycle could bring another ice age in about 50,000 years. Periodic glaciations are part of the natural cycle of climatic change, and any future glaciation would lock much of the Earth's liquid water into ice, so the sea level would fall.

Any climatic change in the distant future could be exaggerated by a shift in the Earth's orbit—which is possible over a period of around 41,000 years. And, over a period of about 22,000 years, a likely variation in the Earth's axis could complicate matters. Such changes would alter the alignment of the seas and land masses to incoming solar radiation, and bring changes to ocean currents and the position of climatic belts.

An icy world with smaller oceans is only one possible outcome. A fall in sea level as the ice caps increased would uncover areas of shallow sea as smooth new coastal plains. In places, "bridges" of freshly dry land might link continents, block ocean currents, and so trigger further climatic changes.

OCEANS AND MAJOR SEAS

Name	Area SQUARE MILES/SQUARE KM	Average depth FEET/METERS
Pacific Ocean	64,000,000/165,753,600	13,740/4,190
Atlantic Ocean	33,420,000/865,544,458	12,250/3,730
Indian Ocean	28,350,000/73,423,665	12,700/3,870
Arctic Ocean	5,110,000/13,234,389	3,960/1,210
South China Sea	1,148,500/2,974,500	4,000/1,220
Caribbean Sea	1,063,000/2,753,063	8,600/2,620
Mediterranean Sea	966,750/2,503,786	4,690/1,430
Bering Sea	875,750/2,268,105	4,700/1,435

WATER WORLD

Some 4.5 billion years ago, not long after the planets of the solar system were formed, our planet had a very different appearance. Instead of being an azure gem, Earth was an airless, bleak, asteroid-blasted chunk of hot rock, little different from nearby Venus and Mars. Luckily, Earth did not stay desolate. Our planet soon began to produce and accumulate the stuff of life: water.

Volcanic activity from the planet's fiercely dynamic interior released steam. Emerging into the cold, this steam condensed and fell back as the first rain. The Earth's new and growing atmosphere prevented this water from boiling into space, and it began to accumulate—first in pools and then in huge expanses.

At that time, the solar system was a more crowded place, and Earth was subjected to countless impacts. Many of the intruders must have been icy bodies, such as comets, that added to the young planet's store of liquid water. Both Venus and Mars may have also possessed considerable amounts of liquid water. But on Venus—which is closer to the Sun and hotter—the water simply boiled away, while on Mars—which is farther from the Sun and cooler—any seas would have been frozen and evaporated. The Earth was just the right size and temperature to retain its water.

It is impossible to exaggerate the importance of the oceans to life on Earth. Without them, we would not be here. The most favorable place for the beginning of life, some 3 billion years ago, was the

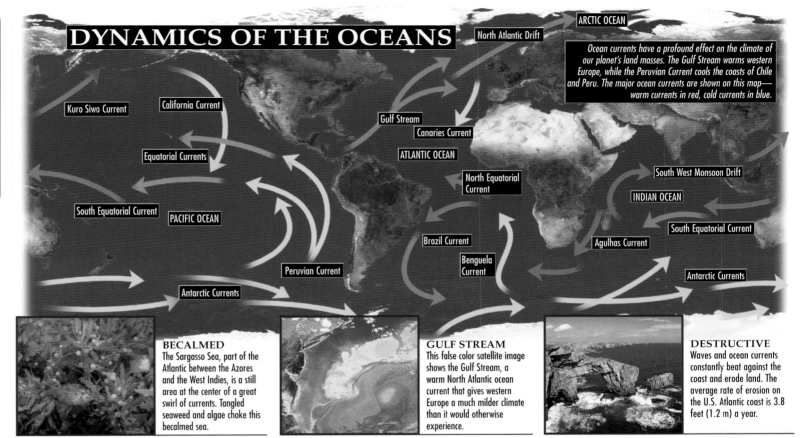

DYNAMICS OF THE OCEANS

ARCTIC OCEAN

North Atlantic Drift

Ocean currents have a profound effect on the climate of our planet's land masses. The Gulf Stream warms western Europe, while the Peruvian Current cools the coasts of Chile and Peru. The major ocean currents are shown on this map— warm currents in red, cold currents in blue.

Kuro Siwo Current

California Current

Gulf Stream

Canaries Current

ATLANTIC OCEAN

Equatorial Currents

North Equatorial Current

South West Monsoon Drift

INDIAN OCEAN

South Equatorial Current

PACIFIC OCEAN

Brazil Current

Agulhas Current

South Equatorial Current

Peruvian Current

Benguela Current

Antarctic Currents

Antarctic Currents

BECALMED
The Sargasso Sea, part of the Atlantic between the Azores and the West Indies, is a still area at the center of a great swirl of currents. Tangled seaweed and algae choke this becalmed sea.

GULF STREAM
This false color satellite image shows the Gulf Stream, a warm North Atlantic ocean current that gives western Europe a much milder climate than it would otherwise experience.

DESTRUCTIVE
Waves and ocean currents constantly beat against the coast and erode land. The average rate of erosion on the U.S. Atlantic coast is 3.8 feet (1.2 m) a year.

warm ocean. Only very much later did living things move to dry land.

LIQUID LIFELINE

The waters continue to aid life. Not only do the oceans provide people and animals with abundant food in the form of fish, but they also keep the world habitable as minute sea-living algae, called phytoplankton, produce atmospheric oxygen.

The oceans are the world's "thermostat": They prevent the planet from getting too hot or too cold. Because water is very good at absorbing and retaining heat, the seas moderate the climate by serving as a heat sink. The excess heat that oceans absorb and store during the summer is released in

winter, maintaining a relatively constant temperature on dry land.

The activities of modern humankind put the oceans in peril. Sewage and chemicals are thoughtlessly dumped into the oceans, jeopardizing the sea creatures that much of humanity depends on for food. And there is considerable evidence to suggest that global warming may be melting the polar ice caps at an increasing rate, causing the levels of the oceans to rise and threatening major coastal cities and entire low-lying island-nations—such as the Maldives in the Indian Ocean and Tuvalu in the Pacific—with inundation.

Of all the planets we know, Earth alone has liquid water. To this resource we owe our existence. And we depend upon it for our future.

EARTH'S TIDES

G ravity, the force that binds the universe, is also the key to the Earth's rising and falling tides. The combined gravitational effects of the Sun and the Moon constantly pull the world's oceans in different directions and create tidal effects. But there are many other factors that complicate this basic process. Friction, the Earth's spin, the tilt of its axis and the waning power of the Moon's attraction on the far side of the Earth all conspire to make our planet's oceans a complex battleground of forces.

WHAT IF...

...THE MOON CAME CLOSER?

D uring the 1850s, French astronomer Edouard Roche (1823–63) tried to work out what would happen if the Moon were nearer the Earth. He calculated that the tidal force would increase by the inverse cube of the distance. So if the Moon were just a third of its present distance, the force would be 27 times greater.

This implies that ocean tides would be so enormous that they would regularly swamp many of the world's lowland regions. Most of New York and London, for example, would be flooded at every high tide. The tidal force would also be strong enough to cause parts of the Earth's crust to flex up and down with the tides, resulting in catastrophic earthquakes. The Moon would be affected even more. According to Roche's calculations, if the Earth and the Moon were as little as 9,670 miles (15,560 km) apart, the tidal effect would be strong enough to pull the Moon to pieces.

...VOLCANOES WERE TIDAL?

B ack in 1979, the Voyager space probes flew past Jupiter's moon Io and revealed it to be the most volcanically active world in the solar system. Io is so close to Jupiter that it is subjected to huge tidal forces from the giant planet's massive gravitational pull. These forces are so powerful that they continually squeeze and stretch Io, heating it up like a paper clip that is bent back and forth.

In recent years, scientists have begun to speculate whether the Earth's volcanoes may be affected by the tides. The Moon's pull is far too weak to provide the internal heating effect that drives Io's volcanoes, but it may be enough to trigger imminent eruptions: When magma wells up beneath a volcano and threatens to erupt, it may be the passage of the Moon overhead that tips the balance. There is evidence to suggest that this was the case during the century's biggest eruption, that of Mount Pinatubo in the Philippines in 1991.

U.S. TIDAL VARIATIONS

	CITY	MAXIMUM TIDAL VARIATION
1	SEATTLE, WASHINGTON	14.9 FT/4.5 M
2	CRESCENT CITY, CALIFORNIA	10.2 FT/3.1 M
3	SAN DIEGO, CALIFORNIA	8.9 FT/2.7 M
4	GALVESTON, TEXAS	2.4 FT/0.7 M
5	MOBILE, ALABAMA	2.3 FT/0.7 M
6	CHARLESTON, SOUTH CAROLINA	8.0 FT/2.4 M
7	WASHINGTON, D.C.	4.1 FT/1.2 M
8	NEW YORK, NEW YORK	7.0 FT/2.1 M
9	BOSTON, MASSACHUSETTS	14.7 FT/4.5 M
10	PORTLAND, MAINE	13.5 FT/4.1 M

MAKING WAVES

The water of the oceans is bound to our planet by the force of the Earth's gravity. But the Earth is not alone in space, and both the Moon and the Sun throw their own gravitational pulls into the equation. The combined effect is to tug the oceans this way and that around the globe.

The Moon's gravity stretches the Earth into an oval. The effect is so tiny that the land masses of the planet are distorted by little more than eight inches. But because of water's fluidity, the effect on the oceans is more noticeable. At the point on the Earth directly beneath the Moon, the ocean is tugged into a bulge of high water. At the same time, a second tidal bulge forms on the opposite side of the planet. This is partly a result of the centrifugal force created by the Moon and Earth's combined rotation around their common center of mass, a theoretical point called the barycenter.

Because the Earth spins on its axis once every 24 hours, the two bulges sweep around the planet in waves, creating two high tides per day at every point on the globe. But the twice-daily cycle is complicated by the fact that the Earth is tilted, which puts the Moon alternately to the north and south of the Equator. This creates slight differences between the two tides each day and adds a daily set of local variations to the twice-daily rhythm.

IMPERIAL LESSONS

When the Roman Emperor Julius Caesar set out to invade Britain in 55 B.C., his knowledge of tides was confined to the land-locked Mediterranean where tidal effects are minimal. On his arrival in Britain, Caesar landed his ships on a sloping beach at low tide—and nearly lost them all when the tide came in!

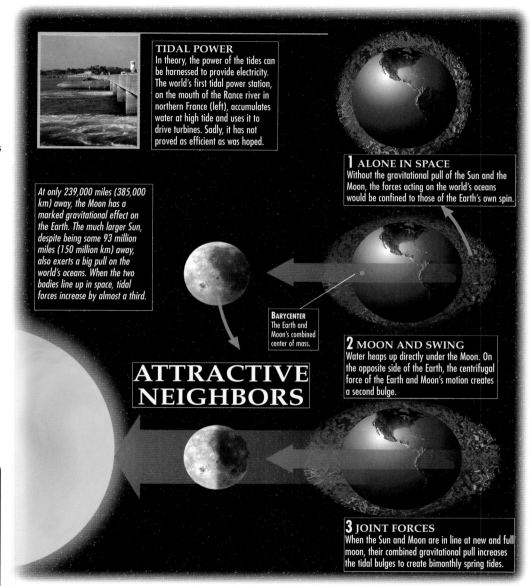

TIDAL POWER
In theory, the power of the tides can be harnessed to provide electricity. The world's first tidal power station, on the mouth of the Rance river in northern France (left), accumulates water at high tide and uses it to drive turbines. Sadly, it has not proved as efficient as was hoped.

At only 239,000 miles (385,000 km) away, the Moon has a marked gravitational effect on the Earth. The much larger Sun, despite being some 93 million miles (150 million km) away, also exerts a big pull on the world's oceans. When the two bodies line up in space, tidal forces increase by almost a third.

ATTRACTIVE NEIGHBORS

1 ALONE IN SPACE
Without the gravitational pull of the Sun and the Moon, the forces acting on the world's oceans would be confined to those of the Earth's own spin.

BARYCENTER
The Earth and Moon's combined center of mass.

2 MOON AND SWING
Water heaps up directly under the Moon. On the opposite side of the Earth, the centrifugal force of the Earth and Moon's motion creates a second bulge.

3 JOINT FORCES
When the Sun and Moon are in line at new and full moon, their combined gravitational pull increases the tidal bulges to create bimonthly spring tides.

THE PLOT THICKENS

A further complication is added by the Sun, whose gravitational pull on the Earth also affects the tides. The tidal force of the Sun and Moon together is almost a third more than that of the Moon alone, with the Sun imposing a solar rhythm on the lunar rhythm. At the new and full moons, when the two bodies are in line, they combine to create extra-high spring tides. When the Moon is in its first and last quarters, the Sun is at right angles to it, and

PREDICTING TIDES

As late as the 1960s, tide machines were used to predict local tidal patterns. The first one (right) was created by William Thomson (later Lord Kelvin, of temperature scale fame) in 1873. An early form of computer, it used cogs and pulleys to calculate the various tidal factors.

their gravitational pulls work against each other to create extra-low neap tides.

If the Earth were completely smooth, this would probably be the end of the story. But in reality, the tidal forces are weakened by friction between the ocean and the seabed to the point where the twice-daily tidal waves get slightly left behind the orbiting Moon. The waves are also continually disrupted by areas of land as they sweep around the Earth, creating yet more local variations.

At the same time, the Earth spinning on its axis causes the tidal waves to oscillate around the world's ocean basins like water in a bath. This means that high tides do not necessarily occur when the Moon is overhead, but when the oscillations accumulate to their greatest height. Each ocean basin is a different shape and so has its own pattern of oscillations. In the South Atlantic, for example, the tides oscillate from south to north and take around 12 hours to sweep from the tip of South Africa to the Equator. In the North Atlantic, they sweep in a counterclockwise direction.

Until recently, the sheer complexity of the tidal forces acting on the Earth meant that the only way to predict tides was by years of patient observation. Now, computer programs do the job—a development for which oceanographers are extremely grateful.

EARTH'S ATMOSPHERE

The atmosphere is wrapped around our planet like a protective shield. Without it, life as we know it could never have evolved. Compared with the Earth itself, the atmosphere is no thicker than the skin on an orange. Yet this wafer-thin layer keeps us supplied with air to breathe and water to drink, filters out harmful rays and maintains our mild climate. The atmosphere has taken millions of years to evolve and is still evolving today. A crucial factor in that evolution is the very life that it helps to support.

WHAT IF...

...PLANTS MADE LESS OXYGEN?

The twin problems of deforestation and air pollution have forced scientists to look closely at the link between those parts of the Earth covered by plants and the balance of gases in the atmosphere.

In recent years, we have learned that the interaction between the atmosphere, the hydrosphere (oceans and lakes) and the biosphere (all living organisms) is far more important and complex than anyone had thought. We know for certain that the Earth's oxygen supply comes entirely from plants. If there were a lot fewer plants—whether through human destruction or from natural causes—there would be less oxygen in the atmosphere. Not only would we find it hard to breathe, but the ozone layer would thin out further, exposing us to harmful ultraviolet radiation from the Sun. We might even find it hard to light fires and run engines.

...THE GREENHOUSE RAN WILD?

Our nearest planetary neighbor stands as a warning of what could happen if our atmosphere changes too much. Venus is only a little nearer to the Sun than we are and is roughly the same size as the Earth. But it is the hottest planet in the solar system, with surface temperatures that soar to an infernal 840°F (450°C). Why?

Venus was always slightly too warm for the water vapor in its atmosphere to condense, so oceans never formed. Water vapor is also a major greenhouse gas, so the surface of Venus became hotter still. With no oceans to absorb carbon dioxide, huge quantities of this, too, remained in the Venusian atmosphere, where it increased the greenhouse effect. The result is runaway global warming that has turned Venus into an oven.

The future may see the need to manufacture ozone artificially. A single factory could meet present requirements. The big problem would be how to release it into the stratosphere.

IN THE AIR TONIGHT

COMPOSITION

NITROGEN	78.08%
OXYGEN	20.95%
ARGON	0.93%
CARBON DIOXIDE	0.03%

PLUS TRACES OF NEON, KRYPTON, XENON, HELIUM, NITROUS OXIDE, METHANE, CARBON MONOXIDE

LAYER	HEIGHT	TEMPERATURE RANGE
TROPOSPHERE	UP TO 7.5 MILES (12 KM)	65°F TO −75°F (18°C TO −24°C)
STRATOSPHERE	7.5 TO 31 MILES (12 TO 50 KM)	−75°F TO 32°F (−24°C TO 0°C)
MESOSPHERE	31 TO 50 MILES (50 TO 80 KM)	32°F TO −100°F (0°C TO −38°C)
THERMOSPHERE	50 TO 440 MILES (80 TO 708 KM)	−100°F TO 3,600°F (−38°C TO 1,982°C)
EXOSPHERE	440 TO 625 MILES (708 TO 1,006 KM)	3,600°F (1,982°C)

LIVING IN AN OCEAN OF AIR

The atmosphere is the most taken-for-granted of our natural resources. Water, food, warmth and even light may be in short supply, but air has always been free and plentiful. Now, pollution and climatic changes are forcing us to treat our atmosphere with more respect.

We live at the bottom of an ocean of air that has several distinct layers. The troposphere—the lowest layer, and the only one in which we can survive unprotected—gets its warmth from the ground and from the oceans, which are heated by the Sun. Temperatures in the troposphere fall as we leave the warm ground behind. The rate of cooling is remarkably even, at about 10°F (6°C) per 3,000 feet (915 m).

The troposphere is perfectly suited to life on Earth: Nowhere else in the solar system possesses a combination of gases quite like it— but neither did the Earth when the planet was young and lifeless.

Scientists still argue about to what extent life

HIGH DRAMA

AN ATTEMPT TO MONITOR THE ATMOSPHERE IN 1869 NEARLY ENDED IN TRAGEDY. JAMES GLAISHER OF THE GREENWICH ROYAL OBSERVATORY ASCENDED IN A BALLOON FROM WOLVERHAMPTON, ENGLAND. AT 29,000 FEET (8,840 M) HE LOST CONSCIOUSNESS. LUCKILY, GLAISHER'S COMPANION RELEASED A GAS VALVE JUST BEFORE HE, TOO, FAINTED. THEY DESCENDED SAFELY.

THE LOWER ATMOSPHERE

39,000 FEET (12,000 METERS)
On the edge of the stratosphere, there is little turbulence. The air, made poisonous by ozone, contains almost no water vapor. Pilots love the stratosphere for its smooth, thin air and fast jet-stream winds that help to speed airplanes on their way.

10,000 FEET (3,000 METERS)
This is as high as most people get with their feet still on the ground. In atmospheric terms, it is only just above sea level. Even so, there is already a significant drop in temperature and oxygen levels and a marked increase in UV radiation.

SEA LEVEL
The lowest layer of the atmosphere, the troposphere, accounts for 75% of the atmosphere's total mass. It also contains vast amounts of dust and water vapor. As the Sun warms the ground, it keeps the mixture churning and creates the phenomenon that we call weather.

has shaped the atmosphere, or has been shaped by it. But samples of primeval air taken from ice cores and amber (fossilized tree resin) reveal that the balance of gases in the air has remained more or less steady over the last few hundred million years, with an oxygen level of around 21%.

If the level were any lower, breathing would be difficult; if it were higher, plants would burst into flames. It remains to be seen whether 250 years of industrialized society will have any long-term effects on this balance.

ABOVE THE CLOUDS

As we leave the troposphere behind, the atmosphere becomes less hospitable. The next layer, the stratosphere, starts at around 39,000 feet (12,000 m). At this point the air thins noticeably and temperatures begin to rise, rather than fall, as exposure to the full force of the Sun increases.

Even so, without the stratosphere, life below would be impossible. Much of the oxygen in the stratosphere is in the form of ozone, which is

not breathable, but which provides an excellent filter for the harmful ultraviolet radiation contained in sunlight. The lower stratosphere also contains significant amounts of water vapor and carbon dioxide. These two gases work like the panes of glass in a greenhouse, trapping the heat that is radiated from the sun-warmed ground below and preventing it from being lost to space. It is this "greenhouse effect" that keeps us comfortably warm.

At about 30 miles (50 km) up, the mesosphere begins. Here the atmosphere is far too thin to breathe, but it still acts as a shield against space debris. Most of the debris, consisting of meteoroids and dust from comet tails, is burned up by friction as it hurtles through the rarefied air at speeds of up to 44,000 miles (70,000 km) per second. Each year, the mesosphere accounts for around 50,000 tons (45,000 tonnes) of debris. Most of the pieces are no bigger than sand grains.

Still farther up, starting at about 50 miles, is the thermosphere. At this height there is scant protection from the glare of the Sun and temperatures soar to 3,600°F (1,980°C). The "air," by now mostly oxygen, is only one ten-thousandth of its sea-level density, yet it is still thick enough to drag satellites down from their orbits within a few years.

The final layer, more than 400 miles (640 km) above the Earth, is called the exosphere. Consisting mostly of helium and hydrogen, this becomes steadily more rarefied until, at about 600 miles, it is no denser than the wind of solar particles that fills interplanetary space.

93

THE AURORAE

...THE SUN BECAME EVEN MORE ACTIVE?

In the past decade or so, scientists have become increasingly interested in the links between the aurorae and solar activity. It has become clear that surges in the solar wind are not just linked to spectacular auroral displays, but can affect electrical and communications equipment, too.

Inhabitants of North America and Canada have suffered frequent power cuts due to sudden increases in solar activity. Satellites are even more vulnerable: In March 1991, a surge played havoc with the Earth's magnetosphere, completely knocking out the MARECS-1 communications satellite and badly damaging the NOAA's weather satellite GOES-7. So it could be that although a more active Sun might result in auroral displays to rival a New Year's fireworks display, our increasingly technologically dependent society pays a heavy price.

...THE AURORAE MOVED?

One of the most fascinating features of the Earth's magnetic field is that it is known to have reversed direction again and again over geological time. There are some scientists who believe that it is doing so at this moment. They point to the fact that the field has been getting steadily weaker since measurements of it were first taken in the early 19th century.

It is thought that when the magnetic field reverses, it does not simply switch off and then start again in the other direction; instead, it becomes distorted and grows weaker. If this happened, the Earth would be stripped of its protective cocoon and the solar wind would be able to penetrate a far greater proportion of the upper atmosphere. Under these circumstances, it is likely that aurorae would begin to "creep" toward the equator. Instead of the rarity in subtropical latitudes that they are today, they could become as frequent a sight in Florida and Spain as they are in Alaska and Lapland.

That's show business: If the Earth's magnetic field begins to reverse and the solar wind penetrates more of the atmosphere, the aurorae could become a regular sight in the night sky of some unlikely places.

The "northern lights" of the aurora borealis are among nature's most beautiful sights. With shimmering curtains of colors, dazzling white streamers and bright green rays flashing with red, the aurorae regularly stage spectacular displays in the polar night skies. For centuries, the cause of this apparently supernatural light show remained a mystery. Only recently have we come to realize that they are created by high-energy particles that stream from the Sun and collide with the Earth's atmosphere.

THE AURORAE IN HISTORY

37 A.D.	A DISPLAY SEEN FROM ROME TRICKS THE EMPEROR TIBERIUS INTO SENDING SOLDIERS TO PUT OUT WHAT IS SAID TO BE A "FIRE" IN THE PORT OF OSTIA.
93–839 A.D.	DISPLAYS SEEN IN SCOTLAND ARE VARIOUSLY DESCRIBED AS "ARMIES FIGHTING IN THE HEAVENS" AND "POOLS OF BLOOD IN THE FIRMAMENT."
MARCH 16, 1716	AN AURORA OVER LONDON, ENGLAND, IS NOTED BY THE ASTRONOMER SIR EDMUND HALLEY.
SEPTEMBER 2, 1859	DISPLAYS REPORTED IN HAWAII AND THE TROPICS ARE LINKED TO THE ACTIVITY OF SOLAR FLARES.
FEBRUARY 4, 1872	DISPLAYS ARE VISIBLE IN INDIA, EGYPT AND THE CARIBBEAN.
SEPTEMBER 25, 1909	DISPLAYS ARE VISIBLE IN SINGAPORE (LATITUDE 1°25' NORTH).
MAY 15, 1921	DISPLAYS ARE VISIBLE IN SAMOA (LATITUDE 14° SOUTH).
JANUARY 25, 1938	A BLOOD-RED DISPLAY IS SEEN FROM ENGLAND, SPAIN AND PORTUGAL.
MARCH 30, 2000	SPECTACULAR DISPLAY VISIBLE AS FAR SOUTH AS TEXAS AND FLORIDA.

LIGHT SHOW

The aurorae (pronounced "or-ror-ree") are not occasional freaks of nature: They are a permanent feature of the Earth's upper atmosphere. Auroral displays vary in intensity, and sometimes they fade away almost to nothing. But they are always there.

There are aurorae in both hemispheres of the Earth—the aurora borealis around the North Pole and the aurora australis around the South Pole. On satellite pictures, they show up as oval bands that encircle the Earth's magnetic poles like giant halos. The size and shape of these halos change continuously, but they never completely disappear. The aurorae are also gigantic and stretch high into the atmosphere. The lowest fringes hang about 50 miles (80 km) above the ground; the upper rays extend more than 600 miles (960 km) into space—three times farther than the Space Shuttle's orbit.

An auroral display resembles a giant television. In a TV, streams of electrically charged particles (electrons) from the tube are guided by a magnetic field onto the lines of the screen, causing the lines to glow. In an auroral display, charged particles from the Sun strike the atoms and molecules of the Earth's atmosphere, causing them to glow in a similar way. This stream of particles is called the solar wind: It radiates continuously from the Sun's corona at over 300 miles (480 km) per second.

Fortunately for us, most of the Earth is shielded from this hurricane of charged particles by its magnetic field, which surrounds the planet like a cocoon. But there are two holes in the magnetic field, one above each magnetic pole. It is through these holes, called the polar clefts, that the solar wind pours, giving rise to the aurorae.

MIXING COLORS

Aurorae appear in so many colors because each gas in the atmosphere glows a different hue when struck by solar particles. The color also varies according to both the electrical state and concentration of the gas.

THE AURORA

MAKING TRACKS
Aurorae follow the angle of the Earth's magnetic field lines. This angle varies depending on longitude.

LOW OXYGEN
When solar particles strike oxygen atoms, the atoms glow green or red. In the lower atmosphere, there is mostly a green glow because the concentration of atoms is high.

HIGH OXYGEN
In the upper atmosphere there are fewer oxygen atoms. The rate of collision between the oxygen atoms is less, so red auroral displays predominate.

BLUE NITRO
Particle collisions with nitrogen are responsible for blue, purple and red displays, depending on whether the nitrogen atoms are charged (ionized) or not.

For example, oxygen, when struck at low altitudes (about 60 miles/100 km up), glows a brilliant green—the most common of auroral colors; at higher altitudes (around 200 miles/320 km), it results in the vivid red aurorae that are seen during major disturbances. Nitrogen atoms, by contrast, glow blue when electrically charged (ionized) and red when neutral. Nitrogen can also emit purple light, which happens when the atoms are struck by the ultraviolet radiation contained in sunlight.

Although aurorae are ever-present, they are also ever-changing. Satellites such as the IMP-8

(Interplanetary Monitoring Platform) have monitored the solar wind. From their observations we know that aurorae are at their most spectacular when the solar wind blows fiercely enough to create magnetic storms—disturbances in Earth's magnetic field. We also know that mirror-image aurorae flare simultaneously around the North and South poles. Yet it may still be some time before we fully understand the complex relationship between aurorae, the Sun and the Earth's magnetic field. For now, as they have done for centuries, the "northern lights" remain a beguiling mystery.

LIGHT NAMES

THE AURORAE HAVE BEEN GIVEN VARIOUS NAMES IN DIFFERENT PARTS OF THE WORLD. IN SCOTLAND YOU WOULD HEAR THEM REFERRED TO AS THE "MERRY DANCERS" OR THE "HEAVENLY DANCERS." THE NAME "AURORA" ITSELF COMES FROM THE LATIN WORD MEANING "DAWN." ITS FIRST KNOWN USE WAS IN A BOOK WRITTEN IN 1649 BY THE FRENCH ASTRONOMER PIERRE GASSENDI.

EARTH'S MOON

The Earth's giant satellite is a constant feature in our skies, changing its appearance as its orientation to the Sun shifts throughout each lunar orbit, but always keeping the same face turned toward us—trapped this way in aeons past by the same tidal forces that raise and lower the Earth's own seas. Earth is the only terrestrial planet with such a large satellite, and its origin was a mystery until relatively recently. Rocks returned from the Apollo lunar landings showed that it has a mixture of Earth-like and alien rocks, and it now seems that the Moon was born in a "big splash" collision between the still-molten Earth and a small rogue planet early in its history. Due to its lack of an atmosphere and geological activity, the Moon acts as an astronomical time capsule—it preserves details of meteor impacts from billions of years ago, and has been used to piece together a history of the inner Solar System. Indeed, some astronomers think that the Moon's influence has helped stabilise Earth's rotation and protect it from some meteor impacts, perhaps fostering the development of life.

An arresting image of the moon setting behind the Earth, taken from the space shuttle Discovery. The hugely ambitious and successful Apollo program of the late 1960s and 1970s allowed scientists a remarkable insight into the nature and history of our only natural satellite.

THE MOON

The Moon is the most familiar sight in the night sky. This is because it is the Earth's closest companion and travels with the Earth in space around the Sun. The Moon orbits the Earth at a distance of just 238,850 miles (384,392 km); yet, despite being so near, it is very different from our own planet—a gray desert, dotted with craters from ancient asteroid collisions. So far the Moon is the only body in space on which humans have landed. The recent discovery of water there has stimulated plans for an eventual return.

WHAT IF...

...WE RETURNED TO THE MOON?

Twelve American astronauts landed on the Moon between 1969 and 1972 during the Apollo program. No one has been back since, although NASA plans to send astronauts back to the satellite before the year 2020. NASA is also developing plans for a Mars mission that may be launched from the Moon. Analysis of the rocks brought back by the Apollo astronauts show that the Moon contains useful amounts of metals, such as aluminum, iron, titanium and magnesium, which could be used to build spacecraft. And since the Moon's gravity is weaker than the Earth's, launching rockets from there would be far easier.

...WE ESTABLISHED A MOON BASE?

The discovery of water ice at the Moon's poles has stimulated plans to set up bases using pressurized cylinders, like those used in present-day space stations. The cylinders will be covered with lunar soil to protect them from cosmic rays and meteorites.

At first, scientists will use the bases to explore the mountains, craters and valleys of the Moon to find out more about its history. Observatories will be established to obtain a clearer view of the sky than is possible from Earth, where our atmosphere gets in the way. It will also be desirable to set up radio telescopes, to study the cosmos free of interference from radio transmissions on Earth.

Energy to power the lunar bases will come from sunlight. Solar power will also be used to convert the water ice at the poles into hydrogen and oxygen for fuel; oxygen for breathing will be extracted from the deposits currently "locked away" as oxides in the Moon's rocks. The same rocks could be used to extend the bases. Plants for food will be grown in Moon soil, with added water and fertilizer; farm animals and fish will be kept in pressurized domes.

Eventually, we will mine the Moon for the valuable minerals that it contains. Instead of rockets, magnetic ramps called mass drivers may be used to propel containers of Moon rocks into space. The rocks will then be ferried to Earth or processed in space factories. One day it may also be possible for tourists to take vacations on the Moon, living in lunar hotels and visiting sites such as Tranquillity Base, where the first Apollo astronauts landed in 1969.

MOON FACTS & FIGURES

DIAMETER	2,160 MILES (3,480 KM)
AXIS TILT	6° 41' RELATIVE TO ITS ORBIT
TIME TO ORBIT EARTH	27.3 EARTH DAYS
LENGTH OF DAY	27.3 EARTH DAYS
DISTANCE FROM EARTH	238,850 MILES (384,392 KM)
SURFACE TEMPERATURE	253°F/123°C (DAY) TO −387°F/−197°C (NIGHT)
SURFACE GRAVITY	0.17 OF THE EARTH'S
MASS	0.012 OF THE EARTH'S
VOLUME	0.02 OF THE EARTH'S
DENSITY	3.34 TIMES THAT OF WATER

MOONSCAPE

Since the Moon is so close, we can see its most prominent features with the naked eye. Most obvious are the dark areas that form the familiar "Man-in-the-Moon" pattern. In reality these are lowlands, formed by giant meteorites that smashed into the Moon long ago, which were then filled by dark lava. They are called maria, a Latin word meaning "seas" (singular: mare); although there has never been any water in them, that is what they looked like to early observers. Many of the maria are given fanciful names, such as Mare Tranquillitatis (Sea of Tranquillity) where Apollo 11 astronauts landed in 1969. The bright areas on the Moon are highlands.

The rugged, colorless appearance of the Moon is in stark contrast to the surface of the Earth. There is no air on the Moon, so there are no clouds to spoil our view. A look through a pair of binoculars reveals that the Moon is pitted with craters, the largest of which can engulf a fair-sized city. These, too, were formed long ago by meteorite impacts. Some of the craters are bordered by bright streaks, called rays, which consist of crushed rock thrown out by the crater-forming impacts. Lunar craters are named after famous people, mostly astronomers. The most notable example is Tycho, in the Moon's southern hemisphere, where the formation is particularly noticeable around the time of a full moon.

BIRTH OF THE MOON

Astronomers believe that the Moon was born about 4.6 billion years ago when the youthful Earth was hit by another body, larger than the present Moon. The colliding body shattered completely under the force of the impact, which also melted part of the Earth's outer layers. The debris from the collision then went into orbit around the Earth, where it collected together to form the Moon. The lack of air and liquid water means that there is no erosion, with the result that the Moon's surface features have remained virtually unchanged for millions of years.

Whenever we look at the Moon, we always see the same side. This is because the Moon turns on its axis in exactly the same time (27.3 days) that it takes to orbit Earth—a phenomenon known as "synchronous rotation." Until space probes orbited the Moon, no one knew what the far side looked like. The first probe to photograph the Moon's far side was Russia's Luna 3 in October 1959; since then, it has been fully mapped. The main difference is that the far side is mostly covered with bright, crater-marked highlands, and has fewer large, dark mare areas.

MARIA
The darker areas of the Moon's surface. They are depressions created by the impact of giant meteorites which were filled with dark lava when the Moon experienced volcanic activity.

HIGHLANDS
The bright areas on the Moon are craggy highlands, whose light rocks reflect the sunlight. This contrasts with the dark, sunlight-absorbing rock in the lava-filled maria.

CRATERS
The surface of the Moon is pitted with craters, the largest of which can engulf a city. They, too, were formed long ago by meteorite impacts. Some craters at the poles never receive sunlight, and in 1998 were found to contain traces of water ice.

RAYS
Some craters are surrounded by bright streaks called rays. These consist of crushed rock thrown out of the crater at the time of impact.

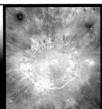

FAR HORIZONS

BECAUSE THE MOON IS ONLY ONE-QUARTER THE SIZE OF THE EARTH, ITS SURFACE IS MORE NOTICEABLY CURVED. THIS ALSO MEANS THAT THE MOON'S HORIZON IS MUCH CLOSER—SO MUCH SO THAT IF YOU WERE STANDING INSIDE A LARGE CRATER ON THE MOON, YOU WOULD BE UNABLE TO SEE THE SURROUNDING WALLS BECAUSE THEY WOULD BE BELOW THE HORIZON!

MOON DUST
The Moon is covered in very fine dust called regolith. It is made from rock pulverized by eons of meteor impacts. Some pieces of rock are shaped like droplets, where the rock has been melted and splashed across the surface.

ANATOMY OF OUR MOON

CHANGING VIEWS OF THE MOON

Ever since humans first looked up to the night sky, the Moon has inspired our imagination. The face of night has—over thousands of years—represented countless gods to numerous civilizations. But with the invention of the telescope, the Moon quickly took on a different complexion. Eyes, nose and mouth gave way to rocks, craters and ancient volcanic flows. Today the Moon holds a special place in human history for another reason—it is the only world, other than the Earth, that bears the prints of human feet.

WHAT IF...

...THE MOON COULD TELL US MORE ABOUT THE EARTH'S BIRTH?

More than 50 spacecraft have successfully gone to the Moon since the first—Luna 1, in 1959. With thousands of photographs taken of the lunar surface every year, astronomers know more about our Moon than any body—except the Earth—in the universe. But despite all the missions and photographs, the Moon still has plenty to teach astronomers. Further exploration of the lunar surface may reveal clues, not only to the formation of the Moon, but also the birth of our own planet.

Today, the most popular theory of the creation of the Moon is the "great impact" theory. This idea suggests that both Moon and Earth were born from the same molten mixture around 4.5 billion years ago. The two bodies were separated when a massive planetoid crashed into the newly formed Earth. A large cloud of debris was thrown up into orbit around our infant planet, and eventually this material coalesced and become the Moon.

Since the cataclysmic impact, the Earth and her sister Moon have evolved in very different ways.

On Earth, weather and geological activity continually reshapes the landscape. Rocks have been broken and reformed in an endless cycle, so evidence of the planet's geological past is constantly being erased: Throughout the Earth's long history, its surface has never been older than a few hundred million years. On the Moon, the story is very different. Geological activity there ceased long ago—relatively soon after the Moon's formation. This means lunar rocks are evidence of what the Earth was like soon after the impact.

Of the 50 missions that have reached the Moon, nine have brought lunar rock back to Earth. The oldest samples returned were formed 4 billion years ago—just 500 million years after the supposed collision.

The possibility of finding rocks older than these—perhaps even from before the impact—is something of a holy grail for astronomers. An opportunity to analyze such ancient rock would fill many of the gaps that currently exist in our understanding of the Earth's history and planetary formation in general.

To learn the Earth's complete history, we must go back to the Moon. Our distant sister's scarred, beaten face can teach us far more about planetary formation than the Earth's apparently youthful skin.

ILLUMINATING THE MOON

c. 350 B.C.	ARISTOTLE BELIEVES THE MOON IS AN INHABITED WORLD
1609	GALILEO GALILEI PRODUCES SOME OF THE FIRST LUNAR DRAWINGS
1651	ITALIAN ASTRONOMER GIOVANNI RICCIOLI NAMES MOST FEATURES ON THE MOON
1959	SOVIET SPACE PROBE LUNA 3 IS THE FIRST TO PHOTOGRAPH THE FAR SIDE OF THE MOON
1969	APOLLO 11 ASTRONAUTS RETURN TO EARTH AFTER THEIR SUCCESSFUL TRIP TO THE MOON, BRINGING SAMPLES OF LUNAR ROCK
1972	HARRISON SCHMITT, APOLLO 17 ASTRONAUT AND GEOLOGIST, TRAVELS TO THE MOON
1994	CLEMENTINE PROBE MAPS GRAVITATIONAL VARIATIONS TO REVEAL PREVIOUSLY UNKNOWN CHARACTERISTICS OF THE MOON'S INTERIOR
1998	LUNAR PROSPECTOR MAPS MINERAL DISTRIBUTION ON THE MOON'S SURFACE AND SEARCHES FOR WATER ICE

MOON STRUCK

To the ancient Egyptians, the Moon represented the god Thoth, a symbol of wisdom and justice. Other ancient civilizations also placed great significance on our shining neighbor, plotting its path across the sky or conducting special ceremonies during lunar eclipses. Around the third century b.c., Greek philosophers began to discuss the Moon as an inhabited world very distinct from our own.

The invention of the telescope did not instantly discredit the idea that the Moon was a fertile land. In the early 1600s, Italian astronomer Galileo Galilei pointed one of his telescopes toward the Moon. Although Galileo's telescope could—at best—magnify just 20 times, he could still see that the lunar surface was "full of inequalities, uneven, full of hollows and protuberances, just like the surface of the Earth itself." Galilei suggested that darker regions were aqua (water) and the rest of the surface was terra (land)—a notion that led to the dark areas receiving the name maria, or "seas."

With larger telescopes, the idea of land and seas began to evaporate. In 1837, German astronomer Wilhelm Beer and his colleague Johann Mädler mapped the visible side of the Moon and concluded that the lunar surface was a cratered desert. Other astronomers commented on how clear the lunar surface always appeared—a sign that it could not have a significant atmosphere. With no air and no visible water on the ground, most astronomers agreed that life could not exist on the Moon.

CRATERS GALORE

One thing the Moon clearly possessed was craters—and plenty of them. By the turn of the 20th century, the question of how this scarred landscape had formed was hotly debated. A few astronomers argued that they were formed by impacts. But many others believed they were the result of long-extinct volcanoes. The argument could not be settled without closer observation.

When Apollo astronauts brought lunar rocks to Earth, the samples were discovered to consist of a conglomeration of different materials—suggesting that the surface of the Moon had, for millions of years, been pummeled by meteoroids.

NEIGHBORHOOD WATCH

GIRL POWER
In a 15th-century painting (above) the Moon is depicted as a woman watching over the Earth's oceans. As early as the second century A.D., Egyptian astronomer and geographer Ptolemy (below) suggested a link between the Moon and the seas.

MOON MAP
On August 5, 1609, English scientist and mathematician Thomas Harriot (right) drew the first-ever map of the Moon (left). Harriot beat Italian astronomer Galileo Galilei, who also studied the Moon, by a few months.

FAR OUT
In late 1959, the Soviet probe Luna 3 (right) began to orbit the Moon. It gave the human race a first look—albeit blurry—at the lunar far side (left).

From an oil painting to a photograph: As telescopes—and later space probes—have improved, the surface of the Moon has come under increasing scrutiny.

MAN ON THE MOON

HUMANS FINALLY PUT A FOOTPRINT ON THE MOON IN 1969 (RIGHT). BUZZ ALDRIN AND NEIL ARMSTRONG SPENT 22 HOURS IN THE MARE TRANQUILLITATIS REGION STUDYING THE LUNAR SURFACE, AND COLLECTING ROCK SAMPLES. DESPITE MANY YEARS OF EXPECTATION, HUMANS HAVE YET TO PUT THEIR FOOT DOWN ON ANOTHER BODY IN THE SOLAR SYSTEM.

INNER COLOR
A false-color map of the Moon's gravity (right) was produced using data from the probe Clementine (below). Launched in 1994, Clementine allowed astronomers to map the Moon more thoroughly than ever before.

MAN IN THE MOON

MANY SHAPES CAN BE FOUND IN THE MARKINGS OF THE MOON. THE FAMILIAR "MAN IN THE MOON" IS FORMED FROM A NUMBER OF DARK REGIONS THAT—WITH A LITTLE IMAGINATION—BECOME EYES, NOSE AND A MOUTH. BUT THIS FACE IS NOT THE ONLY MAN ON THE MOON. THERE IS ANOTHER MAN, WHO CAN BE SEEN CARRYING A BUNDLE OF STICKS ON HIS BACK (LEFT SIDE OF MOON, ABOVE). ACCORDING TO FOLKLORE, HE WAS BANISHED TO THE MOON AS PUNISHMENT FOR GATHERING THORNS ON A SUNDAY.

PHASES OF THE MOON

At one time, the phases of the Moon controlled peoples' lives. The full moon made it possible to travel at night, the new moon was thought to bring good luck, and coastal communities knew that the tides depended on the position of the Moon. Today, few of us know what phase the Moon will be in tonight, and many people are unsure what causes its apparent changes in shape. Yet the truth is that the lunar cycle affects not only astronomers, but each of us on the planet Earth.

WHAT IF...

...THE MOON SHRINKS?

In the future, the Moon will move farther away from Earth as it has done throughout history—in fact, if there had been people around a billion years ago, they would have seen a giant moon hanging in the sky.

In the far distant future, a month (the time that the Moon takes to go around the Earth) will be about twice as long as it is now and the Moon will appear much smaller in the sky. But the Earth's day will have slowed, too, and will be as long as this month.

At this point the Moon will only be visible from one hemisphere of the Earth—probably the side with the greatest landmass—and its only movement will be up and down in the sky. The Moon will continue to go through its cycle just as it does today, although it will take about 58 of our present days to do so. At sunset the tiny Moon will be at first quarter, then it will take another 27 present days to go through full to last quarter, when the Sun will lazily lift itself above the horizon for another extended Earth day.

The cause of these changes is the gravitational pull of the Moon on the Earth's oceans which, as well as raising tides, slows down the Earth's rate of rotation. This in turn causes the Moon's orbit to spiral outward into space—though at the present rate of just under an inch a year, the familiar phases of the Moon will be with us for millions of years to come. Whether any humans will be around to watch the tiny, unmoving Moon during the month-long nights is another matter!

LUNAR CYCLE STATISTICS

PERIOD OF ROTATION AROUND THE EARTH (SIDEREAL MONTH)	.27.3 DAYS
INTERVAL BETWEEN NEW MOONS (SYNODIC MONTH)	.29.5 DAYS
NUMBER OF NEW MOONS IN A YEAR	.13
BEST MONTHS FOR OBSERVING DIFFERENT PHASES	
CRESCENT NEW MOON	.APRIL (END)
FIRST QUARTER	.MARCH
FULL MOON	.DECEMBER
LAST QUARTER	.SEPTEMBER
CRESCENT OLD MOON	.JULY (END)

MOONLIGHT SERENADE

The Moon's orbit around Earth is counterclockwise, like most of the orbits in our solar system. In the course of each orbit—or lunar cycle—its appearance changes from a thin crescent, through half phase to full, then back to a thin crescent again. These changes are entirely due to the angle at which the Sun's light strikes it: The Moon has no light of its own.

When the Moon is growing in size, up to full moon, it is said to be waxing; after full, it is waning. The cycle begins at new moon. Strictly speaking this takes place when the Moon is almost in line with the Sun, which means that it is in the daytime sky and can't normally be seen. But most people consider the Moon's first appearance as a thin crescent to be the new moon, which occurs a day or two after the true new moon.

The new moon is a crescent because it is almost completely backlit. It can be seen in the western sky just after sunset, and sets soon after the Sun itself. The earliest sighting is some 14 hours after the true new moon, but many people regard even a 3- or 4-day-old moon as still new. By the time the Moon is seven days old (that is, seven days after new) it has grown to a half moon. Astronomers call this the first quarter, because the Moon is now one quarter of the way around its orbit.

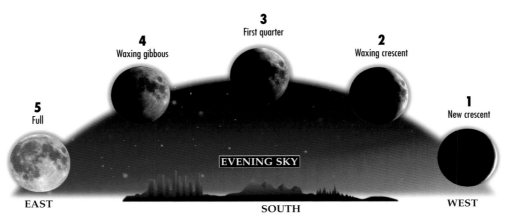

3 First quarter
4 Waxing gibbous
2 Waxing crescent
5 Full
1 New crescent

EVENING SKY

EAST SOUTH WEST

THE LUNAR MONTH

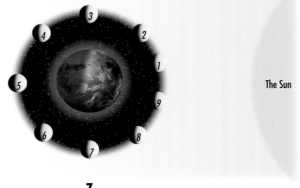

As the Moon orbits the Earth (right), its appearance changes according to how much of the Sun's light it reflects. Each numbered position in the orbit matches the view in the sky above and below. At the start of the lunar month, the Moon is only visible in the evening sky (above, views 1–5); by the time it has reached full moon it can be seen all night, rising in the evening sky at sunset in the east and setting in the morning sky at sunrise in the west. After full moon, the Moon is visible in the morning sky (views 5–9). Everywhere on Earth sees virtually the same phase of the Moon on any one day.

The Sun

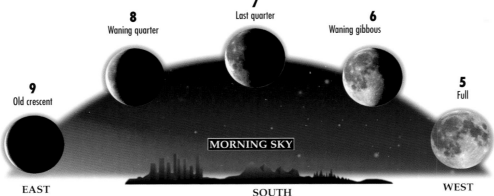

7 Last quarter
8 Waning quarter
6 Waning gibbous
9 Old crescent
5 Full

MORNING SKY

EAST SOUTH WEST

IN THE CRADLE

QUITE OFTEN WHEN THE MOON IS A CRESCENT THE WHOLE OF ITS DISK IS FAINTLY VISIBLE—SOMETIMES CALLED "THE OLD MOON IN THE NEW MOON'S ARMS." THE REASON FOR THIS PHENOMENON WOULD BE OBVIOUS IF YOU WERE STANDING ON THE MOON—ABOVE YOU WOULD BE THE NEARLY FULL EARTH WITH ITS BRILLIANT WHITE CLOUDS, CASTING LIGHT ON THE LANDSCAPE MANY TIMES BRIGHTER THAN THE FULL MOON DOES IN OUR SKY.

WANING MOON

The half moon can be seen more or less due south at sunset, and follows the sun down to the west just a few hours later. After another seven days, the Moon has grown to full. Now it rises at around sunset, but is more or less directly opposite the Sun in the sky—roughly due east. The phases just before and after full moon are called gibbous, a word which comes from the Latin gibbosus meaning "hunchbacked."

About a week after full, the Moon is at half phase again but with the opposite side to the first quarter illuminated. It rises roughly an hour later each night, so you will probably only see this phase early in the morning. A few days after last quarter, the Moon is back to a crescent again—this time in the pre-dawn sky, rising an hour or two before the Sun in the east. Then, for a few days, the Moon "disappears" as it passes directly between the Earth and the Sun, signaling the start of a new cycle.

TRUE STORY

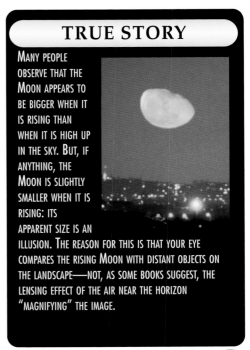

MANY PEOPLE OBSERVE THAT THE MOON APPEARS TO BE BIGGER WHEN IT IS RISING THAN WHEN IT IS HIGH UP IN THE SKY. BUT, IF ANYTHING, THE MOON IS SLIGHTLY SMALLER WHEN IT IS RISING: ITS APPARENT SIZE IS AN ILLUSION. THE REASON FOR THIS IS THAT YOUR EYE COMPARES THE RISING MOON WITH DISTANT OBJECTS ON THE LANDSCAPE—NOT, AS SOME BOOKS SUGGEST, THE LENSING EFFECT OF THE AIR NEAR THE HORIZON "MAGNIFYING" THE IMAGE.

GEOLOGY OF THE MOON

The Earth and the Moon are both made from similar rocky material—but there the comparison ends. Almost devoid of water and lacking an atmosphere, there are no erosive forces of wind and rain to wear away the lunar surface. With no active tectonic plates, the Moon also escapes the periodic volcanic upheavals that rearrange the surface of the Earth. As a result, the Moon is effectively frozen in time. With its ancient craters and primordial lava flows, our nearest neighbor is like a geological museum of the distant past.

WHAT IF...

...THE MOON WERE HIT AGAIN?

A 10-mile-wide (16-km-wide) asteroid is detected hurtling toward the Earth. People begin to panic, and are only slightly reassured when they hear that it is expected to miss us—because, they are told, it will hit the Moon instead in a 30-mile-per-second (50-km-per-second) collision.

As the asteroid speeds inward, those with a clear line of sight see a brilliant flash near the Moon's western rim. The asteroid is destroyed in a terrible impact. The place where it struck glows dimly for days. But the site is hard to see from Earth because it is close to the edge of the disk. The public breathes a sigh of relief and believes nothing has changed, other than the new 100-mile-wide (160-km-wide) crater on pictures sent back by lunar-orbiting spacecraft.

Then, in the following months, seismometers on the Moon—which recorded the shock of the impact—begin to register a series of moonquakes near the impact site. At first, the shock waves are hundreds of miles deep, but gradually they approach the surface.

Suddenly one night, people on Earth notice a glowing red streak rising above the Moon's western rim. The lunar orbiter has a better view: It transmits pictures of a new volcano from which incandescent magma jets erupt 100 miles (160 km) into the sky, driven by the force of escaping gas. As the magma splashes back, it collects into lava flows that work their way downhill and spread across the floor of the 600-mile-wide (960-km-wide) Orientale Basin.

The fire fountain spurts for many weeks but becomes weaker as the fissure begins to clog up. Lava continues to gush quietly from isolated vents along the fissure for years. It carves mile-wide channels and flows out into the basin beneath a shell of magma that has already cooled and hardened. Years later, when the last glows from the cracks in the lava have faded, all that remains to mark these titanic events is a giant dark smudge on the floor of the Orientale Basin.

To observers on Earth, the spectacle is over—but what kind of volcano might be left behind? The chances are that it would be smaller than Mount St. Helens, one of the largest on Earth. More likely it would resemble a small cinder cone or volcanic dome. Strangely, despite the Moon's volcanic past, only a scattering of small volcanoes—no more than a few thousand feet high and 6–10 miles (10 to 16 km) across—have been found on the lunar surface. The fact is that the Moon cooled billions of years ago, unlike the Earth, and is now a dead world. It would take a cataclysm to reawaken it.

THE MOON'S CRUST

ELEMENT	% IN HIGHLAND CRUST	% IN MARE CRUST
OXYGEN	45.0	45.0
SILICON	21.0	20.0
ALUMINUM	12.0	4.0
CALCIUM	11.0	6.0
IRON	5.0	13.0
MAGNESIUM	4.0	8.0
TITANIUM	0.4	4.0
SODIUM	0.4	0.1
POTASSIUM	0.1	0.1
CHROMIUM	0.1	0.3

ROCKY MOON

Even a glance at the Moon immediately reveals two very different types of lunar surface—the light-colored, rugged terrain of the highlands and the darker, flatter maria—the Moon's dark "seas." Both are scarred by craters, the result of ancient asteroid and comet impacts. But there are fewer craters on the maria because of their relative youth.

The older highlands formed as minerals from the interior floated to the surface, cooling as our neighbor evolved beyond its initial molten state. In the Moon's second, volcanic stage of development, eruptions of molten rock flooded the impact basins to form the maria, which are characterized by much darker basalt rock.

The Moon's youngest deposits are ejecta—blankets of fragmented material that were flung out from the most recent impacts to cover all of the older regions. In fact, it is rare to find solid rock anywhere on the lunar surface, since impacts over the ages have shattered every inch of the original terrain. Most of the existing surface is made up of scattered fragments known as regolith, which in places meld to form a rock type called breccia.

We also know that the Moon, like the Earth, is made up of layers. The outer layer, the crust, is 40–60 miles thick (60–100 km). Below it lies the mantle, now mostly solid, but once the source of ancient eruptions. Although both the crust and the mantle are made of rock, the mantle has more magnesium and iron, while the crust contains more aluminum. The Moon may have a solid core of iron or iron sulfide, no more than about 450 miles (720 km) across.

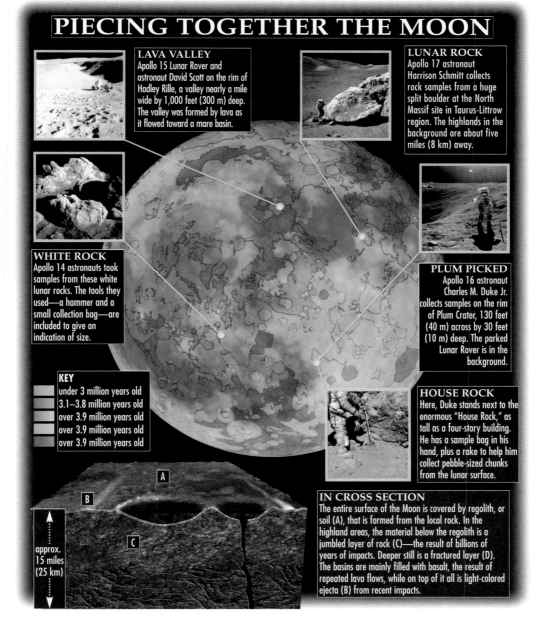

PIECING TOGETHER THE MOON

LAVA VALLEY
Apollo 15 Lunar Rover and astronaut David Scott on the rim of Hadley Rille, a valley nearly a mile wide by 1,000 feet (300 m) deep. The valley was formed by lava as it flowed toward a mare basin.

LUNAR ROCK
Apollo 17 astronaut Harrison Schmitt collects rock samples from a huge split boulder at the North Massif site in Taurus-Littrow region. The highlands in the background are about five miles (8 km) away.

WHITE ROCK
Apollo 14 astronauts took samples from these white lunar rocks. The tools they used—a hammer and a small collection bag—are included to give an indication of size.

PLUM PICKED
Apollo 16 astronaut Charles M. Duke Jr. collects samples on the rim of Plum Crater, 130 feet (40 m) across by 30 feet (10 m) deep. The parked Lunar Rover is in the background.

KEY
under 3 million years old
3.1–3.8 million years old
over 3.9 million years old
over 3.9 million years old
over 3.9 million years old

HOUSE ROCK
Here, Duke stands next to the enormous "House Rock," as tall as a four-story building. He has a sample bag in his hand, plus a rake to help him collect pebble-sized chunks from the lunar surface.

IN CROSS SECTION
The entire surface of the Moon is covered by regolith, or soil (A), that is formed from the local rock. In the highland areas, the material below the regolith is a jumbled layer of rock (C)—the result of billions of years of impacts. Deeper still is a fractured layer (D). The basins are mainly filled with basalt, the result of repeated lava flows, while on top of it all is light-colored ejecta (B) from recent impacts.

approx. 15 miles (25 km)

LUNAR EVOLUTION

The geological history of the Moon has been worked out using the same principles that apply on Earth. For example, we can be sure that one feature that overlies or cuts across another must be the younger of the two, and that the more impact craters there are on a particular surface, the older that surface must be.

The Apollo landings between 1969 and 1972 provided an opportunity to learn even more. NASA was careful to choose landing sites in both highland and maria regions, and photographs taken from the orbiting spacecraft were used to add to our existing maps of the lunar surface. Exact locations were chosen on the basis of where rocks from two or more sources could be collected—that is, where there was a junction between two rock types, or where nearby impacts were expected to have excavated and flung out a mixture of material.

Studies of the rock samples returned by the Apollo missions have allowed the lunar surface to be fitted into a more precise relative time scale. Dating Moon rocks by measuring the products of radioactive decay has revealed that the highlands suffered intense meteoritic bombardment between about 3.8 and 4 billion years ago. Unfortunately, this flurry of activity has also obscured the age of the original rock: It is probably about 4.5 billion years old, whereas samples of basalt brought back from the maria range in age from 3.1 to 3.8 billion years.

Only one scientist, Harrison Schmitt, has visited the Moon. But, thanks to the samples brought back by the Apollo astronauts, geologists know more about our satellite's history than they know about any other body in the solar system except the Earth.

SEAS OF THE MOON

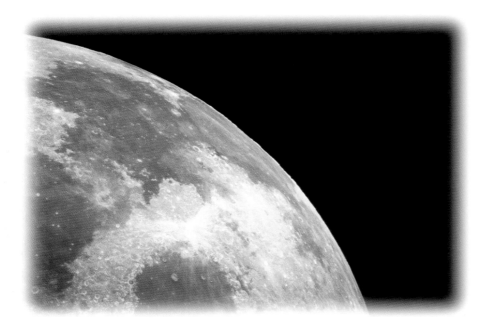

Around 350 years ago, when European astronomers first began to observe the large, dark areas on the lunar surface through telescopes, they thought they saw water. They christened these lunar oceans maria (the plural of mare, Latin for "sea"). We now know that the maria are the solid remnants of gigantic lava flows that oozed from the Moon's once-molten interior over three billion years ago. Remarkably, they have changed very little since then.

WHAT IF...

...A NEW MARE FORMED?

January, 2130: Lunary geologists working at Armstrong Base on the edge of the Moon's Mare Imbrium bring their 10-year Project Moonhole to a successful conclusion—on time and under budget. A deep borehole, probing hundreds of miles through the lunar mantle, has provided conclusive proof that the Moon is geologically dead. "There isn't enough molten material inside the Moon to fill a coffee cup," says the project leader. "There can never be any new maria."

February, 2130: Members of the Senate Appropriations Committee are paying a surprise visit to Nixon Station on the lunar farside. It is the most secret installation on the Pentagon's books. Out of sight of earthbound telescopes, a select group of scientists and engineers are working on Project Nemesis—the creation and storage of large quantities of antimatter.

The project promises an interstellar space drive, cheap transport around the solar system and—almost as an afterthought—pocket-sized weapons of unimaginable power. But it is way over budget. In the course of an angry meeting, the Project Director explains why. "We have almost 10 pounds [4.5 kg] of antimatter here," he says. "That's enough energy for a couple of gigaton bombs—or to send a ship to Alpha Centauri."

But the Appropriations Committee is adamant. Money is tight: Nemesis must be shut down. "So what do you expect me to do with all this antimatter?" yells the Director. Beside himself with rage, he slams the budget report down and inadvertently thumps a big red button.

As the confinement field shuts down, the antimatter touches the ordinary matter of the base. Twenty pounds of mass turns itself into a colossal surge of pure energy. A huge chunk of the Moon's surface is vaporized and rock for hundreds of miles around the base glows orange-red and flows like water.

As seen from Earth, a new Moon undergoes a startling transformation. For a full 20 minutes, its dark disk is ringed by fire. Experienced observers find it uncannily similar to a total eclipse of the Sun—except that it is taking place in the middle of the night. For a week afterward, a trace of the farside glow is still visible. Then the molten surface cools and settles.

A new mare has formed on the Moon.

THE MOON'S MAJOR SEAS

Latin Name	Common Name	Latitude	Longitude	Approx. Dimensions miles/km
Mare Frigoris	Sea of Cold	56°N	7°E	700 by 95/1,130 by 150
Mare Imbrium	Sea of Showers	33°N	16°W	720 by 720/1,160 by 1,160
Mare Fecunditatis	Sea of Fecundity	9°S	51°E	550 by 375/890 by 605
Mare Nubium	Sea of Clouds	20°S	16°W	400 by 400/645 by 645
Mare Serenitatis	Sea of Serenity	27°N	18°E	400 by 375/645 by 605
Mare Tranquillitatis	Sea of Tranquility	8°N	30°E	340 by 490/545 by 790
Mare Australe	Southern Sea	47°S	92°E	610 by 610/980 by 980
Oceanus Procellarum	Sea of Storms	20°N	58°W	1,250 by 810/2,010 by 1,300

OCEAN OF BASALT

The lunar maria are the result of vast outpourings of dense, molten basaltic lava from the molten mantle that once seethed below the Moon's crust. They cover 16% of the lunar surface, to a thickness of between 1,500 and 5,000 feet (450 and 1,520 m). Most of the maria date back some 3.2 to 3.8 billion years, although the youngest are only 1 billion years old. They are typically circular in outline and are rimmed by mountains. Most of them fill the bottoms of very large, very old impact craters.

The maria appear dark because of the low reflectivity (albedo) of their basaltic rock—it reflects only 7% of the sunlight that shines on it, which gives it an albedo of 0.07, compared with the Earth's 0.39. Even so, small variations in the type of light reflected from different maria have helped to define 13 types of maria basalt.

One of the most striking things about the maria is their uneven distribution: They cover nearly one-third of the side facing the Earth, but just 2% of the far side. This is because the Moon's crust is much thinner on the Earth-facing side—38 miles (61 km) compared with 63 miles (101 km)—which made it much easier for the lava to breach it.

Many of the maria also display a wealth of volcanic features that are unique to the Moon.

Among them are the rilles, with their distinctive flat floors and steep, parallel walls. Often found around the edges of maria, they show where the crust to either side has slipped. Sinuous rilles, by contrast, are winding channels formed by ancient lava flows, or by collapsed lava tubes—caves cut by underground lava streams that later fell in after the source of the lava had dried up.

OCEAN EXPLORER

ITALIAN ASTRONOMER AND PHILOSOPHER GIOVANNI BATTISTA RICCIOLI (1598–1671) MADE ONE OF THE FIRST DETAILED TELESCOPIC STUDIES OF THE MOON AND NAMED MOST OF THE LARGEST MARIA AND CRATERS, INCLUDING ONE NAMED FOR RICCIOLI HIMSELF (CIRCLED). THE NAMES WERE PUBLISHED IN HIS BOOK ALMAGESTUM NOVUM IN 1651 AND ARE STILL USED TO THIS DAY. RICCIOLI, A JESUIT PRIEST, REMAINED A STAUNCH BELIEVER IN THE PTOLEMAIC SYSTEM, WHICH PLACED THE EARTH AT THE CENTER OF THE UNIVERSE, AS OPPOSED TO THE NEWLY EMERGING COPERNICAN THEORY, WHICH PLACED THE SUN AT THE CENTER OF THE SOLAR SYSTEM.

FEATURES OF THE MOON'S SEAS

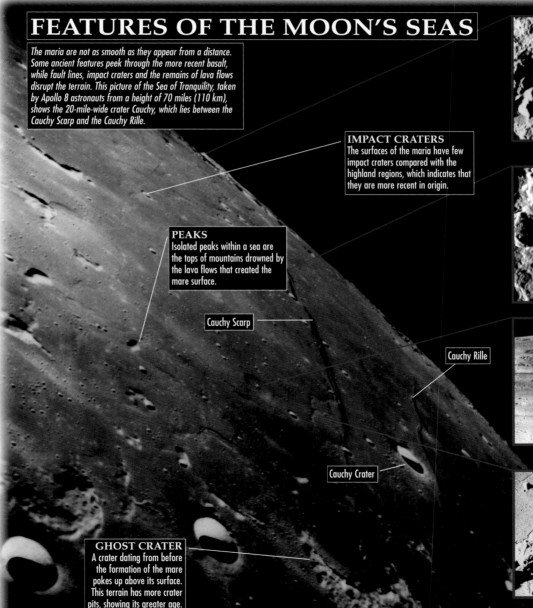

The maria are not as smooth as they appear from a distance. Some ancient features peek through the more recent basalt, while fault lines, impact craters and the remains of lava flows disrupt the terrain. This picture of the Sea of Tranquility, taken by Apollo 8 astronauts from a height of 70 miles (110 km), shows the 20-mile-wide crater Cauchy, which lies between the Cauchy Scarp and the Cauchy Rille.

IMPACT CRATERS
The surfaces of the maria have few impact craters compared with the highland regions, which indicates that they are more recent in origin.

PEAKS
Isolated peaks within a sea are the tops of mountains drowned by the lava flows that created the mare surface.

Cauchy Scarp

Cauchy Rille

Cauchy Crater

GHOST CRATER
A crater dating from before the formation of the mare pokes up above its surface. This terrain has more crater pits, showing its greater age.

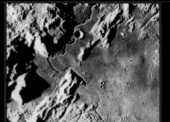

SINUOUS RILLES
These deep, snaking fissures were mostly created by lava flows, or by the collapse of underground channels, after the formation of the maria. The most famous sinuous rille is Hadley Rille (left) in the Mare Imbrium.

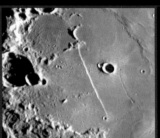

SCARPS
Scarps like the 72-mile (116-km) long , 1,300-foot (396-m) high Straight Wall in the Mare Nubium were once thought to be the work of aliens. Now we know that they are simply fault lines. The slope of the "Wall" is in fact only about 20°, but looks steeper under a low Sun.

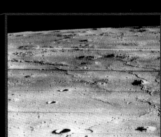

DOMES
Domes are volcanic plugs that are thought to mark the locations of minor eruptions that took place after the formation of the maria. The domes shown at left, in the Oceanus Procellarum, are just a few hundred feet high.

WRINKLE RIDGES
The wrinkled ridges that traverse many of the maria were probably formed by tectonic movements of the lunar surface after the lava flows that formed the maria had cooled and solidified.

HIGHLANDS OF THE MOON

The Moon is actually a mountainous satellite, on which 80% of the surface is made up of rounded hills. To the naked eye, these highland areas shine silvery white against the lowland "seas," their distinctive color the result of their unique geological makeup. Highland rock samples gathered by Apollo 16 in 1972 provided the first solid evidence of the Moon's enigmatic past. But most of the highlands are on the far side of the Moon, so geologists have turned to the so-called dark face in order to find out more.

WHAT IF...

...WE COULD TAKE A VACATION IN THE HIGHLANDS?

The discovery that the surface of the Moon could supply all the basic requirements for a living environment is good news for adventurous tourists. It means that one day, before long, the ancient, pitted lunar highlands could boast the long-held dream of a lunar theme park or a domed biosphere.

The greatest economic problem facing potential lunar vacation developments has always been the huge cost of lifting sizeable payloads out of the Earth's deep gravity well. Now, we know that we do not need to export bulky raw materials to the Moon: Recent evidence seems to confirm that there is even water there, concentrated at the lunar poles.

Although there is no breathable atmosphere on the Moon, most lunar rocks contain about 40% oxygen. Since the 1980s, scientists have been working on ingenious, and increasingly effective, electrochemical processes to extract this precious resource. Meanwhile, up in the polar regions above the Mare Frigoris, the eternal sunshine enjoyed by the highlands makes them the perfect location to install solar-powered generators.

Down in the bottom of the polar craters the Sun never shines, making these dark holes the coldest places on the Moon. But they offer a stability of temperature and light that would make them the ideal sites for the construction of artificial worlds. Faced with building a "polar adventure complex," engineers and scientists would know exactly the amount of heat required to create the perfect climate, and the never-setting Sun on the polar cap could be relied upon to provide a consistent source of energy.

Meanwhile, from the safety of a domed world built in the crater Gagarin—between the Mare Moscoviense and the Mare Ingenii—visitors would experience the longest days and nights of their lives. Each night lasts for two weeks, so a solar-powered society would probably require considerable backup power. But the solar wind has deposited hydrogen, helium and other elements in the lunar soil—elements that could be released by placing the soil in a vacuum and then heating it just a few degrees. It might even be possible to use the abundant isotope helium-3 to fuel "clean" nuclear fusion reactors. With so many resources potentially at our disposal, vacations could be the last thing on lunar visitors' minds.

EARTH AND MOON

	THE ROCKY MOUNTAINS	THE LUNAR HIGHLANDS
APPROXIMATE AGE	65–35 MILLION YEARS OLD	4.1–4.4 BILLION YEARS OLD
HUMAN PRESENCE FROM	10–8,000 YEARS B.C.	1969
MAX. HEIGHT ABOVE SEA/MEAN LEVEL	14,433 FEET (4,399 M)	c.20,000 FEET (6,096 M)
AVERAGE BULK DENSITY	344 POUNDS PER CUBIC FOOT (156 KG PER 0.03 CUBIC M)	206 POUNDS PER CUBIC FOOT (93 KG PER 0.03 CUBIC M)
PREDOMINANT ROCK TYPE	SHALE AND SANDSTONE	ANORTHOSITES AND BRECCIAS
MINERAL RESOURCES	COPPER, SILVER, GOLD, LEAD, ZINC, COAL, MOLYBDENUM, BERYLLIUM AND URANIUM	TITANIUM, IRON, ALUMINUM, MAGNESIUM, CALCIUM AND URANIUM
WATER	RIVERS, LAKES AND GLACIERS	NONE
PLANT LIFE	FOREST AND ALPINE WOODLAND	NONE
ANIMAL LIFE	DIVERSE WILDLIFE	NONE
FORMED BY	COLLISION OF TECTONIC PLATES	SOLIDIFICATION OF BOILING CRUST

ORIGINAL CRUST

To discover the true age of the Moon, scientists needed to find samples that had remained undisturbed since they first solidified in the primeval lunar crust. Most of the original crust had either been buried by lava or repeatedly disrupted by impacts. But when the lunar probe Ranger 9 crashed into the Alphonsus crater in 1965, it sent back images of a highland area (above) that was composed entirely of anorthosites and free of maria. Geologists later identified the pale highlands as the only areas in which primeval evidence could have survived. So the later Apollo (15, 16 and 17) targeted these areas for the retrieval of rock samples.

ANCIENT HILLS

The ancient lunar highlands hold a wealth of scientific history on and beneath their rocky surface. They are pitted with meteorite craters of all sizes and littered with the rocky debris that rained down after the impacts. Similar, but larger, impacts also created the great basins, which filled with lava to form the maria—the lunar "seas"— that show up so clearly as dark patches today. But most of this activity took place a very long time ago. Scientists estimate that, apart from the occasional cosmic collision, the Moon has been a relatively quiet place for the last 3.9 billion years.

The highlands appear distinctively pale, even to the naked eye, thanks to their unique geology. They are largely formed from anorthosites—rocks rich in calcium and silicon, with a high aluminum content that produces the pale coloring. These rocks are so old that even the youngest anorthosite is far older than the most ancient rocks found on Earth. They are also much older than the other two types of rock found on the Moon—the volcanic basalts that fill the dark maria, and the breccias, formed when different types of rock are compounded by the heat and pressure of impacts.

The lunar probes and crewed missions of 1966 to 1972—which sent back the first photographs of

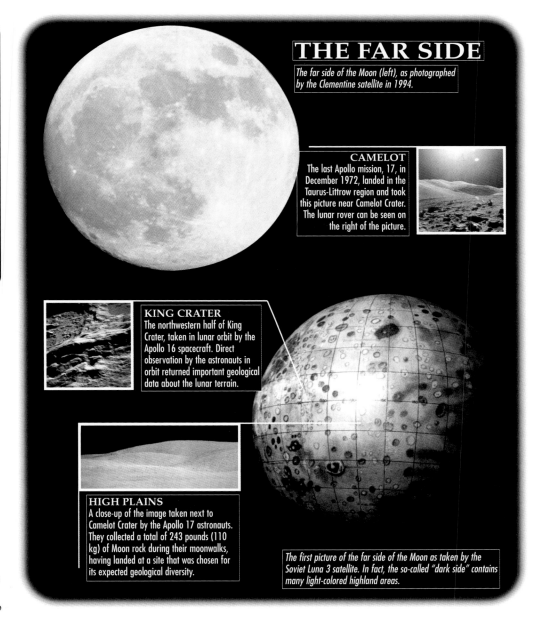

THE FAR SIDE

The far side of the Moon (left), as photographed by the Clementine satellite in 1994.

CAMELOT
The last Apollo mission, 17, in December 1972, landed in the Taurus-Littrow region and took this picture near Camelot Crater. The lunar rover can be seen on the right of the picture.

KING CRATER
The northwestern half of King Crater, taken in lunar orbit by the Apollo 16 spacecraft. Direct observation by the astronauts in orbit returned important geological data about the lunar terrain.

HIGH PLAINS
A close-up of the image taken next to Camelot Crater by the Apollo 17 astronauts. They collected a total of 243 pounds (110 kg) of Moon rock during their moonwalks, having landed at a site that was chosen for its expected geological diversity.

The first picture of the far side of the Moon as taken by the Soviet Luna 3 satellite. In fact, the so-called "dark side" contains many light-colored highland areas.

WEIGHTY MISSION

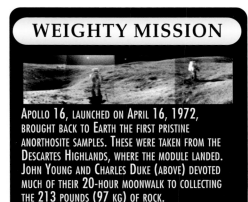

Apollo 16, launched on April 16, 1972, brought back to Earth the first pristine anorthosite samples. These were taken from the Descartes Highlands, where the module landed. John Young and Charles Duke (above) devoted much of their 20-hour moonwalk to collecting the 213 pounds (97 kg) of rock.

SECRETS OF THE HIGHLANDS

Even after the early Apollo landings, there was much that scientists didn't understand about the Moon. They knew that it had once been volcanically active, but there were competing theories about how it had formed. Lunar geologists suspected that only samples from the ancient highlands would provide the answer, and with the samples provided by Apollo 16 in 1972, they were able to identify the age of the rocks. Many of the samples gathered from the highlands proved to be over 4 billion years old.

With primordial Moon rocks in their hands, scientists at last felt able to explain how the Moon had evolved. After the accretion of small chunks of material in Earth orbit some 4.6 billion years ago, the Moon's surface heated up under continuing bombardment over the next 200 million years. It eventually melted into a giant ocean of lava, causing volatile material to be lost to space, and less dense material—such as the aluminum and calcium found in the highlands—to float to the top.

Between 4.4 and 4.1 billion years ago, the crust at last began to solidify. The resulting ripples created extensive highlands that were only disturbed by the showers of space debris that piled crater upon crater over the next million or so years. And without wind, water or organic processes to erode or change the surface, that is the way things have stayed ever since—a perfect and lasting record of a geological age that existed before time on Earth effectively began.

the Moon, and then recovered samples from the lunar surface—were crucial in establishing such basic geological data. Apollo missions 15, 16 and 17 between them gathered a huge amount of material, and Apollo 16 was the first to land in the highlands to collect its valuable ancient cargo. Even today, these rock samples are still being analyzed. What they tell us continues to form the basis of our understanding of lunar geology and history.

109

CRATERS OF THE MOON

The most abundant features on the Moon's surface are its distinctive craters. Most of these pockmarks were formed by impacts early in the history of the solar system, when space was littered with debris left over from the formation of the planets. Since the invention of the telescope, a bewildering number of crater shapes and sizes have been catalogued, from ancient ghostly remnants almost obliterated from the surface to young ray craters with bright streaks radiating hundreds of miles. Each has a unique story to tell.

WHAT IF...

...WE COULD WITNESS A CRATER BEING FORMED?

What would we be able to see if the Moon were struck by an asteroid, and what implications would it have for Earth? It so happens that during the last thousand years just such an event was recorded.

It was a warm June evening in England, in the year 1178. In the western sky hung a serene crescent Moon. A group of five monks were casually admiring our celestial neighbor when, according to the Chronicle of Gervase of Canterbury, "...suddenly the upper horn split in two. From the midpoint of this diversion a flaming torch sprang...spewing out, over a considerable distance, fire, hot coals and sparks." The monks had witnessed a lunar impact. Much later, orbiting spacecraft found a fresh-looking impact site just on the edge of the Moon's far side. Now called Giordano Bruno, the crater almost certainly marks the spot where, on June 18, 1178, the monks saw the "flaming torch of fire."

Giordano Bruno is about 13.5 miles (22 km) in diameter, so the meteoroid that caused it was relatively small. Yet the impact was clearly visible on Earth. A larger impact would be much more spectacular. As long as the object struck the near side of the Moon (the only side of the Moon that is

ever visible from Earth), we could be sure of a good view. The effects of the impact would probably be visible for hours, even to the naked eye. Any ensuing eruptions would glow an orange-red, best seen if the impact took place on the part of the Moon in darkness at the time. The site could well be visible with infrared telescopes for days or even weeks afterwards.

Even a large impact on the Moon would pose only a minor threat to the Earth. Although pieces of the Moon have been found on Earth in the form of meteorites, only the largest chunks of rock are able to make it through the Earth's shielding atmosphere. While we cannot rule out the Moon being struck by an object large enough to hurl a shower of rocks in our direction, the chances are astronomically small—that is, compared with those of 4 billion years ago, when the solar system was a far more violent place.

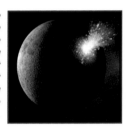

A massive meteor impact on the Moon might look like a raging bonfire from Earth. Witnessing this spectacle could be as popular as viewing a solar eclipse—if we knew when it was coming. A large-enough impact would form a complex crater (a crater with a central peak) because of the huge rebound from the impact.

CRATER TYPES

CRATER CLASSIFICATION	NAME	SIZE (MILES/KM)	LOCATION
OLD	PRINZ	33/53	25.5°N, 44.1°W
YOUNG	TYCHO	53/85	43.4°S, 11.1°W
RAYED	COPERNICUS	58/93	09.7°N, 20.1°W
SIMPLE/BOWL	LINNE	7/11	27.7°N, 11.8°E
CENTRAL PEAK	THEOPHILUS	69/111	11.4°N, 26.4°E
TERRACED	LANGRENUS	83/136	08.9°S, 61.1°E
MULTI-RINGED	ORIENTALE	580/933	19.4°S, 92.8°W

DEEP IMPACT

Anyone who glances through a small telescope at the surface of the Moon can't fail to notice that it is littered with pockmarks. These walled circular enclosures are craters. For 200 years a fierce debate raged as to whether the Moon's craters were formed by impacts or by lunar volcanoes. It is now accepted that most of them resulted from a relentless bombardment of rock and ice chunks of various sizes between 4.6 and 3.8 billion years ago. Since then, impacts have continued to form craters, but at a much lower rate.

A closer look at the lunar surface reveals a huge range of crater shapes and sizes. Generally, the shape of a lunar crater becomes more complex the larger it is in size. Small craters—with diameters less than 5 miles (8 km)—are bowl-shaped with rims that stand above the surrounding land. Around the rim lies a roughly circular fan of fragmented rock called an ejecta blanket, made of material thrown out during the impact. Strings of smaller irregular craters are often seen beyond the ejecta blanket. These are called secondary craters: They were formed by larger chunks of rocks that hurled from the surface by the original impact. Craters that are 7–20 miles (11–32 km) across have flat rather than bowl-shaped floors. Their walls often slump where the rock has slipped inward, giving them a scalloped appearance.

Larger craters are called complex craters. They have flat floors, terraced walls, and intriguing central peaks—huge mountains of material thrown upward in the last stages of the crater's formation. Two good examples of complex craters are Copernicus and Tycho. These have a distinctive pattern of bright rays emanating from their rims. The rays are made of material flung thousands of miles upward and outward. Rays are typical of young craters. The bright far-flung ejecta has not had time to darken from the constant milling action of the cosmic rays and microscopic dust that rain down on the lunar surface. Craters up to 200 miles (320 km) in diameter have features that are similar to

Copernicus and Tycho, although not all of these have kept their rays.

Craters with diameters larger than about 200 miles (320 km) are called impact basins. All lack central peaks and the largest form multiple rings. Multi-ringed basins are the biggest impact structures on the Moon. Most are over 500 miles (800 km) in diameter and are the legacies of truly cataclysmic collisions. Impact basins are often flooded by smooth dark lava, which welled up though fractures in the lunar crust or formed as the crust melted from the intense heat released when a giant impactor struck. The rings are similar to ripples on a pond, now frozen forever on the Moon's crust.

EFFECTS OF AGE

COPERNICUS (BOTTOM LEFT IN THE IMAGE AT RIGHT), IS CONSIDERED A YOUNG CRATER AT 1 BILLION YEARS OLD. YOUNG CRATERS LIKE THIS

ONE OFTEN HAVE BRIGHT RADIATING STREAKS CALLED RAYS. BUT OVER MANY MILLIONS OF YEARS, THEY LOSE THEIR YOUTHFUL LOOKS. THE RAYS ARE SLOWLY CHURNED AND BLACKENED BY THE CONSTANT BOMBARDMENT OF COSMIC PARTICLES AND MICROMETEORITES.

A TRIP TO THE FAR SIDE

THE FAR SIDE OF THE MOON REMAINED A MYSTERY UNTIL THE RUSSIAN SPACECRAFT LUNA 3 IMAGED IT IN 1959. THE SPACECRAFT DISCOVERED AND MAPPED A NUMBER OF CRATERS, INCLUDING THE GIORDANO BRUNO (BELIEVED TO BE THE CRATER FORMED IN AN IMPACT SEEN FROM THE EARTH IN 1178), TSIOLKOVSKY, LOMONOSOV, JOLIOT-CURIE, PASTEUR AND JULES VERNE CRATERS.

A CATALOG OF CRATERS

LAVA-FILLED PEAK CRATER

On the Moon's far side, out of sight from Earth, the crater Tsiolkovsky is about 110 miles (180 km) in diameter. It is a complex crater with a prominent central peak and a flooded floor of dark basaltic lava. This lava probably surged upward from beneath the lunar crust after the impact, which occurred about 3.8 billion years ago. This image was taken by Apollo 13 in 1970.

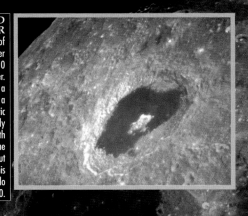

OVERLAPPING CRATERS

Apollo 15 brought back this image of the crater, Aristarchus (top left) whose bright ejecta blanket overlaps the crater Herodotus (top right). Aristarchus is the brightest crater on the moon. The two craters can almost be said to be twins, as they both have a diameter of 22 miles (35 km). But Herodotus is older and duller.

CRATER CHAIN

This chain of small craters was photographed by Apollo 17 in 1972—the last time humans visited the Moon. It lies close to the large crater Copernicus, and it was formed by material ejected in the impact that gave birth to Copernicus. Crater chains such as these are also referred to as secondary craters.

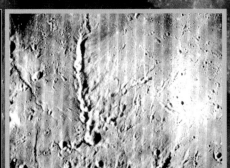

RAY CRATER

Dionysus, imaged by the 1994 Clementine probe, is a small ray crater with a diameter of 12 miles (19 km). Near the crater's rim are bright rays, which gradually darken toward the further reaches of the ejecta blanket. The location of the Dionysus crater is unusual—it lies at a precise junction between the maria and the highlands.

FAR SIDE OF THE MOON

From Earth, the Moon presents an unchanging face. This is the Moon's near side. But the opposite hemisphere is a much less familiar place. The far side of the Moon is always turned away from the Earth and cannot be observed directly. Before the Space Age, it was unseen by human eyes—but now space probes have furnished astronomers with stunning photographs and accurate maps of the far side's rugged surface. Some scientists have suggested that it would make the ideal site for radio and optical telescopes.

WHAT IF...

...WE COULD PUT A TELESCOPE ON THE FAR SIDE?

Many scientists have suggested that the Moon's far side might be an ideal place to put an astronomical observatory. It has no atmosphere to block or distort the light from distant stars, and it is screened from light and radio emissions from the Earth—in fact, it is probably the most radio-quiet place in the solar system, since it is the only place never reached by the radio racket from Earth. But could the idea really work?

A 1991 study by the Texas Space Grant Consortium concluded that it could. The study found that the best site for the observatory would be Hertzsprung, a large crater on the eastern edge of the Moon's far side. Crater floors are smooth and easy to build on. And because Hertzsprung is situated on the Moon's equator, a telescope there would be able to view both the northern and southern hemispheres of the sky.

The observatory would be equipped with radio, optical and infrared telescopes, enabling astronomers to survey a large portion of the electromagnetic spectrum. Electrical power could be provided either by nuclear reactors, or upright solar panels situated at one of the Moon's poles—where they would receive constant sunshine.

Heavy-duty power cables would carry the electricity from the lunar pole to the observatory. Similarly, a fiber-optic link would connect the observatory to a radio relay station on the Moon's near side, which would then beam data and pictures back to Earth. An orbiting relay would not be appropriate—it would pollute the far side with radio noise.

Because the Moon rotates once every 28 days, the lunar night lasts for 14 days, allowing marathon observation sessions. Some astronomical objects are so faint that they can only be seen after a telescope has stared at them, gathering light, for a very long time. The maximum exposure time at an Earth-based observatory is about 8 hours. For telescopes in low Earth orbit, such as the Hubble Space Telescope, it is much less—about 45 minutes. The long lunar night, however, would allow 14-day exposure times, revealing extremely faint objects.

And that is not the only way a lunar observatory could outperform an orbiting telescope. The density of the Moon's atmosphere is tiny—as low as one-hundredth of the particle density in low Earth orbit, where Hubble is located. This means that a Moon telescope would not only have a much clearer view of the universe, but it would also suffer much less erosion than its Earth-orbiting counterparts—so much less that it could outlast them by a century.

FAR SIDE FEATURES

Name	Latitude	Longitude	Name	Latitude	Longitude
Apollo crater	37° S	153W	Mendeléev crater	6° N	141° E
Birkhoff crater	59° N	148W	Planck crater	57° S	135° E
Gagarin crater	20° S	150E	Poincaré crater	57° S	161° E
Hertzsprung crater	0°	130W	Schrödinger crater	75° S	133° E
Korolev crater	5° S	157W	Tsiolkovsky crater	21° S	129° E
Lorentz crater	34° N	100W	Van de Graaff	27° S	172° E
Mare Moscoviense	27° N	147E	Mare Orientale	19° S	93° W

FAR SIDE FANTASY

Before humans first ventured into space, the far side of the Moon was a total mystery: No one knew what it looked like or what lay there. Then, in 1959, the Russian Luna 3 spacecraft returned the first-ever pictures from the far side. Luna flew behind the Moon during its orbit of Earth. The Luna 3 photos revealed a rugged surface, peppered with craters and almost completely lacking the smooth lunar seas, or maria, that feature so abundantly on the near side.

Since Luna 3, other space missions have returned sharper images, allowing the far side to be accurately mapped. Lunar Orbiters mapped the whole Moon from orbit, and the Apollo space missions returned breath-taking imagery when they swung around the far side of the Moon. More recently, in 1994, the uncrewed Clementine spacecraft studied the Moon from lunar orbit, making a number of discoveries about its structure and geology. And in 1998, Lunar Prospector surveyed the mineral content of the far side.

Perhaps most significantly, Clementine confirmed the reason why the far side has so few maria. The maria are areas where collisions

NEAR AND FAR

THE MOON HAS A FAR SIDE BECAUSE OF TIDAL FRICTION. THIS WAS FIRST SUGGESTED BY GEORGE DARWIN (1845–1912; RIGHT), SON OF CHARLES DARWIN, THE PIONEER OF EVOLUTION THEORY. THE PULL OF THE EARTH AND MOON'S GRAVITY ON EACH OTHER PRODUCES A TIDAL BULGE IN EACH OBJECT. IN THE PAST, AS THE MOON ROTATED, IT WOULD TRY TO CARRY ITS TIDAL BULGE AROUND WITH IT, WHILE EARTH'S GRAVITY TRIED TO PULL IT BACK. THIS FRICTION EVENTUALLY SLOWED THE MOON'S SPIN UNTIL ITS ROTATIONAL AND ORBITAL PERIODS WERE THE SAME, LEAVING ONE SIDE ALWAYS POINTING TOWARD EARTH.

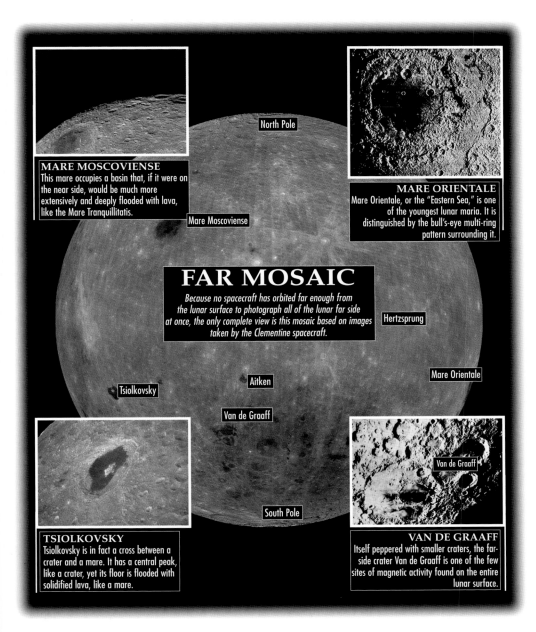

MARE MOSCOVIENSE
This mare occupies a basin that, if it were on the near side, would be much more extensively and deeply flooded with lava, like the Mare Tranquillitatis.

Mare Moscoviense

North Pole

MARE ORIENTALE
Mare Orientale, or the "Eastern Sea," is one of the youngest lunar maria. It is distinguished by the bull's-eye multi-ring pattern surrounding it.

FAR MOSAIC

Because no spacecraft has orbited far enough from the lunar surface to photograph all of the lunar far side at once, the only complete view is this mosaic based on images taken by the Clementine spacecraft.

Hertzsprung

Aitken

Mare Orientale

Tsiolkovsky

Van de Graaff

Van de Graaff

South Pole

TSIOLKOVSKY
Tsiolkovsky is in fact a cross between a crater and a mare. It has a central peak, like a crater, yet its floor is flooded with solidified lava, like a mare.

VAN DE GRAAFF
Itself peppered with smaller craters, the far-side crater Van de Graaff is one of the few sites of magnetic activity found on the entire lunar surface.

LUMPY TERRAIN

Clementine allowed much more detailed estimates of the thickness of the Moon's crust by mapping tiny variations in the Moon's gravitational field from point to point across its surface.

The crust was found to be thickest in an elevated region surrounding the South Pole–Aitken basin—a large depression in the far side's southern hemisphere. The elevation is believed to be debris ejected during the violent impact that formed the basin.

Clementine also made the first comprehensive, three-dimensional topographic maps of the far side. The probe carried a laser altimeter that fired pulses of laser light down to the ground as the probe orbited. Measurements were made of the time it took for each pulse to return. Combining this with the probe's known orbital altitude gave the height of the terrain below.

Clementine found the lunar terrain to be much lumpier on the far side than the near side. It also showed that the heights of the Moon's surface features vary by as much as 10 miles (16 km)—about the same as the Earth's surface features, even though the Moon is only a quarter the size of the Earth. The 1,550-mile-diameter (2,500-km-diameter) South Pole–Aitken Basin alone is seven-and-a-half miles deep.

Although the far side is described as unobservable from Earth, it is in fact possible to see roughly 18% of it, thanks to an effect called libration. This effect is caused by the Moon speeding up and slowing down in its elliptical orbit around the Earth. This makes the Moon appear to rock from side to side about its rotation axis. As it rocks, observers on Earth can peek around the back slightly. The effect makes prominent features on the edge of the far side clearly visible to amateur astronomers. The Mare Orientale was one such feature—named "Eastern Sea" because it was on the eastern edge as seen from Earth. When astronauts visited the Moon, this became the west, but the sea retains its old name.

with asteroids or comets have split open the Moon's crust. This allowed molten lava to well up from beneath and solidify, forming a smooth cap over the impact region. However,

Clementine found that the lunar crust is thicker on the far side. This makes it harder for lava to break through to the surface after an impact and generate the lava flows that produce the maria.

FORMATION OF THE MOON

The Moon is an oddity—a satellite so large that the Earth-Moon system is almost a double planet. For generations, rival theories strove to explain its origin. But evidence brought back by the Apollo landings may have settled the argument. Most scientists now believe that 4.5 billion years ago, the molten, new-made Earth was struck a glancing blow by another planet. The impact destroyed the wanderer and hurled huge chunks of the Earth into space. From this ancient collision, our nearest neighbor was born.

WHAT IF...

...MORE THAN ONE MOON HAD FORMED?

Like the Moon, the solar system formed out of the accumulation of material in an orbiting, disk-like structure. But whereas the disk orbiting the Sun gave birth to at least nine planets and countless smaller astronomical objects, the debris circling the Earth was only sufficient to form a single large satellite. If circumstances had been different, could the Earth, too, have had more than one moon?

Computer simulations show that when the Earth is subject to a large impact, it is indeed possible for two moons to form. But this poses the question of whether a pair of satellites could remain in stable orbits for any significant length of time. The answer seems to be no. The simulations imply that a two-moon system would be a short-lived phenomenon: The smaller of the satellites would be flung back into the Earth or ejected into deep space thanks to the gravitational pull of the larger satellite, or else the two moons would collide.

This aside, if our planet had acquired two satellites, things on Earth would certainly be different. Solar eclipses, for example, would be more frequent, but total eclipses might never occur. And the tides, which at the moment repeat on a roughly 13-hour cycle, would become almost unfathomably difficult to predict—as well as more extreme than they are in reality. Another effect would be a difference in the length of days and nights. The Moon's gravity influences the rotation of the Earth, speeding it up or slowing it down according to where it is in its orbit. When the Moon first formed, some 4.5 billion years ago, it was much closer to the Earth and so the Earth spun faster—perhaps as rapidly as once every five hours. Two moons would have prolonged this effect, resulting in much shorter days and nights today.

More controversially, two moons might have led to increased emotional instability among the Earth's human population. There is statistical evidence, for example, that the incidence of both accidents and violent assaults increases around the time of a full moon in a cycle that resembles that of the tides. With two moons contributing to people's mood swings, their behavior might be even more unpredictable than it already is. And that hardly bears thinking about.

Two moons dominate the night sky of an alternate Earth. Such an arrangement might well have diminished the appeal of an alternate Apollo program. As an epoch-making slogan, "First Man on A Moon" does not quite make it.

THEORIES THAT FAILED

BEFORE THE APOLLO MISSIONS OF 1969–72, THERE WERE THREE COMPETING THEORIES ON THE MOON'S FORMATION. DATA FROM THE MOON LANDINGS NOW SUPPORT A FOURTH THEORY—THAT THE MOON WAS FORMED IN A GIGANTIC COSMIC COLLISION.

CO-ACCRETION HYPOTHESIS

STATED THAT THE EARTH AND MOON FORMED INDEPENDENTLY, AS A DOUBLE PLANET, FROM THE ORIGINAL NEBULA (CLOUD OF GAS AND DUST) THAT CREATED THE SOLAR SYSTEM.

PROBLEM IF THE EARTH AND MOON HAD EVOLVED FROM THE SAME EMBRYONIC ENVIRONMENT, THE GEOLOGICAL COMPOSITION OF THE BODIES WOULD BE MORE SIMILAR.

FISSION HYPOTHESIS

PROPOSED THAT THE MOON "SPUN OFF" ITS PARENT PLANET WHEN THE EARTH WAS NEWLY FORMED AND ROTATING MUCH MORE RAPIDLY ON ITS AXIS.

PROBLEM IF THE EARTH AND MOON FORMED OUT OF THE SAME FULLY FORMED PLANET, THE OUTER LAYERS, AT LEAST, WOULD BE MORE ALIKE. THE MOON SHOULD ALSO ORBIT IN THE PLANE OF THE EARTH'S EQUATOR, WHICH IT DOES NOT.

CAPTURE HYPOTHESIS

SUGGESTED THAT THE MOON WAS ORIGINALLY A SEPARATE ASTRONOMICAL OBJECT THAT WAS DRAWN IN BY THE EARTH'S GRAVITY.

PROBLEM THE MOON IS PROBABLY TOO MASSIVE EVER TO HAVE BEEN CAPTURED IN THIS WAY. NOR DOES THE THEORY ACCOUNT FOR THE GEOLOGICAL SIMILARITIES BETWEEN THE TWO BODIES.

COSMIC COLLISION

Astronomers may never know for sure how the Moon was formed, but most today favor the so-called "giant impact" theory. The early solar system was a chaotic place—a whirling disk of spinning debris that slowly clumped together to form the planets. The giant impact theory proposes that in the midst of this maelstrom, the infant Earth was hit by another planet-sized body. The catastrophic collision sent a vast cloud of debris swirling around what remained of the Earth, and this debris later coalesced to form the Moon.

The beauty of the giant impact theory is that it explains both the differences and the similarities between the two bodies. The composition of the Moon and Earth are similar, but by no means identical. Some differences can be explained by gravity—the Earth's core has more iron than the Moon's, for instance, because the Earth's greater mass would naturally have attracted the heaviest material. But some lunar rock is quite unlike any on Earth—perhaps because it came from the debris of a shattered impacting planet.

The theory evolved in the early 1970s, when data from the Apollo missions began to cast doubt over earlier ideas on how the Moon was formed. Two astronomers, William Hartmann and Donald Davis, from the Planetary Sciences Institute in Tucson, Arizona, had been estimating the sizes of mini-planets, or planetesimals, that might have formed near to the Earth in the early days of the solar system.

THE MAKING OF OUR MOON

IMPACT
A wandering planet, traveling at a speed of several miles per second, plunges into the young Earth, still hot from its recent formation. Within four hours, the core of the impacting planet merges with the Earth's mantle and the outermost layers of both planets are blasted into space.

DEBRIS
Under the influence of the Earth's gravity and centrifugal force (the force that pulls an orbiting body away from the center of its orbit), debris from the collision swirls into a gigantic disk. Much of the material in the disk falls back to Earth, but the rest begins to coalesce into steadily larger objects.

MOONBIRTH
Within a few years, the largest bodies in the disk have swept up most of the free debris and gathered together to form a single large satellite. The Earth–Moon system is now complete, but still young and hot. In time, the two bodies will cool and drift apart to form the system that exists today.

At over one quarter the diameter of the Earth, the Moon is the fifth-largest natural satellite in the solar system. Only Pluto's moon Charon—which is nearly half Pluto's diameter—is bigger in relation to its parent planet, but the Pluto system may have formed in a very different way.

REFORMED?

MIRANDA, ONE OF THE SATELLITES OF URANUS, WAS ONCE THOUGHT TO HAVE A HISTORY AS VIOLENT AS THE MOON'S. MIRANDA'S TERRAIN INCLUDES THREE RELATIVELY YOUNG AND HEAVILY RIDGED AREAS, WHILE THE REST OF THE LANDSCAPE IS OLDER AND CRATERED, IMPLYING THAT IT REFORMED AFTER BEING SHATTERED IN A GIANT IMPACT. BUT NOWADAYS MOST ASTRONOMERS THINK THAT THE GIANT RIDGES ARE DUE TO TIDAL VOLCANIC ACTIVITY OF THE TYPE THAT ALSO FLEXES JUPITER'S MOON IO.

MOTHER MOON

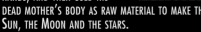

THE IROQUOIS BELIEVE, THE MOON WAS FORMED BY HAHGWEHDIYU, THE "GOOD CREATOR" (RIGHT). HE SHAPED THE SKY WITH THE PALM OF HIS HANDS, AND THEN USED HIS DEAD MOTHER'S BODY AS RAW MATERIAL TO MAKE THE SUN, THE MOON AND THE STARS.

IMPACT SIMULATIONS

Davis and Hartmann's research showed that several objects, each much larger than the present-day Moon, could have coalesced near enough to Earth to pose an impact threat, and that such an impact would create the right quantity of debris to form a satellite like the Moon.

At first, astronomers could only guess at the size of impacting body needed to release such a vast amount of material. But now, with the help of computer simulations of impacts, they can follow the progress of between 1,000 and 3,000 interacting fragments that vary in size from a few dozen miles to a few hundred miles across. The simulations appear to confirm that a giant impact at a certain angle could have led to the formation of the Earth-Moon system within just a few years. And in July 1997, a team of scientists from the University of Colorado presented the latest estimate for the mass of the impacting body. They showed that around 60–85% of the debris in an orbiting disk simply falls back to the parent planet's surface. This implied that for the remainder to coalesce into a satellite the size of the present-day Moon, the Earth must have been hit by an object 2½–3 times the size of Mars.

Even so, the giant impact theory is not conclusively proved. To date, all of the computer simulations that lead from giant impact to the formation of a Moon-sized satellite leave the Earth rotating about twice as fast as geological evidence suggests that it should have been at the time. Our nearest neighbor is not about to give up its secrets so easily—if it ever does so.

WATER ON THE MOON

On March 5, 1998, NASA excitedly declared that its uncrewed Lunar Prospector probe had detected water on the Moon—tons of it, in fact—in the guise of ice deposits at the lunar poles. If NASA's estimates are correct, the Moon could well be our stepping stone to the stars. A plentiful water supply in the solar system is essential for human space colonization—not only to provide us with water to drink, but with oxygen to breathe and fuel for our rockets in the form of liquid hydrogen and oxygen.

WHAT IF...

...WE USED LUNAR WATER?

Thanks to the discovery of water on the lunar surface, the Moon could become a very busy place in the 21st century. It is almost inevitable that it will eventually host a permanently crewed space base and act as a platform for powerful space telescopes. It could also become the joint launchpad and gas station for the exploration of the rest of the solar system.

A Moon colony would probably be located on top of a crater rim near one of the poles—to give it easy access to the ice and a ready supply of solar energy. Water would be extracted from the soil to provide drinking water and irrigation for growing plants and crops inside a sealed and environmentally controlled biosphere. One potentially energy-saving way of extracting the water would be to place the soil in a near-vacuum and then heat it at 34°F (1°C)—just two degrees above freezing. The resulting water vapor could then be condensed.

According to experts, concrete could also be produced on the Moon from raw materials already found there. Concrete structures should be capable of withstanding the harsh surface conditions, including radiation, the solar wind, micrometeoroids and extreme temperatures. Industrial plants would be built to break down the water into breathable oxygen and into liquid hydrogen and oxygen—the propellants used in the Space Shuttle's main engine—to fuel chemical rockets.

Because of its thin atmosphere and low gravity, launching a payload from the Moon requires less than 17% of the energy of an equivalent launch from Earth. So in theory a lunar rocket assembled from components sent from Earth would not only have ready supplies of fuel on its own doorstep, but would require less of it to travel farther, faster.

Other researchers are looking at new means of rocket propulsion that would take advantage of the ease of Moon launches. One idea is the nuclear-heated steam rocket, which would use a small nuclear reactor to boil water and propel itself on a steady jet of steam.

According to Alan Binder of the Lunar Research Center, "The Moon is going to enable Mars, it's going to enable Venus and Mercury...and everything else. I liken the moon to a natural harbor. It may well be that interplanetary missions won't start on Earth, they'll start on the Moon."

THE LUNAR PROSPECTOR

MANUFACTURER LOCKHEED MARTIN MISSILES AND SPACE, SUNNYVALE, CALIFORNIA

PROJECT COST $63 MILLION (INCLUDES $34M DEVELOPMENT)

PROBE DIMENSIONS HEIGHT 4.25 FEET (1.3 M); DIAMETER 4.5 FEET (1.4 M)

WEIGHT 650 POUNDS (295 KG) (FULLY FUELED)

LAUNCHED JANUARY 6, 1998, 9:28:44 PM EST FROM PAD 46, KENNEDY SPACE CENTER, CAPE CANAVERAL, FLORIDA

LAUNCH VEHICLE LOCKHEED MARTIN ATHENA II, 3-STAGE SOLID-FUEL ROCKET

POWER SOURCE BODY-MOUNTED 202W SOLAR CELLS CHARGING 15-AMP-PER-HOUR NICKEL-HYDROGEN BATTERIES

FLIGHT TIME TO LUNAR ORBIT 105 HOURS

POLAR ORBITS
- JANUARY 11–DECEMBER 19, 1998; HEIGHT 63 MILES (101 KM); PERIOD 118 MINUTES; ORBITAL VELOCITY 3,668 MPH (5,903 KM/H)
- DECEMBER 19, 1998–JANUARY 28, 1999; HEIGHT 25 MILES (40 KM)
- JANUARY 28; HEIGHT 15 MILES (24 KM)

CRASH DATE JULY 31, 1999

ONBOARD EQUIPMENT
- NEUTRON/GAMMA RAY SPECTROMETERS TO ANALYZE LUNAR SURFACE COMPOSITION
- MAGNETOMETER AND ELECTRON REFLECTOMETER TO ANALYZE AND MAP THE MOON'S MAGNETIC FIELD
- ALPHA PARTICLE SPECTROMETER TO IDENTIFY ALPHA PARTICLES IN RADON GAS PRODUCTS EMITTED FROM CRATERS
- DOPPLER GRAVITY EXPERIMENT TO ASSESS FIELD STRENGTH AND VARIATION BY S-BAND RADIATION SIGNALS

HIDDEN DEPTHS

After the Apollo 17 mission in 1972, NASA did not return to the Moon for more than 25 years. The next obvious step was to establish a moonbase, but one of the biggest problems was water—or the lack of it. Water is vital for living in space, not only for drinking, but as a source of oxygen. The cost of transporting large amounts from Earth is prohibitive, not least because the weight puts too much of a demand on the launcher. Unfortunately, the lunar rock samples brought back by the Apollo astronauts were bone dry, implying that the Moon was waterless—around the landing sites, at least.

The rock samples were all from near the lunar equator, where temperatures reach 250°F (121°C). Some people thought that there might be a better chance of finding water at the lunar poles. Their hopes were raised in November 1996 when NASA revealed that the probe Clementine, which had been testing radar instrumentation for the Pentagon during a mission in 1994, had picked up the radar signature of hydrogen at the Moon's south pole—an indicator of the presence of ice.

In January 1998, NASA launched the uncrewed Lunar Prospector with a mission profile that included searching for further evidence of water. Less than two months after the probe went into into lunar orbit, NASA proudly announced that one of the instruments on board—a neutron spectrometer—had confirmed

PROBING THE MOON'S POLAR DEPTHS

POLAR PUZZLE
NASA's probes detected twice as much water ice at the north pole as at the south pole. This was surprising, given that the permanently shadowed area near the north pole is considerably smaller than its south polar equivalent. Perpetually shadowed areas inside deep craters, known as "cold traps," were considered to be the most likely places to harbor water ice.

RADAR CLUE
Radar mapping beams bounced off the lunar surface by the probe Clementine first alerted NASA scientists to the presence of water.

NORTH POLE
Water ice detected in area covering between 3,600 and 18,000 square miles (9,300 and 46,600 square km).

SOUTH POLE
Water ice detected in area covering between 1,800 and 7,200 square miles (4,660 and 18,650 square km).

COLD TRAP
With no internal heat source and hardly any atmosphere to conserve radiant heat, temperatures on those parts of the Moon's surface that receive no direct sunlight an hover below −280°F (−138°C). Areas of perpetual shadow within deep polar craters known as "cold traps" are most likely to contain water ice.

OH, MY DARLING

THE FIRST HINT OF LUNAR WATER CAME FROM THE CLEMENTINE PROBE, A JOINT PROJECT BETWEEN NASA AND THE STRATEGIC DEFENSE INITIATIVE (WHO PROPOSED THE STAR WARS SATELLITE NETWORK). LAUNCHED ON JANUARY 25, 1994, CLEMENTINE TESTED SENSORS AND SPACECRAFT COMPONENTS UNDER LONG EXPOSURE TO SPACE. IT CIRCLED THE MOON 348 TIMES, MAPPING THE SURFACE AND BOUNCING RADAR BEAMS OFF SOME OF THE DEEPEST CRATERS. THE DATA FROM THE SOUTH POLE INDICATED THE EXISTENCE OF REFLECTIVE WATER ICE.

that hydrogen was present in rocks at both lunar poles and that this hydrogen was bound up in frozen water. "Our expectation is that we have areas at both poles with layers of near-pure water ice," says Alan Binder of the Lunar Research Center in Gilroy, California.

Some NASA scientists have estimated that there is between one and six billion tons (5.4 billion tonnes) of ice, buried beneath 18 inches (46 cm) of dry lunar soil and concentrated in a 650-square-mile (1,685-square km) region at each pole. One billion tons of water is enough to fill around 300,000 Olympic-sized swimming pools. It is also thought to be enough to support a community of 2,000 people on the Moon for over a century.

As a final attempt to prove the presence of water on the Moon beyond argument, the Lunar

Prospector team planned a spectacular finale to the mission, sending the probe hurtling to its doom in a suspected ice-filled crater. Telescopes on Earth hoped to detect the ice and water flung up by the impact, but they saw nothing—perhaps because the probe was simply unlucky. So the case for water on the Moon is perhaps not yet quite closed.

LOCKED IN THE SHADOWS

Where did the Moon's water come from? Most scientists believe that it arrived there as ice, brought by countless meteorite and comet impacts over billions of years. Over time, most of this water evaporated in the intense heat of the lunar days and, in the absence of a substantial atmosphere, was lost to outer space. Yet some of

the deep craters that are found at the lunar poles lie in perpetual shadow, where temperatures may never rise above −280°F (−138°C). This creates "cold traps" in which some of the water could have remained frozen for billions of years. The water NASA found could have been deposited by direct impact, or by individual water molecules that randomly migrated over the lunar surface and froze when they reached the polar craters.

The discovery of the water has been particularly inspiring for planetary scientists. Probing the layers of polar ice on the Moon could yield fresh insights into what comets are made of and how often they have hit both the Moon and the Earth. It could also help to reveal whether the rate of impacts changes over time—and if so, whether it is likely to increase.

LUNAR TRANSIENT PHENOMENA

Until recently, most planetary scientists would have said that the Moon is a totally dead and unchanging world. Despite reports of strange glows and clouds on the Moon stretching back some 450 years, the assumption has been that nothing much has happened on the Moon for over a billion years. But new evidence from the Clementine spacecraft indicates that the odd glows, obscurations and other lunar transient phenomena (LTP) reported by earthbound observers may be the results of real lunar events.

WHAT IF...

...YOU COULD WATCH LTP FROM YOUR BACK YARD?

It is perfectly possible that it will be an amateur who spots the next important instance of the mysterious lunar transient phenomena. Incidents of LTP continue to be reported by both professionals and amateurs, just as they have for hundreds of years. While there have reportedly been a few occurrences of lunar glows bright enough to be seen with the naked eye, you would probably need a small telescope in order to see anything. But even a telescope with a mirror as small as three inches in diameter will give you a real chance of witnessing these strange Moon-changes with your own eyes—if you happen to be looking in just the right place at the just right time.

Although the Moon is some 240,000 miles (386,200 km) away, amateur-size telescopes are surprisingly capable of revealing small details on its surface. A telescope with an objective mirror 3 in (75 mm) in diameter, for example, can theoretically detect a feature a mere 1.76 miles (2.83 km) across. Larger instruments can do even better, with a 6-inch (15-cm) telescope revealing details 0.88 miles (1.4 km) in diameter and one with an 8-inch (20-cm) mirror scoping out lunar landmarks as small as 0.65 miles (1

km). It would seem fairly realistic to expect a major gas cloud emission to be at least this big. Generally, reflecting telescopes are considered preferable to refractors for LTP hunting, as they do not share the refractor's problem with false color.

Contrast is also an important factor in LTP visibility from Earth. A brightly glowing mass—even a relatively small one—should be visible to most telescopes. On the other hand, LTP consisting of vapor clouds may be completely invisible even with very large instruments.

But the real trick to detecting lunar transient phenomena is in being able to look at the right spot at the right time. The Moon is a big place, and spotting LTP is highly dependent on luck. The events often last for only a few minutes and usually involve small areas of the lunar landscape. You can improve your chances of seeing something by observing known LTP hot spots. Prime hunting grounds are the perimeters of the maria, or lunar seas. The areas where the seas give way to the rough lunar highlands are renowned as prime LTP hunting ground.

Fresh-appearing craters are also good places to look. In the past, the region immediately surrounding the bright crater Aristarchus has been the site of numerous glows and obscurations.

MOON MYSTERIES

Date	Observers	Event/Comments
November 26, 1540	Monks	Bright star-like point on unilluminated part of Moon
December 10, 1685	Bianchini	Red streak on floor of crater Plato during lunar eclipse
April 19, 1787	William Herschel	Three glowing red points
April 20, 1787	William Herschel	Single, glowing "volcano"
March–April–May 1789	Johann Bode	Very bright spot near crater Aristarchus
September 16, 1891	W.H. Pickering	"Dense clouds of white vapor" in the lunar feature Shröter's Valley
May 12, 1927	H.P. Wilkins	Complete obscuration of the floor of the crater Peirce
August 20, 1951	Patrick Moore	Brilliant white patch seen on the floor of the crater Messier A
November 2, 1958	N.A. Kozyrev	Reddish glow in the floor of crater Alphonsus
July 19, 1969	Neil Armstrong	Brightening of surface near Aristarchus

DEAD OR ALIVE?

Early one winter's morning in November 1540, a monk at a monastery near Worms, Germany, happened to glance up at the crescent moon. He was astonished to see a bright light shining on the dark part of its globe. The starlike point lasted just a few minutes, but the monk made a detailed record of what he had seen.

This story is the first reliable record of an occurrence of lunar transient phenomena (LTP): short-lived glows, flashes, bright spots and obscurations (hazes) on the surface of the Moon. Many observers since then have made similar reports. But the source of the strange apparitions has remained a mystery. By the 19th century it was known that the Moon was incapable of supporting life that could be responsible for the LTP effects. But astronomers still felt that the sightings must reflect real events, theorizing that the Moon might still be slightly volcanically active. The bright lights, small red glows and obscurations could be the result of minor volcanic eruptions. At the time, this seemed logical, since it was mistakenly believed that the Moon's craters were volcanic.

The dawn of the space age brought increased scrutiny of the Moon, resulting in the first hard evidence for lunar transient phenomena. Early in the morning of November 3, 1958, Soviet astronomer Nikolai Kozyrev was examining the lunar crater Alphonsus. Kozyrev was surprised to see that the central peak of the crater appeared

brighter than usual. He obtained a spectrogram — a reading of light coming from the area—that suggested traces of carbon emission. Many astronomers accepted Kozyrev's observation as proof that something was happening on the lunar surface. But when no further evidence appeared, even after the Apollo landings, belief in LTP began to wane. It was also increasingly clear that the lunar craters were formed from asteroid impacts—not by volcanic activity.

In 1999, Jet Propulsion Laboratory scientist Bonnie Buratti decided to put the LTP mystery to rest. She and her colleagues examined lunar images returned by the Clementine spacecraft in 1994. The skeptical scientists were amazed to find that the images seemed to verify the presence of glows and outgassings in some places. These areas appeared to have suffered recent landslides that may have thrown dust into the atmosphere or perhaps allowed the escape of volatile gases. Much more work needs to be done before lunar transient phenomena can be considered proven, but it seems clear that our sleepy old Moon is not as dormant as we thought.

Lunar transient phenomena have been witnessed all over the Earth-facing side of the Moon, but are thought to be most common on the fringes of the maria (dark lowlands). Recent analysis suggests that the release of gas trapped under the surface may play a role in the light shows.

Labels on Moon: Plato, Linné, Peirce, Aristarchus, Schröter's Valley, Ltp Hot spots, Messier A, Alphonsus, Cavendish, Liebig

FIBBER

THE 19TH CENTURY WAS A TIME OF FREQUENT HOAXES. ONE EXAMPLE WAS A LETTER FROM ONE JOHN HAMMES OF IOWA TO SCIENTIFIC AMERICAN MAGAZINE IN 1878 DESCRIBING OBSERVATIONS OF A "LUNAR VOLCANO." BUT ON CLOSE EXAMINATION, THE ENTIRE TALE, FROM DETAILED DRAWINGS OF THE VOLCANO TO TESTIMONIALS OF THE WRITER'S INTEGRITY MADE BY HIS TOWN MAYOR, FALLS APART.

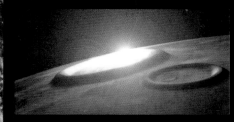

LUNAR FLARE
January 24, 1956 Amateur astronomer R. Houghton is observing the region around the lunar crater Liebig—in the southwest corner of the Moon—when a bright flare flashes from the wall of the nearby crater Cavendish. Alerted by Houghton, professional U.S. astronomers also witness and confirm the event.

RED MOON

DURING ITS 1994 MOON-MAPPING MISSION, NASA'S CLEMENTINE SPACE-CRAFT GATHERED MORE EVIDENCE OF LTP. THIS IMAGE OF THE ARISTARCHUS REGION WAS TAKEN BY CLEMENTINE AFTER A REPORTED OBSCURATION OF THE AREA. THE IMAGE SHOWED THAT THE REGION WAS REDDER THAN USUAL. SCIENTISTS HAVE SPECULATED THAT ESCAPING GAS FROM THE COLLAPSED LAVA TUBE RUNNING THROUGH THE AREA CAUSED THE HAZE.

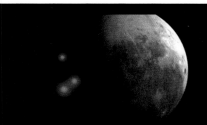

LUNAR "VOLCANOES"
April 19, 1787 William Herschel (1738–1822), the German-born astronomer famous as the discoverer of the planet Uranus, is surveying the Moon with a large telescope when he sees three glowing red points on the unilluminated portion of the disk. Herschel calls these lights "volcanoes" and is convinced that he has witnessed a real event.

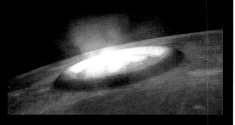

CRATER HALO
July 19, 1969 Orbiting the Moon just before their scheduled landing, Apollo 11 astronauts receive a NASA message that amateur astronomers are reporting a bright glowing area near Aristarchus. Checking the crater from orbit, Neil Armstrong reports that it does indeed seem brighter than normal.

PYRAMIDS OF LIGHT
July 3, 1882 Several residents of Lebanon, Connecticut, observing the nearly full Moon with their unaided eyes, see what appear to be two pyramid-shape protuberances glowing faintly and extending from the upper limb of the Moon. This is one of the less reliable sightings—probably caused by optical effects in Earth atmosphere.

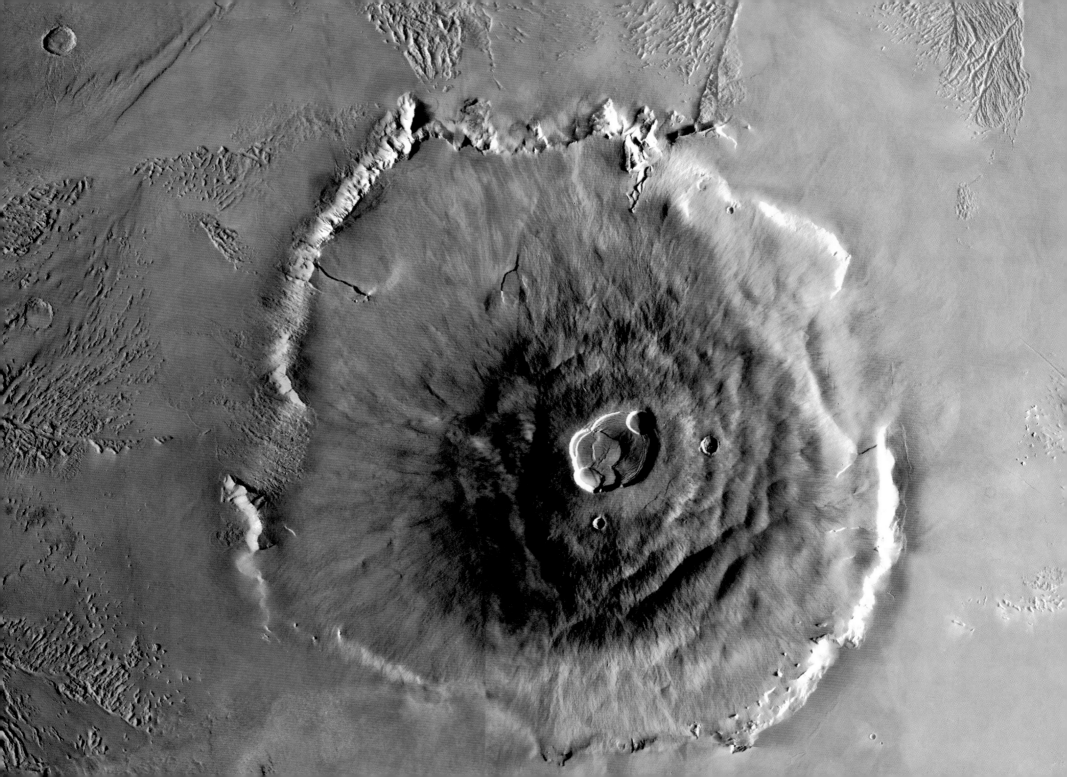

MARS

The famous "Red Planet," Mars is one of the most conspicuous objects in Earth's skies, and has fascinated people since ancient times. Like Venus, it was thought that it might be a hospitable environment for advanced life, and the discovery of polar caps and changing dark patches on the surface—as well as the infamous "canals"—encouraged the idea. When the first space probes flew past, there was huge disappointment at what they revealed—a dry, moonlike, cratered world. It was not until the first orbiters surveyed the planet in detail that Mars's reputation as a fascinating planet was restored. Although they found no signs of life or water, the photographs revealed giant volcanoes, a canyon dwarfing any on Earth, and channels that looked suspiciously like dried-up rivers. Later probes have discovered even more: huge quantities of ice beneath the surface, complex weather systems, and conclusive evidence that Mars had oceans for a long period of its early history. Even the question of past and present life on Mars, once dismissed out of hand, is now taken seriously once more.

A color mosaic image of Olympus Mons, the largest volcano in the solar system and one of the most striking features of the Martian landscape. Olympus Mons is about 370 miles (600 km) in diameter and the summit is 15 miles (24 km) above the surrounding plains.

MARS

Mars is the fourth planet from the Sun and one of Earth's nearest neighbors. It is also the only other planet in the solar system where humans may one day live. Mars is smaller and colder than Earth, but is otherwise remarkably similar: It has days and seasons, a thin atmosphere and, probably, significant reserves of water buried as ice beneath the surface. There is even a chance that Mars once played host to simple life-forms, and that the fossilized remains of long-extinct creatures are buried there.

WHAT IF...

...WE COLONIZED MARS?

The most likely future for Mars from a human point of view is as a mining colony, supplying us with minerals that have been exhausted on Earth. The writers of Star Trek got there first: The great shipyards that built the starship Enterprise-D and other Galaxy class starships are located in synchronous orbit (orbiting at the same speed as Mars itself) some 100,000 miles (160,000 km) above the plain of Utopia Planitia.

This is not as far-fetched as it sounds. Mars is rich in minerals—especially iron and aluminum, which, even in the 24th century, will still be the basis of engineering construction. The other elements required to make complex starship alloys would be needed in much smaller quantities, so they could be shipped in from other sources such as Mercury or the Asteroid Belt.

In Star Trek, the shipyards are fed with materials by a largely automated mining operation at Utopia Planitia itself. The alloys are partially created on the surface, then transferred to the shipyards by a Magnetic Pulse Transfer Generator (MPTG), which shoots the material skyward on a beam of electromagnetic energy. Again, Mars is ideal for such an operation. Its surface gravity is only a third that of Earth's, so the energy needed to transfer materials in this way would be much reduced.

...WE MADE MARS LIKE EARTH?

A more ambitious plan for colonizing Mars involves melting the reserves of ice that are thought to lie trapped below the surface. In time, this water would evaporate into the atmosphere, making it denser and creating a greenhouse effect similar to the global warming that currently threatens life on Earth.

As the icy surface of Mars began to warm up, it would melt more ice reserves, causing seas to form in the lower-lying regions. At this point, plant life could be introduced. The plants would feed off the carbon dioxide-rich atmosphere, releasing the nitrogen and oxygen that are necessary for humans to breathe.

MARS AND EARTH

MARS		EARTH
4,228 miles (6,804 km)	Diameter	7,973 miles (12,831 km)
25° 11'	Axis tilt	23° 27'
687 Earth days	Length of year	365 days
24 hours 37 minutes	Length of day	24 hours
141.6 million miles (227.9 million km)	Distance from Sun	93.5 million miles (150 million km)
−9°F (−13°C)	Surface temperature	59°F (15°C)
0.379 g	Surface gravity	1 g
Carbon Dioxide (90%)	Atmosphere	Nitrogen (78%), Oxygen (21%), Argon (1%)
10 millibars	Atmospheric pressure	1,000 millibars
Silicon, Iron, Oxygen	Composition	Silicon (60%), Aluminum (15%)

A COLD, ROCKY DESERT

Of all the planets in the solar system, Mars is the most like ours. Its axis is tilted like the Earth's, which gives it seasons. Mars has a relatively warm summer, when temperatures in the southern hemisphere can reach 68°F (20°C), but a long, cold winter that sees them plunge to −284°F (−140°C).

Over 4 billion years ago, Mars was covered with massive volcanoes—just like the Earth—and had surface water, which occasionally gathered in flash floods, carving water channels into the surface. There may even have been standing oceans over long periods of time. Like the Earth, too, Mars has a cloudy atmosphere; although the Martian "air" is much thinner, and the wispy clouds are made of carbon dioxide rather than water vapor.

So, despite its many similarities to Earth,

CANAL MYTH

THE BIGGEST MYTH ABOUT MARS—THAT A RACE OF MARTIANS ONCE BUILT CANALS TO CARRY WATER FROM THE POLES TO THE EQUATOR—IS BASED ON A TRANSLATION ERROR. IN 1877, THE ITALIAN ASTRONOMER GIOVANNI SCHIAPARELLI SAW WHAT HE THOUGHT WERE "CANALI," MEANING "CHANNELS," THROUGH HIS TELESCOPE. IN ENGLISH THIS WAS TRANSLATED AS "CANALS" SO PEOPLE ASSUMED THAT THEY MUST BE ARTIFICIAL WATERWAYS. IN FACT, THEY WERE AN OPTICAL ILLUSION.

Mars is a far from hospitable place. If you landed there without a spacesuit, not only would you suffocate but, due to the much lower atmospheric pressure, your blood would boil within minutes.

LIVE FROM MARS
On July 4, 1997, the uncrewed U.S. probe Pathfinder became the first spacecraft to land on Mars since Viking in 1976. After a short delay due to technical glitches, Pathfinder released the Sojourner remote-controlled Mars rover, and the world held its breath as the probe's first pictures of the Red Planet were broadcast back to Earth.

NATURAL FEATURES

North Pole

Arcadia Planitia region

Acidalia Planitia region

Amazonis Planitia region

Ascraeus Mons

Mangala Vallis

Pavonis Mons

Valles Marineris region

Arsia Mons

Noctis Labyrinthus

Magaritifer Sinus region

South Pole

When the first truly detailed maps of Mars were made in the 1970s, scientists added descriptive Latin words such as Planitia ("plain") or Mons ("mountain") to the original place names.

NORTH POLE
Made of frozen carbon dioxide (dry ice) and water ice. The southern ice cap shrinks away to almost nothing in the summer.

OLYMPUS MONS
The largest volcano in the solar system, it towers 17 miles (27 km) above the plains and is 375 miles (600 km) across. It is probably extinct.

VALLES MARINERIS CANYON
A giant canyon across one side of Mars, 3,125 miles (5,030 km) long and so big that the Rocky Mountains would fit comfortably inside it.

ARGYRE PLANITIA
One of many basins on Mars created billions of years ago by asteroid impacts. The crater Galle is about 125 miles (200 km) across.

DUST STORMS

Apart from the lack of oxygen and the low atmospheric pressure, you would also have to withstand the continuous winds that blow across the Martian surface at speeds of over 125 miles per hour (200 km/h), whipping up giant clouds of fine orange-brown dust in their wake. It is this dust that has earned Mars the nickname "The Red Planet," although "Rusty Planet" might be more appropriate, since the color is explained by the high proportion of iron in the planet's rocks—on average, almost twice as much as on Earth. Mars is also very dry and cold. Even on a warm summer's day the ground rarely gets above freezing point, and on winter mornings the rocks are often coated with a fine layer of carbon dioxide "frost."

The two tiny moons of Mars, Phobos and Deimos, race around the planet in about eight

hours and 30 hours respectively. They are thought to be asteroids that strayed too close to the planet in the distant past and were captured by its gravity. If Phobos, the closer of the two, maintains its present orbit, it is likely to crash into Mars in about 100,000 years' time. There is also some evidence to suggest that the planet has suffered similar collisions in the past.

SAND DUNES

THE SURFACE OF MARS HAS THE BIGGEST FIELD OF SAND DUNES IN THE SOLAR SYSTEM, COVERING AN AREA LARGER THAN EARTH'S SAHARA AND ARABIAN DESERTS COMBINED. BUT THERE IS ONE BIG DIFFERENCE—THE DUNES ON MARS ARE SITUATED NEAR THE PLANET'S NORTH POLE, AND DURING WINTER THEY ARE IN PERMANENT DARKNESS.

CHANGING VIEWS OF MARS

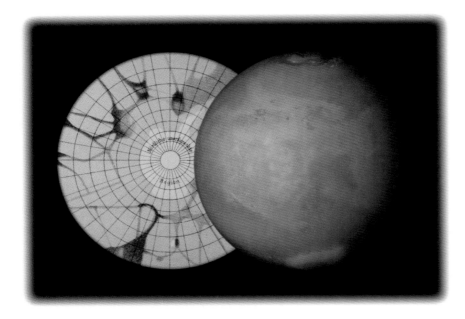

Ever since the 17th century, when Christiaan Huygens first sketched the face of Mars, people have wondered what the planet is like—and if it is home to life. To the eyes of early astronomers, Mars appeared to be a watery world, lush with vegetation. Advanced civilizations—builders of canals and cities—were thought to exist on the planet. Today, we know Mars is a cratered, desolate world. Space probes have mapped all of its barren surface. But Mars still has plenty of secrets. One of them may yet be life.

WHAT IF...

...WE UNCOVERED NEW FEATURES ON MARS?

Despite NASA's best efforts, Mars has yet to reveal all of its secrets to us. Knowledge is only as good as the instruments used to obtain it, and as increasingly sophisticated space probes have sent back data on the red planet, astronomers have had to redraw the face of Mars. Future advanced probes will undoubtedly uncover new features and cast the planet in a different light.

"Is there life on Mars?" remains the most crucial unanswered question. To date, five landers have looked for signs of Martian life. Although none have returned conclusive results, evidence is mounting that Mars was once a hospitable environment in which simple life might have evolved. And Mars is a big place: Its surface area roughly corresponds to the dry land area of the Earth. Future missions to Mars may discover life huddled in protected oases—in or around geothermal vents, for example, or perhaps beneath the barren surface.

Or we may find evidence of past life on Mars in the shape of fossils. In its watery past, Mars may have harbored microbial life. On Earth, enormous colonies of tiny microbes have been buried beneath sediment and formed distinctive large fossil structures. These stromatolites can measure over a yard (0.9 m) across.

But at the moment, the chances of finding Martian fossils—even large ones similar to stromatolites—are slim. Aerial photographs of Mars, however clear, will never be able to reveal fossils or small microbial life on the planet's surface. Some things can only be seen from the ground. Just as you would learn a lot more about the Sahara desert by walking on it than by flying over the Earth, large tracts of the Martian surface must be studied closely to find evidence of life, past or present.

Another problem is that fossils form within rocks. Only by living on, combing over and digging into the Earth's surface have we discovered terrestrial fossils. Mars landers have neither broken Martian rocks nor brought samples back to Earth for analysis.

Many questions will only be settled when a future Mars rover tunnels beneath the desolate surface. Here, caves, lava tubes and unusual rock structures could greet the robot explorer. Perhaps we will find huge reserves of water on the planet—not on the surface, but under it as permafrost. Such a discovery would give Mars a more complex identity, one that is in sharp contrast to its plain exterior, and more like the intriguing planet imagined by early astronomers.

AN AUDIENCE WITH MARS

Date	Name	Work
1659	Christiaan Huygens	drew first sketches of the light and dark areas of Mars
1780s	William Herschel	noted thin Martian atmosphere
1877	Giovanni Schiaparelli	drew first detailed maps of Martian surface
1965	Mariner 4	beamed back 20 photos from the first flyby of Mars
1971	Mariner 9	sent back 7,300 images from first-ever orbital mission
1976	Viking 1 & 2	first probes to land on Martian surface and photograph terrain
1998	Mars Global Surveyor	began to map the whole surface of Mars
2003	Mars Express	discovers widespread subsurface ice and maps surface with stereo camera

SEEING RED

From the first telescopic observations to the latest probe pictures, our view of Mars has been under constant review. The Dutch astronomer Christiaan Huygens (1629–93) made the first sketch of the Red Planet in 1659. He believed that there was life on Mars—as well as on other planets—and the view through his crude telescope led him to think that Mars was similar to the Earth. Huygens could make out dark spots that he mistook for seas, and assumed the bright regions he observed were fields of ice.

Two centuries later, in 1877, Italian astronomer Giovanni Schiaparelli (1835–1910) drew the first good maps of Mars. He used a telescope that could resolve objects down to 50 miles (80 km) in diameter to fill in the planet's surface features. The maps—characterized by dark tracks that Schiaparelli named canali—were controversial: Some astronomers could see his dark bands, others could not.

The wealthy U.S. businessman Percival Lowell (1855–1916) mistranslated Schiaparelli's Italian term, and "canali" became "canals." Excited by what he thought was evidence of extraterrestrial life, Lowell built himself a telescope to investigate further. High up in Arizona, Lowell's 24-inch (60 cm) refractor allowed him to make out Martian surface features just 35 miles (56 km) across. Despite Lowell's conviction, most scientists remained skeptical and dismissed the idea of water canals—although they all continued to believe that dark regions on the planet were patches of vegetation. Not until 1965 were their ideas finally disproved.

UNCOVERED TRUTH

In 1965, images from NASA's first Martian probe—Mariner 4—gave Mars its most dramatic costume change since Huygens first sketched the planet. Disappointingly for some, a dry, rocky world was clearly visible, with no sign of life. Over the next seven years, three more spacecraft paid visits. It was the last of these probes—Mariner 9—that discovered Mars' most spectacular features. As Mariner 9 orbited the

planet, a huge dust storm masked the surface. When the storm subsided, the tops of volcanoes and the outline of huge chasms became clear

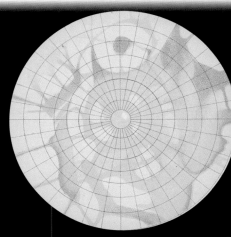

PLANET PLANTS
Like his contemporaries, Giovanni Schiaparelli (left) believed that vegetation flourished on Mars. Just under 100 years later, the Mariner probes flew past Mars, and dark areas like those in Schiaparellli's sketch (above) were revealed to be nothing more than patches of rock uncovered by dust storms.

DESERT SURPRISE
Mariner 9 (left) flew past Mars in 1971. The probe took the first-ever photo-globe of a body in the solar system (right), giving astronomers a complete photographic map of the Martian surface. Although the probe sent back images of huge volcanoes and deep channels, they were all inactive.

THERMAL BLANKET
Mars Global Surveyor (left) began to map the Red Planet in 1998. Using four different instruments, the probe has sent back images of the planet unlike any seen before. One instrument—the Thermal Emission Spectrometer—used temperature differences to map the planet. The night-time image of Mars (far right) shows the coldest areas in purple and blue (–250°F/–120°C); the warmest regions are shown red and white (–150°F/–65°C).

GENTLE TOUCH

IN 1976, THE FIRST SUCCESSFUL MARS LANDERS TOOK CLOSE-UP PHOTOGRAPHS OF THE MARTIAN SURFACE (RIGHT). THE LANDERS ALSO SENT BACK DATA ON THE MARTIAN SOIL AND MARS' ATMOSPHERE. NEITHER OF THE PROBES—NAMED VIKING 1 AND 2— FOUND EVIDENCE OF LIFE OR ITS BIOCHEMICAL BASIS. BUT THAT DOES NOT MEAN LIFE HAS NEVER EXISTED ON MARS.

LOOKING TO MARS

These three maps of Mars were created by three very different methods. The top map is a drawing of Mars made by the Italian astronomer Giovanni Schiaparelli. The middle photograph is a global mosaic made up of 1,500 individual images. The last map was produced by a spectrometer that measured the heat radiated by the planet.

through the orange haze. The atmospheric pressure of Mars was found to be too low for water to remain liquid, but these newly discovered channels suggested that water had once flowed across the surface.

In the mid-1970s, the red surface of Mars became visible from a completely different angle. Two Viking landers touched down and took images of the rock-strewn landscape that surrounded them.

Though the Mars Global Surveyor satellite has now photographed surface objects only five feet across, our view of Mars since the Viking missions has changed largely through exploration of our own planet. The discovery of life in hostile environments on Earth has reignited the idea that Martians could exist in nooks and crannies on the planet's surface—or perhaps underground. Orbital missions can tell us little more: Only painstaking examination at ground level will coax Mars into giving up its tantalizing secrets.

SHADOWLESS LANDSCAPE

AT MARS' CLOSEST APPROACH TO THE EARTH, SUNLIGHT HITS THE PLANET STRAIGHT ON, ELIMINATING THE SHADOWS THAT REVEAL LANDSCAPES. DURING THIS TIME, THE ONLY VISIBLE MARKS ON MARS ARE SO-CALLED ALBEDO FEATURES. IT WAS THESE PATCHES OF LIGHT AND DARK THAT DUTCH ASTRONOMER CHRISTIAAN HUYGENS (ABOVE RIGHT) OBSERVED IN 1659.

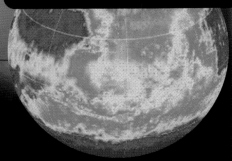

THE SURFACE OF MARS

R ust staining in the soil, imparted by iron oxide, is the simple reason why Mars' surface is so distinctly red. But its color is just one of the planet's unique surface features. In recent years, the Martian surface has been revealed as a rich and varied landscape, shaped by great natural forces such as meteorite impacts, floods, volcanoes, earthquakes and glaciers. Mars' surface is continuing to evolve, due to seasonal freezing and thawing and powerful winds that transport vast clouds of dust across its face.

WHAT IF...

...ASTRONAUTS LANDED ON THE SURFACE OF MARS?

E ven though Mars is the most Earthlike planet in our solar system, the first astronauts to set foot on the Red Planet will experience a totally alien environment. Not only will they have to get used to seeing a pink sky, but the thin Martian atmosphere is made up mostly of unbreathable carbon dioxide, and the air pressure, at a mere 0.6% of Earth's, is little more than in outer space. Mars is also cold. The average recorded temperature there is –80°F (–27°C), with a maximum temperature of 70°F (21°C) and a minimum of –220°F (–104°C).

As on the Moon, insulated and pressurized spacesuits will be essential for life support. Scientists believe that spacesuits similar to those worn by lunar astronauts will be used. Yet the gravity on Mars is about twice that on the Moon, so the suits will have to be half the weight to allow the same freedom of movement.

A major problem for astronauts on Mars will be the Martian soil. Carried by wind speeds of up to 70 mph (110 km/h), the coarsest and most abrasive grains will sandblast and possibly scratch the astronauts' suits and visors. Finer particles might clog up filters and electrical equipment, and if soil particles were to get inside the suits or lander craft, they could irritate or possibly even burn human flesh and corrode the craft's equipment.

To move around the surface of Mars, astronauts will need powered transport. The terrain varies from the mountainous to pancake-flat, and there is soft, fine soil and loose rock almost everywhere. NASA expects that rover vehicles like those used on the Moon, or jeeps like those used by the military in stony deserts on Earth, will be used for the easiest ground. But to get around the harsher landscapes, the astronauts could use helium-filled airships or balloons.

Some lessons can be learned from the experiences of unmanned rovers on the surface. The Spirit and Opportunity rovers, which landed on opposite sides of the planet in 2004, have both experienced difficulties with the surface—even shallow slopes on Mars have proved awkward to traverse due to the slippery sand.

Another problem that astronauts will encounter is that of physical weakness. Some doctors are concerned that after six months of weightlessness, which astronauts will have to go through to get to Mars, their muscles and bones will have become so feeble that even Mars' low gravity—about one-third that of the Earth's—will take weeks to get accustomed to. So, after reaching their destination, the astronauts may not be capable of actually doing very much.

TOP TEN SURFACE FEATURES

NAME OF FEATURE	TYPE OF FEATURE	LOCATION	SIZE MILES/KM
VALLES MARINERIS	TECTONIC CANYON SYSTEM	70°W, 10°S	2,500 MILES (4,023 KM) LONG, 373 MILES (600 KM) AT ITS WIDEST AND 5 MILES (8 KM) DEEP
THARSIS RIDGE	BASALT PLATEAU	100°W, 5°N	6 MILES (10 KM) HIGH BY 2,485 MILES (4,000 KM) WIDE
OLYMPUS MONS	EXTINCT SHIELD VOLCANO	135°W, 20°N	17 MILES (27 KM) HIGH BY 373 MILES (600 KM) WIDE (LARGEST VOLCANO KNOWN)
SYRTIS MAJOR PLANITIA	DARK, ELEVATED FLAT PLAIN	290°E, 10°N	785 MILES (1,265 KM) IN DIAMETER
NORTH POLE ICE CAP	FROZEN H₂0 AND CO₂	EXTENDS TO 65°N IN WINTER	MAXIMUM THICKNESS OF 3 MILES (5 KM); 1,000 MILES (1,600 KM) WIDE IN WINTER, 375 MILES (600 KM) WIDE IN SUMMER
SOUTH POLE	COMPOSED MAINLY OF CO₂	EXTENDS TO 45°S IN WINTER	1,200 MILES (1,930 KM) WIDE IN WINTER, 250 MILES (400 KM) WIDE IN SUMMER
ARGYRE PLANITIA	IMPACT BASIN	42°W, 50°S	870 MILES (1,400 KM) IN DIAMETER
ACIDALIA PLANITIA	IMPACT BASIN	30°E, 50°N	1,625 MILES (2,615 KM) IN DIAMETER
HELLAS BASIN	IMPACT BASIN	294°E, 40°S	1,000 BY 1,250 MILES (1,600 BY 2010 KM) (MARS' LARGEST IMPACT CRATER)
ELYSIUM PLANITIA	IMPACT BASIN	210°E, 20°N	1,554 MILES (2,500 KM) IN DIAMETER

ROCKS OF AGES

We know a great deal about the Martian surface, thanks first to 19th-century astronomers, who identified and named many of its features, and more recently to uncrewed Mars probes that have provided us with stunning closeups of the entire planet.

Much of the surface of Mars is a barren stony desert that looks and behaves much like deserts on Earth. But Mars has a range of distinctive features. Most of them, such as mountains, canyons, extinct volcanoes and craters, were created early in the planet's life. Mars' rivers, and probably its seas, have since shaped and modified many of these landmarks. Some of Mars' most unusual features—like the "Main Pyramid," "City Square" and "The Face"—have even led to claims that they are ruins left by an ancient civilization, although most scientists now believe they are natural features.

But the planet's reddish soil can be found everywhere on Mars—the result of billions of years of rock erosion by wind and water and pulverization by meteorites. This soil, blown around by the wind, covers virtually the entire planet and can be a few inches or many feet deep. It collects to forms drifts, which can be seen from high above the planet's surface as distinctive tapered streaks of soil, often deposited on the leeward sides of craters.

Most of what is known about the composition of Martian soil comes from experiments performed by the Viking Landers. They detected iron-rich clays, calcium carbonate, iron oxides and magnesium sulfate, as well as the silicon dioxide that makes up 50% of Mars' soil. The strange oxidizing agent in the soil, which releases oxygen when wetted, is thought to be a type of peroxide.

The Martian surface can be split into two regions. The highlands, in the southern hemisphere, contain Mars' oldest surface rocks. Many craters and basins (craters more than about 50 miles/80 km wide) are found here. The lowlands, in the northern hemisphere, are less cratered. This is the flattest, smoothest region known in the solar system. Some scientists believe it was shaped by ancient ocean water, as it resembles the heavily sedimented floors of Earth's oceans. Mars' largest volcanoes also exist here in an area known as Tharsis Ridge.

Next to Tharsis Ridge, in the equatorial region, lies the Valles Marineris, a canyon stretching almost the distance from New York to California. But unlike the Earth's Grand Canyon, which was cut by the Colorado River, it is thought to be an ancient tectonic feature, caused by movement in Mars' surface mantle. Other features found here are channels, probably cut by frequent flooding more than a billion years ago, and large sand dunes.

More impressive sand dunes surround the poles. A big dune field entirely circles the northern pole, showing just how important the strong Martian wind is in the carving and shaping of this unique landscape.

CLOSE CALL

THE VIKING LANDERS HAD TO TOUCH DOWN ON EVEN GROUND FOR THEIR EXPERIMENTS TO SUCCEED. VIKING 1 LANDED ON AN AREA THAT LOOKED LIKE A SMOOTH, SPARSELY CRATERED PLAIN—FROM ORBIT. BUT IT HAD COME WITHIN 30 FEET (9 M) OF CATASTROPHE IN THE SHAPE OF A 7-FOOT (2-M) WIDE ROCK, LATER NAMED BIG JOE (ABOVE). VIKING 2 LANDED NEAR SEVERAL CHANNELS UP TO 3 FT (90 CM) WIDE AND FOUR INCHES (10 CM) DEEP. BUT BOTH MADE IT, DESPITE THESE UNSEEN OBSTRUCTIONS.

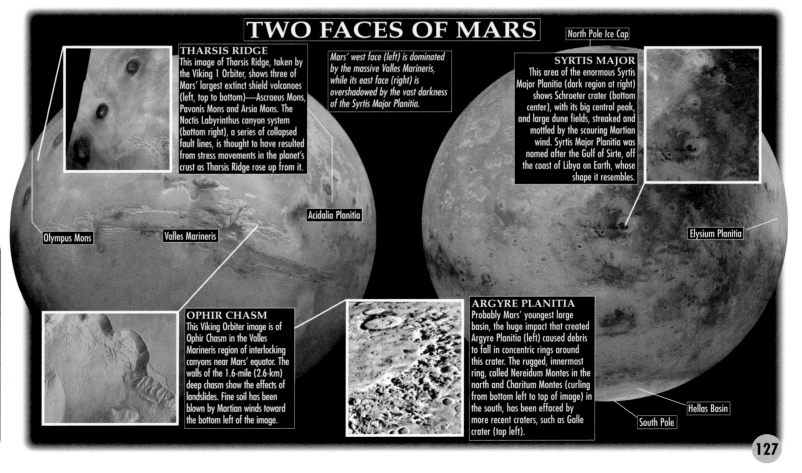

TWO FACES OF MARS

THARSIS RIDGE This image of Tharsis Ridge, taken by the Viking 1 Orbiter, shows three of Mars' largest extinct shield volcanoes (left, top to bottom)—Ascraeus Mons, Pavonis Mons and Arsia Mons. The Noctis Labyrinthus canyon system (bottom right), a series of collapsed fault lines, is thought to have resulted from stress movements in the planet's crust as Tharsis Ridge rose up from it.

Mars' west face (left) is dominated by the massive Valles Marineris, while its east face (right) is overshadowed by the vast darkness of the Syrtis Major Planitia.

SYRTIS MAJOR This area of the enormous Syrtis Major Planitia (dark region at right) shows Schroeter crater (bottom center), with its big central peak, and large dune fields, streaked and mottled by the scouring Martian wind. Syrtis Major Planitia was named after the Gulf of Sirte, off the coast of Libya on Earth, whose shape it resembles.

North Pole Ice Cap

Olympus Mons · Valles Marineris · Acidalia Planitia · Elysium Planitia

OPHIR CHASM This Viking Orbiter image is of Ophir Chasm in the Valles Marineris region of interlocking canyons near Mars' equator. The walls of the 1.6-mile (2.6-km) deep chasm show the effects of landslides. Fine soil has been blown by Martian winds toward the bottom left of the image.

ARGYRE PLANITIA Probably Mars' youngest large basin, the huge impact that created Argyre Planitia (left) caused debris to fall in concentric rings around this crater. The rugged, innermost ring, called Nereidum Montes in the north and Charitum Montes (curling from bottom left to top of image) in the south, has been effaced by more recent craters, such as Galle crater (top left).

South Pole · Hellas Basin

GEOLOGY OF MARS

Mars is one of the smaller planets in the solar system, but its geology is on a grand scale. With a surface sculpted by almost every major geological process known, it has vast chasms, broad lava plains, ancient impact basins, and the largest volcano in the entire solar system—Olympus Mons. But all this has evolved over distinct epochs, with different geological processes dominating at different times. To date these geological periods precisely will probably require a series of crewed trips to the Red Planet.

WHAT IF...

...YOU WERE THE FIRST GEOLOGIST ON MARS?

The year is 2030 and your spaceship is about to touch down on the planet Mars. Just a couple of feet in front of you sits the Commander, who is destined to be the first human being to set foot on the Red Planet. But you have your own moment of glory to consider—you are about to become the first geologist on Mars. Luckily for you, NASA has put a high priority on the science objectives of the mission. The situation was very different with the earlier Apollo Moon program. Then, scientists had to wait until the final mission before they were allowed to set foot on the lunar surface.

As the Mars Module sets down, you see the red, dusty soil billowing around the spacecraft. You expect that the soil will be a mixture of iron-rich clays and *palagonite*, a partially crystallized volcanic rock created by chemical weathering processes. Martian soil is much more complex than the moondust collected by the Apollo astronauts, which was no more than pulverized rock.

As the Martian dust settles, you peer through the window. A remarkable landscape appears—a red, flat flood plain littered with rocks. If the rocks are like those studied by the Mars Pathfinder rover in 1997, they are likely to be a mixture of basalts and other volcanic blocks. Many of them could be rich in silicon—like andesite, a rock found on Earth. But then Mars could be full of surprises.

As the crewmember in charge of science, it will be your job to oversee the experiments and gather as much new data as possible. You will test for seismological activity, analyze soil and rock composition, measure water content, and keep a constant lookout for promising samples. You have brought a fair amount of scientific equipment with you to carry out all your tasks, but the most valuable resource of all will be your expert eyes. No robot can trundle across the Martian landscape with the amount of awareness and scientific vision of a human geologist like you. Bouncing over the surface in the Mars Car, you will be able to visit different geological regions and collect a range of rocks. By studying different rock types, you should be able to pinpoint the geological epochs of Mars. But the time for speculation is over. You are suited up and ready to step out onto a new land.

A geologist on Mars uses a drill while another astronaut looks on. The geological equipment used for the Moon landings will have been adapted to meet the special requirements of Mars' soil.

MARS' GEOLOGICAL TERMS

CATENA	CHAIN OF CRATERS		PLANITIA	PLAIN
CHASMA	CANYON		PATERA	SHALLOW CRATER WITH SCALLOPED EDGES
DORSUM	RIDGE			
FOSSA (PL. FOSSAE)	LONG, NARROW VALLEY		THOLUS	SMALL, DOME-LIKE MOUNTAIN OR HILL
LABYRINTHUS	INTERSECTING VALLEY COMPLEX			
MENSA (PL. MENSAE)	FLAT-TOPPED ELEVATION		VALLIS (PL. VALLES)	VALLEY
MONS (PL. MONTES)	MOUNTAIN		VASTITAS	WIDESPREAD LOWLANDS

ROCKS TO RICHES

About 4.5 billion years ago, at the same time as the Earth was taking shape, another rocky planet began to form slightly farther out from the Sun. It steadily coalesced from the primordial solar nebula—a place brimming with small bodies hurtling around and often crashing together. Mars eventually emerged from this molten turmoil of objects after about 10 million years.

The new planet gradually cooled from its traumatic, fiery birth, separating into a core, a molten mantle and a solid crust. Over the next billion years, new material was added regularly: Asteroids were still numerous in the solar system and continued to strike the young planet. During this period, the massive impact craters scarred Mars for life, covering the planet with deep pits and ejecta—rocky debris created by the explosive impacts. These craters are now best preserved in the ancient southern highlands of Mars.

By the time the solar system had settled down, impacts by asteroids had become rarer—and the impact rate has continued to decline. But then its cooling crust was leading Mars to an entirely new geological era. Volcanoes burst through fractures in the surface, and molten lava poured over the planet. The northern hemisphere, where the crust was thinner, took the brunt of the upheaval. The mountainous Tharsis ridge—the site of Mars' largest volcanoes—was formed.

GEOLOGICAL SCULPTURE ON MARS

These geological maps of the eastern hemisphere (far left) and western hemisphere (below) of Mars show the types of materials in the Martian landscape. The materials that make up the surface features of each area are color-coded. Reddish areas indicate features formed by volcanic materials, greenish and bluish areas indicate features believed to have been formed by water, and brownish areas indicate features formed by impacts.

VALLES MARINERIS CANYON
Formed primarily by faulting in the crust, the Valles Marineris (above) runs for 2,800 miles (4,500 km), from the Noctis Labyrinthus fracture zone (far left of image) to where three parallel canyons merge into a chasm 370 miles (600 km) wide and 5 miles (8 km) deep (center of image), to end in the dark-colored region of Margaritifer Sinus (far right of image).

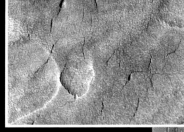

UTOPIA PLANITIA
A close-up of the inside of one of the many craters in the Utopia Planitia (above) shows the cracked surface of the crater's floor, which is filled with a material believed to be sediment.

OLYMPUS MONS
Taken on an angle, this image of Olympus Mons (above) begins to give a sense of the volcano's enormous height. At 15 miles (24 km) high, it is the largest volcano in the solar system. The dark area seen on the top of the volcano and running down its side is the remains of lava flows.

HELLAS BASIN
With a diameter of 1,120 miles (1,800 km), Hellas (right) is the largest basin on Mars. Formed by the ancient impact of a huge asteroid, comet or meteor, the plain inside the basin is the site of many

BELCHING GAS AND VAPOR

At this volcanic time, Mars began to form a denser atmosphere than it has now. This was because volcanoes eject more than just lava—they belch forth gases and water vapor into the air. The thicker atmosphere allowed liquid water to remain on Mars' surface in the form of rivers, lakes and even oceans. But the planet grew colder. When temperatures fell below the freezing point of water, ground ice and possibly glaciers began to form, cutting swaths through the terrain. Landslides tumbled down mountainsides. Eventually the cold may have caused the water to freeze into the soil. Some of the carbon dioxide may have dissolved in the water, or become trapped in the surface rocks as carbon compounds. Much of Mars' atmosphere was also lost, blown away by massive impacts and stripped by the solar wind.

Next came another era of volcanism, covering much of the northern hemisphere with vast lava plains. The Tharsis ridge rose even higher. The Valles Marineris canyon yawned apart, dwarfing the Earth's Grand Canyon. New faults released torrents of water, carving channels and other features. Finally, the Tharsis region gave birth to the enormous volcanoes of Ascraeus Mons, Arsia Mons, Pavonis Mons and Olympus Mons.

Whether Mars still has active volcanoes is open to debate. Geologists have recently discovered volcanic cones at the north pole with no marks from cratering, suggesting that they formed in the very recent past. However, wind is the main force that sculpts Mars' features. Giant sand storms regularly scour the surface, often growing so large that they engulf the entire planet. Mars may not have the tectonic plates and abundant life that constantly reshapes the Earth's surface, but it has enjoyed no shortage of geological activity—and the planet is still changing.

VALLES MARINERIS

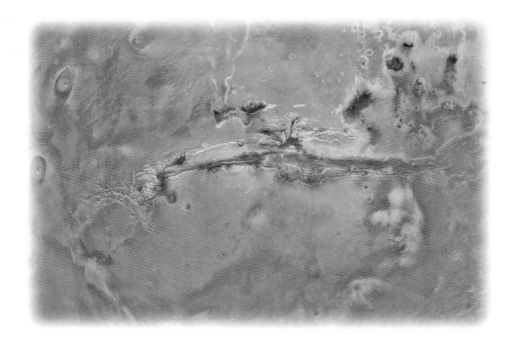

Valles Marineris formed millions of years ago, when immense geological forces split the crust of Mars with a huge fissure at the planet's equator. Four times as deep as the Grand Canyon and roughly as long as the U.S. is wide, Valles Marineris is one of Mars' most impressive surface features. The initial scar was broadened by colossal landslides and erosion from the water that flowed over the surface of ancient Mars. Some channels in the fissure even contain dark sediments that could be the remains of an ancient sea bed.

WHAT IF...

...YOU TOOK A FLIGHT THROUGH VALLES MARINERIS?

It may seem far-fetched, but aircraft are a logical next step in the exploration of Mars. From an altitude of several miles up, cameras mounted on a robot plane could bridge the crucial gap between the highest-resolution satellite images, and the detailed views of rovers and other probes that actually land. It is a sobering thought that even the best orbital photographs of Mars still miss any detail smaller than around 30 feet (9 m) across.

Although Mars's atmosphere is far thinner than Earth's, it is still substantial enough to support an aircraft, provided one can be made that is light enough and has a large enough wingspan to support the weight. NASA did some serious work on the concept in the 1990s, intending to launch an aerial explorer toward Mars around 2006. However, budget cuts and changed priorities meant that the project was shelved, and is now seen as a longer-term option for the next decade.

The first Mars aircraft proposal was actually put forward in the late 1970s—the plane would have been a modified version of the "Mini-Sniffer," a drone aircraft already in use by the U.S. military. When it resurfaced in the 1990s, a variety of models were suggested, including aircraft powered by an onboard fuel supply of hydrazine,

gliders, and solar-powered aircraft that would settle on the ground or glide at night, and charge up during the day. Although none of these concepts was carried forward, for Mars exploration, NASA has pursued the idea of solar-powered aircraft circling the Earth as a cheap alternative to satellites.

The aircraft that came closest to flying on Mars, however, was that selected for the Mars "micro-mission" proposal in 1999. This would have been a small, rocket-powered plane and thus its flying time would have been very limited. The plan was for the plane to deploy from its descent capsule in mid-air, unfold its wings, and then launch itself on a flight over the Martian landscape at speeds of around 250 mph (400 km/h).

Since the cancellation of this micro-mission, NASA has gone back to the drawing board. We have learned so much about the Red Planet in the past decade that they can now plan a more ambitious mission, with a better knowledge of the conditions a Mars aircraft will have to cope with. Prototype "flying wing" designs have been tested on Earth, and there has even been development work on entomopters—small vehicles with flapping wings—which could prove a far more viable option on Mars than on Earth. It is still quite likely that, by the time humans reach Mars, our aircraft will have been circling the planet for some time.

VALLES MARINERIS FACTS

LONGITUDE (OF CENTRAL POINT)	70°
LATITUDE (OF CENTRAL POINT)	11.6°
LENGTH	AROUND 2,500 MILES (4,020 KM)
GREATEST DEPTH	MORE THAN 4 MILES (6 KM)
MAXIMUM WIDTH	370 MILES (600 KM)
AGE	SEVERAL BILLION YEARS
DISCOVERER	NASA SCIENTISTS, USING DATA FROM THE MARINER 9 SPACECRAFT
YEAR DISCOVERED	1971

RED VALLEYS

The Grand Canyon is the most spectacular chasm on Earth, but the sheer scale of Valles Marineris on Mars makes Arizona's greatest landmark look like a scratch. Valles Marineris stretches 2,500 miles (4,020 km) in length, just south of the Martian equator. In places it dips more than four miles beneath the planet's surface and gapes to 370 miles (600 km) wide. By comparison, the Grand Canyon is a paltry 220 miles (350 km) long and less than a mile deep—perhaps equivalent to one of Valles Marineris' smaller tributaries.

Valles Marineris' complex network of canyons and valleys is the result of several different forces. Unlike the Grand Canyon, which was eroded by the Colorado River, Valles Marineris was initially opened up as the crust faulted under tremendous pressures. The canyon walls are long and straight, often with linear ridges along their base—all the defining features of fault scarps. To find the strain that buckled Mars' crust, just look to the western end of Valles Marineris—a maze of fractures that start near the summit of a huge uplift called the Tharsis bulge. This region probably domed upward as lava welled up from below, and Tharsis' immense mass was the pressure that cracked open the surrounding crust.

The new canyons were widened by dust storms and enormous landslides. As parts of Valles Marineris were eroded, thick layers of sedimentary rock built up elsewhere. This was an especially exciting find. These sediments most likely settled out under water, so lakes must have filled some of the canyons in the past.

DAM BURSTERS

The Candor and Ophir Chasmas are two of the canyons that may have been water-filled. Some geologists think that these lakes were once dammed by a ridge that divided Candor from the main canyon. When the rock barrier failed, the waters flooded the canyons.

This would explain the number of teardrop-shape islands in the northern and eastern channels of Valles Marineris, where rapidly

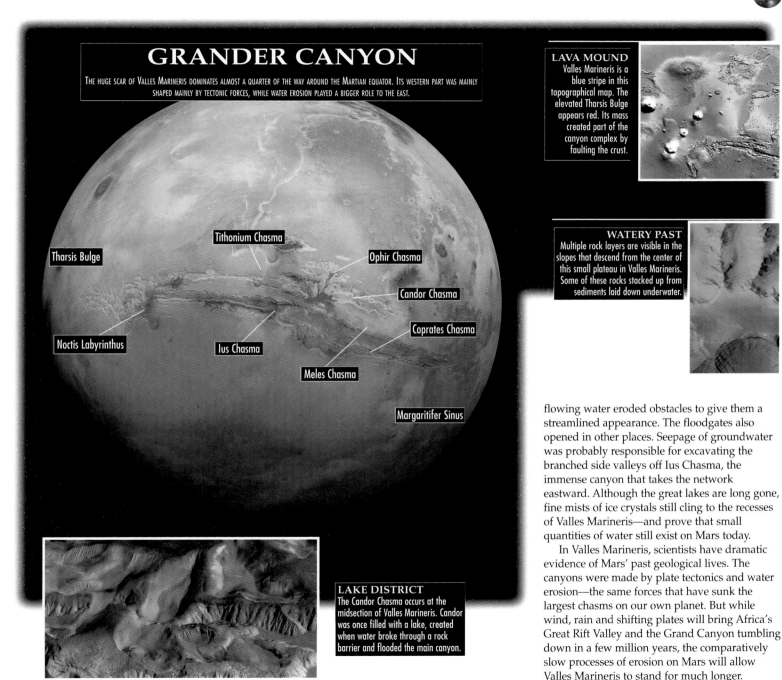

GRANDER CANYON

THE HUGE SCAR OF VALLES MARINERIS DOMINATES ALMOST A QUARTER OF THE WAY AROUND THE MARTIAN EQUATOR. ITS WESTERN PART WAS MAINLY SHAPED MAINLY BY TECTONIC FORCES, WHILE WATER EROSION PLAYED A BIGGER ROLE TO THE EAST.

Tharsis Bulge
Tithonium Chasma
Ophir Chasma
Candor Chasma
Noctis Labyrinthus
Ius Chasma
Coprates Chasma
Meles Chasma
Margaritifer Sinus

LAVA MOUND
Valles Marineris is a blue stripe in this topographical map. The elevated Tharsis Bulge appears red. Its mass created part of the canyon complex by faulting the crust.

WATERY PAST
Multiple rock layers are visible in the slopes that descend from the center of this small plateau in Valles Marineris. Some of these rocks stacked up from sediments laid down underwater.

LAKE DISTRICT
The Candor Chasma occurs at the midsection of Valles Marineris. Candor was once filled with a lake, created when water broke through a rock barrier and flooded the main canyon.

flowing water eroded obstacles to give them a streamlined appearance. The floodgates also opened in other places. Seepage of groundwater was probably responsible for excavating the branched side valleys off Ius Chasma, the immense canyon that takes the network eastward. Although the great lakes are long gone, fine mists of ice crystals still cling to the recesses of Valles Marineris—and prove that small quantities of water still exist on Mars today.

In Valles Marineris, scientists have dramatic evidence of Mars' past geological lives. The canyons were made by plate tectonics and water erosion—the same forces that have sunk the largest chasms on our own planet. But while wind, rain and shifting plates will bring Africa's Great Rift Valley and the Grand Canyon tumbling down in a few million years, the comparatively slow processes of erosion on Mars will allow Valles Marineris to stand for much longer.

VOLCANOES OF MARS

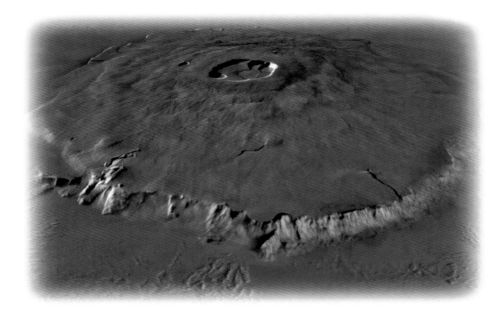

The largest volcanoes in the solar system tower above the surface of Mars. The Earth and Venus both have many volcanoes—as does Jupiter's moon Io—but these are puny compared to those of Mars, which are giants in both height and spread. Even though the largest, Olympus Mons, stands 17 miles (27 km) high—three times the height of the Earth's tallest volcano, Mauna Loa, in Hawaii—it was known only as a dot on the surface until the Mariner 9 space probe visited in 1971. And Olympus Mons is just one of 20 huge volcanoes that dominate the Martian landscape.

WHAT IF...

...THE VOLCANOES WOKE UP?

Although there may still be some occasional activity, astronomers believe the last major volcanic activity on Mars was about 30 million years ago. Venus and the Earth, the larger rocky planets, are still hot enough inside to feed active volcanoes with lava. But Mars is a fairly small planet and its interior cooled more rapidly than those of its more massive neighbors. However, recent discoveries of apparently fresh volcanic activity around the north pole mean it is no longer unthinkable that the Martian volcanoes could erupt once more. If they did become active again, though, they might permit a still-theoretical process called terraforming that would incubate life on the surface of Mars.

As well as lava and ash, erupting volcanoes release gases into the atmosphere. Scientists believe that this is how Venus, the Earth and Mars got their atmospheres in the first place. Billions of years ago, when the solar system was very young, early volcanoes threw out gases such as carbon dioxide, nitrogen and water vapor that collected in a thin layer around these rocky planets, making their first atmospheres. Venus and the Earth were large enough for their gravity

to hold on to these gases. However, Mars is only about half their size, so today it has only a thin atmosphere of mainly carbon dioxide gas, insufficient for either plant or animal life. Farther from the Sun than any other rocky planet, Mars also has a surface that is too cold for the liquid water that permitted life to evolve on Earth. And even if its surface were warmer, liquid water would evaporate immediately under the low atmospheric pressure.

Mars' surface environment has not always been so unfriendly to life. Dramatic photographs of dried-up channels show that at one time water flowed on the surface, proving that Mars was warmer in the past, with a denser atmosphere. If the Martian volcanoes became active again, they would release gases that could thicken up the atmosphere once more.

The atmosphere would still be mainly carbon dioxide, though. To turn it into something suitable for animal life, we would need to introduce primitive plants that would gradually convert the carbon dioxide to oxygen through the process of photosynthesis, just as they do on Earth. It could never be a quick process, but eventually we might create a second planet where humans could live.

MARS VOLCANO RECORDS

NUMBER OF LARGE VOLCANOES ON MARS	20
OLDEST	TEMPE PATERA: 3.4 BILLION YEARS OLD
YOUNGEST	OLYMPUS MONS: 30 MILLION YEARS OLD
TALLEST	OLYMPUS MONS: 16.7 MILES (26.9 KM) ABOVE AVERAGE SURFACE HEIGHT; 14.3 MILES (23 KM) ABOVE SURROUNDING SURFACE
WIDEST	ALBA PATERA: 1,674 MILES (2,694 KM) DIAMETER
LARGEST "CALDERA" OR SUMMIT CRATER	ARSIA MONS: 68.2 MILES (110 KM) DIAMETER

MONSTERS OF FIRE

The surface of Mars bears witness to an active geological history. Wind and water have scoured and sculpted the Martian crust, leaving meandering river beds, broad deltas and smoothed-out rock formations like those familiar to us on Earth. But the most spectacular relics of Mars' past are its huge volcanoes—the biggest in the solar system. Even though Mars is only a little more than half the size of the Earth, these extinct giants dwarf any on our planet. So why did Mars breed such monsters?

Despite the dramatic evidence of eruptions on its surface, Mars' geology has been much more stable than the Earth's. The thick Martian crust stayed immobile above the geological "hot spots" that fueled its volcanoes, and so while the planet's interior was hot enough to drive lava up to the surface, the mountains of lava and ash that piled up in eruption after eruption kept growing.

The same forces are building volcanoes on Earth today. But our planet's crust is fractured into continent-sized plates that are constantly on the move. Because of this, volcano-firing hot spots must find new outlets for their lava every few million years. The Hawaiian Island chain shows this phenomenon in action. The islands are a line of mostly extinct volcanoes that map the progress of the Pacific plate over a hot spot. The Earth's biggest volcano, Mauna Loa, is simply the youngest of the chain to appear above sea level—which is why it is at the end of the line, dominating the southern part of the island of Hawaii.

EVIDENCE FROM ORBIT

All that we know of Mars' volcanoes has come from photographs taken by orbiting spacecraft, starting with Mariner 9 in 1971. Since then, the two Viking probes (1976) and the Mars Global Surveyor (1997–9) have returned more detailed images that tell the life histories of Mars' volcanoes. From studies of impact craters on volcano slopes, astronomers

have calculated that the oldest volcano still visible on the surface rose up over 3 billion years ago, while the youngest may still be active today. The oldest type, called pateras, are shaped like upside-down saucers, rising only a few miles above the surrounding terrain, and were probably formed from a mixture of lava and ash. The youngest and tallest are shield volcanoes—the same type as the Earth's Mauna Loa—with gentle slopes spreading out over huge areas. The lava that made these was relatively thin, and flowed huge distances before cooling and solidifying. A third type of Martian volcano is the dome, with steeper sides made from stickier lava or fewer eruptions.

But not all volcanoes resemble the popular image of a cone-shaped mountain topped by a smoking crater. Lava can also spread out over a wide area, filling up valleys and submerging impact craters to leave smooth plains. When Mars was young, eruptions covered most of the planet's northern hemisphere in just this way—the same process that gave our Moon its so-called "seas."

The great days of Martian volcano-building are long gone. It must have been a very different planet then, with massive outgassings into the atmosphere along with the ejections of lava. But though we may never see Olympus Mons blow its top again, there is mounting evidence for small-scale volcanic activity still disturbing Mars' crust. Only further exploration of the surface will tell.

MARINER 9'S DISCOVERY

WHEN THE MARINER 9 SPACECRAFT ARRIVED AT MARS IN NOVEMBER 1971, IT FOUND THE PLANET COMPLETELY HIDDEN BY A GLOBAL DUST STORM. WHEN THE DUST SETTLED, THE FIRST THINGS MARINER SAW WERE FOUR HIGH MOUNTAINS TOPPED BY HOLLOW CRATERS. IT HAD DISCOVERED THE GIANT MARTIAN VOLCANOES.

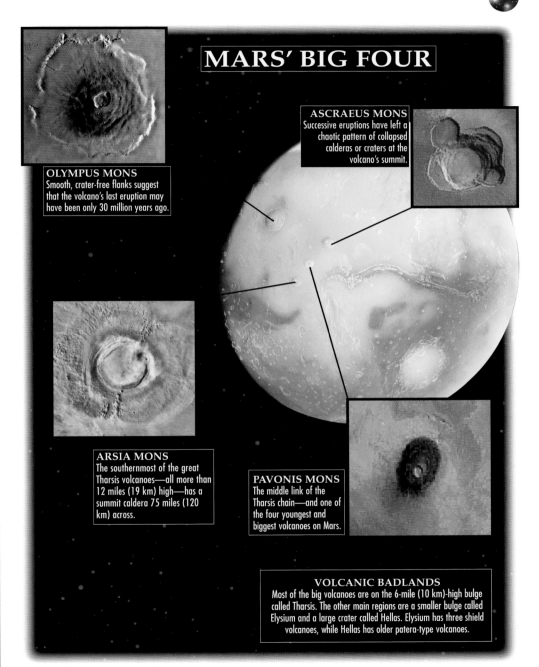

MARS' BIG FOUR

OLYMPUS MONS
Smooth, crater-free flanks suggest that the volcano's last eruption may have been only 30 million years ago.

ASCRAEUS MONS
Successive eruptions have left a chaotic pattern of collapsed calderas or craters at the volcano's summit.

ARSIA MONS
The southernmost of the great Tharsis volcanoes—all more than 12 miles (19 km) high—has a summit caldera 75 miles (120 km) across.

PAVONIS MONS
The middle link of the Tharsis chain—and one of the four youngest and biggest volcanoes on Mars.

VOLCANIC BADLANDS
Most of the big volcanoes are on the 6-mile (10 km)-high bulge called Tharsis. The other main regions are a smaller bulge called Elysium and a large crater called Hellas. Elysium has three shield volcanoes, while Hellas has older patera-type volcanoes.

POLAR CAPS OF MARS

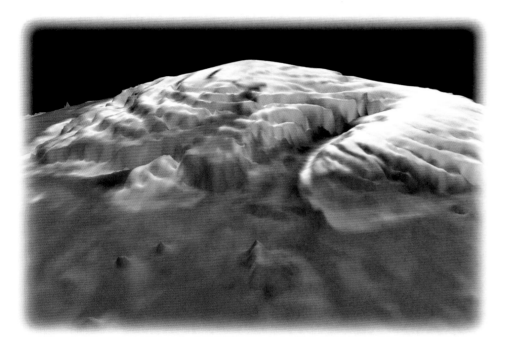

C old, dark and only briefly tipped toward the Sun, the poles of Mars are the planet's deepfreezes. These areas are covered in giant white caps that advance and retreat with the passing seasons. Much of the planet's water is locked in the polar ice, along with vast deposits of solid carbon dioxide—dry ice. Summers on Mars are warm enough to free some of the poles' frozen carbon dioxide. But the water ice is a permanent fixture— and hostile as the poles are, they may be the best place for a human base on Mars.

WHAT IF...

...WE BUILT A POLAR BASE ON MARS?

Mars makes a strong candidate for the location of a space colony—if only because it looks a little like home. Mars is more like the Earth than any other planet around the Sun. Its day is almost the same length as ours, and its axial tilt is nearly identical. This means that the two planets have similar seasonal variations in climate. The first step would be to establish a test base there—but how do we pick the best site?

A source of water will be the most important consideration wherever the space pioneers decide to settle. We cannot transport water to Mars— shipping just one gallon of water to the Red Planet could easily cost a million dollars. So the best place to build a Mars station would be near the most obvious water supply, at one of the polar caps.

Some scientists think that there is enough water ice on Mars to cover the whole planet in an ocean—if we could thaw it out. Not all of that water is at the poles. Much of it lies under the surface in a layer of permafrost. But the polar caps remain the most likely source of running water. Recent results from the Mars Global Surveyor spacecraft have shown that the north cap harbors around 300,000 cubic miles (1.25 million cubic km) of water ice.

An adequate supply of water is not the only reason for establishing a polar base on Mars. Humans need oxygen with at least one-sixth of the Earth's atmospheric pressure to breath comfortably. But the air on Mars is 95% carbon dioxide (CO_2) at a pressure of much less than one-hundredth of an atmosphere. One solution to the oxygen shortage could come from the CO_2 molecule itself. It is made up of two atoms of oxygen and a single atom of carbon, so in theory, it should be possible to extract the necessary oxygen from CO_2. This would take enormous amounts of energy whatever the CO_2 source, but it would be easier to work with the CO_2 ice that freezes over the poles than the very thin atmosphere.

There is a downside to settling at the poles, though. As on Earth, these regions will be in total darkness for half the planet's year. On Mars, that translates to 11 Earth months. Temperatures at the poles are also bitterly cold, at –150°F (–65°C) or less. But the lure of water and air still makes life at the ends of Mars an attractive option—and if we ever possess the technology to make a Mars base reality, we should be able to bring our own light and heat.

POLAR CAP STATISTICS

Northern Cap	
Winter diameter	2,500 miles (4.020 km)
Summer diameter	600 miles (965 km)
Peak height compared with equatorial ground level	–3 miles (–5 km)
Maximum thickness of polar deposit	1.8 miles (2.9 km)
Southern Cap	
Winter diameter	1,000 miles (1,600 km)
Summer diameter	200 miles (320 km)
Peak height compared with equatorial ground level	+1.9 miles (+3 km)
Maximum thickness of polar deposit	1.3 miles (2 km)

SUBLIMATING POLES

Astronomers have been looking at Mars' polar caps for more than three centuries. By the early 1700s, one observer had resolved the Red Planet's lowest latitudes clearly enough to describe them as "white stains." But it was only in 1781 that the British observer Sir William Herschel (1738–1822) suggested that Mars' polar caps, like Earth's, might be made of frozen water.

Herschel was partly correct. Polar temperatures on Mars are far too cold to melt water ice even in summer—yet the ice caps shrink dramatically in this season. The caps do contain water ice, but also large quantities of more volatile frozen carbon dioxide (CO_2), which sublimates—changes from solid to gas—as temperatures rise.

Summer in Mars' southern hemisphere comes when the planet is closest to the Sun. This proximity means that the southern cap shrinks far more than its northern counterpart. But Mars moves fastest at this time of year, so the south pole speeds through its warmest weather. The short-lived thaw, and the fact that the polar water ice does not melt at any time of the year, means that the southern cap never vanishes completely.

Winter is the longest season on southern Mars. As temperatures fall, CO_2 freezes out of the atmosphere and falls as snow that enlarges the polar cap. Similar effects also change the size of the northern polar cap. Each cap is much larger in winter than in summer.

Other changes take place over much longer time scales. Around and under the polar ice, Mars' surface is built up in smooth layers that are each some 100 feet thick (30 m). They are crater-free, so they must have been laid down recently—in astronomical terms, at least.

Mars is a very dusty planet. Over the course of about 50,000 years, global winds blow huge clouds of volcanic ash and dust from the equatorial regions to the poles. As these deposits build up, they are mixed with frozen water and CO_2. If there is then a relatively long warm spell on the planet—caused by a change in axis tilt or orbital eccentricity—the CO_2 sublimates, leaving the dust behind. This causes the new stratum to slump into a terrace with a sheer drop. Over a million years, the cycle of deposit and collapse has built a series of 100-foot (30-m) high steps on the edges of the polar caps.

For all of their instabilities, the polar caps are still water reservoirs, and if we ever establish a base on Mars, it might have to be in these regions.

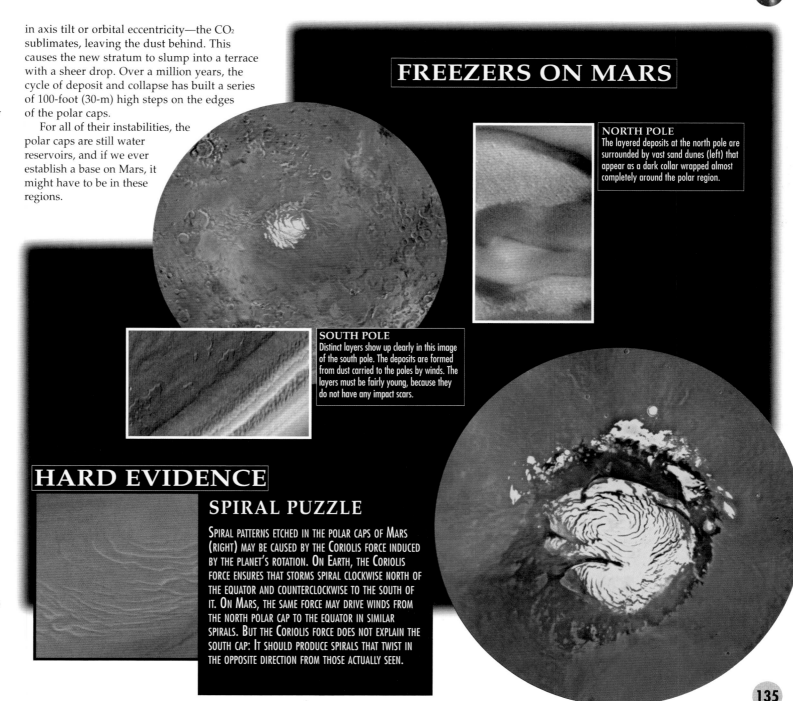

FREEZERS ON MARS

NORTH POLE
The layered deposits at the north pole are surrounded by vast sand dunes (left) that appear as a dark collar wrapped almost completely around the polar region.

SOUTH POLE
Distinct layers show up clearly in this image of the south pole. The deposits are formed from dust carried to the poles by winds. The layers must be fairly young, because they do not have any impact scars.

HARD EVIDENCE

SPIRAL PUZZLE

SPIRAL PATTERNS ETCHED IN THE POLAR CAPS OF MARS (RIGHT) MAY BE CAUSED BY THE CORIOLIS FORCE INDUCED BY THE PLANET'S ROTATION. ON EARTH, THE CORIOLIS FORCE ENSURES THAT STORMS SPIRAL CLOCKWISE NORTH OF THE EQUATOR AND COUNTERCLOCKWISE TO THE SOUTH OF IT. ON MARS, THE SAME FORCE MAY DRIVE WINDS FROM THE NORTH POLAR CAP TO THE EQUATOR IN SIMILAR SPIRALS. BUT THE CORIOLIS FORCE DOES NOT EXPLAIN THE SOUTH CAP: IT SHOULD PRODUCE SPIRALS THAT TWIST IN THE OPPOSITE DIRECTION FROM THOSE ACTUALLY SEEN.

MARTIAN ATMOSPHERE

Before the era of space probes, many astronomers believed that the Martian atmosphere was a relatively dense blanket that might even be able to support life. But from the 1960s onward, orbiting spacecraft and landers beamed back the disappointing news: The air on Mars was thin, desperately cold and composed mainly of life-choking carbon dioxide. It may not always have been so. The same spacecraft have also found evidence that Martian air might once have been as thick as the Earth's.

WHAT IF...

...YOU VISITED MARS WITHOUT A SPACE SUIT?

Margaritifer Sinus, Mars, August 24, 2158: A geologist had a lucky escape today following a freak accident, reports the Marswide Press Agency. Dr. Annabel Carpenter, 27, was today recovering in the Marineris Hospital after surviving an unprecedented five minutes unprotected in the Martian atmosphere.

Carpenter, who arrived on Mars from Earth only two months ago, was working alone in Coprates Chasma. She was driving her All Mars Vehicle, which allows a near shirtsleeve environment. As Carpenter drove through a shadowed area, her vehicle plunged into a deep dust pit. The impact damaged her pressure suit, and Carpenter could not call for help because the radio antenna was submerged by the dust.

In order to clip on the antenna extender—a length of wire that is carried to extend the range of a radio unit in an emergency—it was necessary to work outside the vehicle. Carpenter's oxygen mask was still in order, but the low pressure of the Martian atmosphere would cause her blood to boil within two minutes.

Carpenter knew that it is normal, although discouraged, for personnel to remove gloves for short periods of time in order to manipulate small items, with no harmful effects. She reasoned that if her skin temperature could be kept low enough while working in room clothing, she should survive for some time. Since she would be submerged in the dust pit, at a temperature well below freezing, it should be possible to carry out the emergency antenna repair.

"It felt kind of tingly," she said of her sensations as she fumbled for the stub antenna beneath the dust. "It was really cold, but I knew that I had to call for help. Overall, I was outside the vehicle for just under five minutes."

An ambulance raced to the scene. But the only ill effect that Carpenter suffered was mild frostbite as she touched the outer hull of her vehicle.

Previous cases of exposure to the Martian elements have involved personnel whose pressure suits have become ruptured. In most cases, lack of oxygen causes the victim to black out within half a minute. The only protection afforded by the atmosphere is in reducing the ultraviolet radiation from the Sun compared with deep space, though sunburn is likely after no more than 10 minutes' exposure of the skin to full sunlight.

AIRS ON EARTH AND MARS

MARS' ATMOSPHERE

	ABUNDANCE (IN BARS)	PERCENTAGE
CARBON DIOXIDE	0.0062	95
NITROGEN	0.00018	2.7
ARGON	0.00010	1.6
OXYGEN	0.000002	0.13
WATER	0.00000039	0.03

PRESSURE: LESS THAN 10 MILLIBARS AT THE SURFACE

EARTH'S ATMOSPHERE

	ABUNDANCE (IN BARS)	PERCENTAGE
NITROGEN	0.78	77
OXYGEN	0.21	21
HYDROGEN	0.01	1
ARGON	0.94	0.93
CARBON DIOXIDE	0.000355	0.035

PRESSURE: AROUND 1,000 MILLIBARS AT SEA LEVEL

ABSENT AIR

Mars' atmosphere is laden with red dust, and its pink skies are some of the most scenic in the solar system. But any space tourist would certainly need a pressure suit to survive. At the surface, Mars has an atmospheric pressure of no more than 10 millibars. On Earth, you would have to travel to 120,000 feet (36,600 m), four times the height of Mount Everest, to reach such thin air.

What gas there is on Mars is mainly carbon dioxide, a greenhouse gas. But there is not nearly enough of it to warm up the chilly planet. Mars' atmosphere contributes only about 12°F (6.5°C) to the average temperature of –65°F (–18°C). On Earth, carbon dioxide makes up less than one percent of the air, but raises temperatures by around 63°F (35°C).

Scientists once thought that Mars' atmosphere was mainly made up of nitrogen, like the air on Earth. But the Viking landers uncovered the more inhospitable truth in 1976, when they found that only three out of every hundred atmosphere molecules were nitrogen. Mars was not so short of nitrogen in the past. Atoms come in various isotopes—each with a different weight. Martian air contains a higher proportion of heavy nitrogen atoms than Earth's air: Many of the light isotopes have escaped.

If the young Mars did have a thicker atmosphere, the puzzle of its water-based features would be solved. The very thin, very cold Martian air means that water cannot exist as a liquid now—it appears as ice or vapor. Only flowing water in a thicker, warmer atmosphere could have shaped the channels recorded by the Mars orbiters.

GAS ESCAPE

So where is all the missing air? Some of it may have simply drifted away. The Red Planet has low gravity—only one-third of that on Earth—that is too weak to hang on to many air particles. Other parts of the atmosphere were removed more violently. Mars was bombarded by meteors in its early history, and much of the atmosphere was literally blasted into space. But other planets also came into the line of fire—and still managed to cultivate healthy second atmospheres.

One theory suggests that some of the Martian air was carried away by the solar wind. On Earth, the air is shielded by our planet's magnetic field. On Mars, the magnetic field is too weak to offer much protection. Atoms in the upper atmosphere are *ionized*—given an electrical charge—by sunlight. The solar wind sweeps these charged particles out into space. Lighter atoms are the easiest to pluck away, and the heavyweight molecules that make up the majority of the air now are the only ones left.

The liquid water that excavated rivers and channels on the young planet may also have destroyed much of the atmosphere. Carbon dioxide probably dissolved in the water, to be deposited later as carbonate minerals. The discovery of carbon in Martian meteorites supports this idea. The same process takes place on the Earth, with one essential difference—Earth has volcanoes. Carbon dioxide is absorbed into the Earth's surface in just the same way as it is on Mars, and is eventually blasted back into the atmosphere by lava flows.

But the lock-up of carbon dioxide in minerals is permanent on Mars: There are no plate tectonics to free the trapped carbon. There is evidence that Mars has had violent volcanic epochs in the past, but the planet's surface is still now. Mars' missing atmosphere is irretrievable—and the Red Planet will remain a cold, dry and inhospitable desert.

MARTIAN SKIES

CLOUDY SKY
These stratus clouds are about 10 miles (16 km) above Mars' surface. They consist of water ice condensed on dust particles suspended in the atmosphere. Clouds on Mars have been observed to cover vast regions.

MORNING MISTS
Mist forms over early-morning Mars. Martian water is constantly changing between gas and solid. Particles of ice condense on the surface at night, only to turn into vapor under the first rays of the Sun. The vapor recondenses in the cold atmosphere to form a haze of ice particles.

The daily movements of the Sun create atmospheric havoc on Mars. Solar heat causes early morning mists, and global dust storms are at their most violent when Mars is closest to the Sun.

DUST RAGE
Planetwide dust storms are stirred by winds of more than 60 mph (96 km/h). The sand that covers Mars is very fine, and is easily whipped up to heights of 30 miles (48 km). The dust may shroud the whole planet for months before it settles to the surface.

SOIL SECRETS

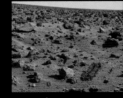

THE ATMOSPHERE ABOVE THE YOUNG PLANET MARS MAY HAVE BEEN AS THICK AS THE EARTH'S AIR IS NOW. THE PROOF IS LOCKED AWAY IN THE RED SOIL THAT BLANKETS THE MARTIAN SURFACE (SHOWN HERE SAMPLED BY PATHFINDER). LIQUID WATER RAN OVER MARS MILLIONS OF YEARS AGO. IN THE PROCESS, IT DISSOLVED MUCH OF THE CARBON DIOXIDE ATMOSPHERE AND TRAPPED IT WITHIN THE ROCKS AND SOIL.

WATER ON MARS

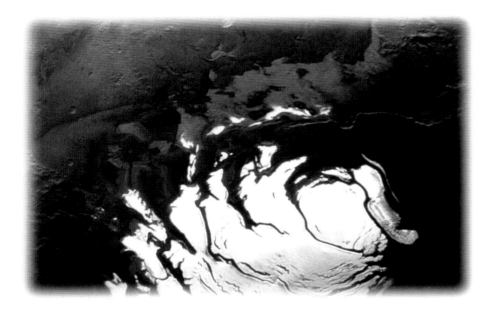

The Italian astronomer Giovanni Schiaparelli (1835–1910) started a myth when he said he'd seen *canali* (channels) on Mars. These were later found to be an optical illusion and a scientific impossibility—spectroscopic analysis of light from Mars showed a dry place with so little atmosphere that water would have boiled away instantly. Yet Mars wasn't always a desert. Space probes have discovered the remains of dried-up riverbeds, flood plains and shallow seas, so what happened in the distant past to freeze-dry the entire planet?

WHAT IF...

...MARS BECAME A WATERY PLANET AGAIN?

The collapsed landscape of broken boulders, known as *chaotic terrain*, suggests that Mars was the victim of a series of floods long after its atmosphere was thinned and its remaining water frozen beneath the surface. A popular theory is that hot plumes of lava periodically well up beneath the layers of ice until the ice melts and bursts through the surface.

Another possibility is that asteroid impacts heat up the water: The more recent impact craters have what look like mudflows around them, suggesting that the asteroids melted ice in the frozen ground as they hit the surface. An alternative hypothesis is that Mars' orbit may vary somewhat over millions of years, producing vast climatic changes. If these orbital fluctuations brought Mars closer to the Sun, the warmer temperatures could result in the melting of ice below the planet's surface.

If we choose to colonize Mars in the future, we may decide to artificially alter planetary conditions so that water can once again flow across its surface. Scientists have worked out a variety of methods to *terraform* Mars (make the Martian environment more like that of the Earth). Most of them begin with a means of thickening the planet's atmosphere. If we

are unable to do this, then recreating rivers and seas on Mars will be an impossible task. Terraforming Mars may involve melting the planet's poles, which are composed of carbon dioxide and some frozen water. Simply scattering a few million tons of black graphite across the dry ice will cause it to absorb more heat. Orbital space mirrors could shine light down on the poles, or solar power stations could be constructed to beam down microwaves. Meanwhile, a network of factories could be built to manufacture extremely powerful greenhouse gases.

A more far-fetched scheme could involve finding a way to alter the course of comets so that they crashed into the Martian surface, bringing both heat and water to the Red Planet.

As air pressure increased and the temperature rose, liquid water would flow across Mars once again. The first sign would probably be a river running through the bottom of the Valles Marineris canyon, which is located in the warm region of the equator and has the lowest elevation on the planet. Water would collect in the planet's vast basins, recreating the Oceanus Borealis in the northern hemisphere, and an "Oceanus Australis" would take the place of the southern icecap. Perhaps 600 years after our first landing on Mars, our descendants will enjoy their first Martian rain shower.

EVIDENCE FOR WATER

1. **THE POLES:** ....THE SOUTH POLE IS ALMOST ENTIRELY FROZEN CO_2 BUT THE NORTH POLE IS MAINLY WATER ICE.
2. **CIRRUS CLOUDS:** ....THESE CLOUDS ARE FORMED BY WATER ICE CRYSTALS AROUND 10 MILES (16 KM) UP.
3. **FROST:** ....FROST APPEARS ON CRATER FLOORS IN THE MORNING ON MARS.
4. **ERODED CRATERS:** ..OLDER CRATERS HAVE BEEN SMOOTHED OUT BY WATER FLOW.
5. **CHAOTIC TERRAIN:** ..SUBSIDENCE AND SCATTERED BOULDERS INDICATE MASSIVE FLOOD CHANNELS.
6. **DRY RIVERS:** ......FORMED BY TRIBUTARIES, THESE DRY RIVER CHANNELS END IN DELTAS.
7. **SALTS:** ....SOILS EXAMINED BY THE VIKING PROBE WERE AS MUCH AS 20% WATER-DEPOSITED SALTS.
8. **SHALLOW SEA:** .....THE NORTHERN HEMISPHERE IS MARKED BY THE POSSIBLE REMAINS OF AN ANCIENT OCEAN.
9. **ROUNDED STONES:** ..THESE WERE SEEN BY THE PATHFINDER PROBE IN THE ARES VALLIS FLOOD PLAIN.
10. **MARS METEORITES:** ..APART FROM SUSPECTED "MICROFOSSILS," MARTIAN METEORITES ALSO SHOW SIGNS OF WATER.

FROZEN IN TIME

Planetary scientists were slow to discover Mars' secret history. The first probes to fly past the Red Planet returned only a handful of images, which showed Moon-like cratered plains. And when Mariner 9 orbited Mars in 1971, it took many months for the spacecraft to begin mapping the planet—almost all of Mars' features were obscured by a global dust storm. But when the storm eventually cleared, Mariner 9's pictures proved to be worth the wait. They showed enormous volcanoes, far bigger than any found on Earth. They also showed a multitude of features that suggested liquid water had once scoured the planet's surface. These included vast canyons, eroded craters, *chaotic terrain* of broken rock caused by sudden flooding, and long, riverlike channels fed by tributaries that run downhill.

Although some researchers tried to dismiss this evidence, suggesting other processes such as lava flows that might have caused erosion, successive spacecraft have only strengthened the evidence for a once-watery Mars. Orbiter spacecraft such as the Vikings, Mars Global Surveyor, and Mars Express have provided ever-clearer images of water-formed features, while Mars rovers—in particular Spirit, which landed in 2003—have discovered minerals in the Martian soil that could only have formed if the surface was submerged for sustained periods of time. Astronomers still argue over the extent of the water, though—some imagine short-lived temporary lakes on the surface, but others suggest Mars was once a blue planet, with a great ocean, the Oceanus Borealis.

Another question is when and how the water disappeared. One suggestion is that radiation from the Sun was able to break up water molecules in the atmosphere, and because of the weak gravity, light hydrogen atoms were then carried away by the solar wind. Another idea is that the water remains in underground reservoirs. In 2002, the Mars Odyssey probe detected the signature from massive amounts of water ice just below the surface around both the north and south poles.

On Earth, the geological process of plate tectonics recycles carbonates from rocks into the air, as continental plate movements redistribute the molten mantle. Mars lacked the energy for this process. If carbon and oxygen from the air got chemically locked into the Martian rocks, they stayed there—shrinking the atmosphere further. In a reverse greenhouse effect, the thinner the atmosphere got, the colder it became. Perhaps 2 billion years ago, much of the remaining atmosphere became frozen carbon dioxide—or dry ice—and the last of the water retreated below the surface, finding refuge at the planet's poles.

DELTA GROOVES

The Mariner 9 spacecraft brought back the first pictures of Mangala Vallis—a 370-mile (595-km), water-carved outflow channel running across Mars' southern hemisphere. The channel is probably the product of massive flooding by water that broke through from beneath the planet's surface crust. Just like a river on earth, it begins with a network of tributaries, which then meet in a narrow main channel. This runs downhill, eventually thickening out at the mouth (see above) like a terrestrial river delta. Struck by its similarity in size and shape to the greatest river on Earth, scientists decided to name its downstream plain Amazonis Planitia.

LAYERED ROCK
This image from the Mars Global Surveyor spacecraft shows layered rock in the Coprates Catena area, which lies at the center of the massive Valles Marineris canyon. Layered rock on Earth, such as that found in Arizona's Grand Canyon, is often the result of sediment deposited by ancient lakes.

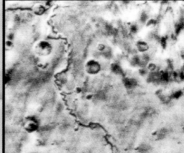

TRIBUTARIES
This Viking image of the Nirgil Vallis canyon shows tributaries off the main channel. They were probably formed by springs located on cliffs overlooking the canyon. As the water weakened the cliffs, they eventually collapsed. Each collapse forged a new tributary, which grew longer as the spring maintained it.

ISLANDS
The water that carved channels to the north and east of the vast Valles Marineris canyon had huge erosive power. One consequence was the formation of streamlined islands where the water encountered obstacles. This image shows islands formed as the water was diverted by large craters in its path.

LIFE ON MARS

A succession of visiting spacecraft have left little doubt: Mars is a frozen ball of arid desert, its thin atmosphere made up almost entirely from unbreathable carbon dioxide. It seems to be no place for living things. But the Red Planet used to be different. Billions of years ago, water flowed on a much warmer surface. Life may have come to Mars, possibly even before it arose on the Earth. And as we know from Earth, life is tenacious. So despite the evidence so far, this apparently sterile planet might still be alive.

WHAT IF...

...WE COULD CLONE ANCIENT MARTIAN LIFE?

So far, the only traces of life on Mars that have been discovered are some possible fossil remnants of ancient microorganisms. These were discovered in Antarctica, inside a meteorite that was blasted from Mars to Earth by a cosmic impact billions of years ago. But many scientists are unconvinced by the fossils and suspect that they may have a non-living origin. In any case, the remains are billions of years old, and there is absolutely no genetic material in them.

Early in the 21st century, though, at least one Mars landing probe should be returning samples from the planet's surface to Earth. And there is also the possibility of a crewed mission or even a Mars base. Scientists will soon have far more to analyze than battered meteorites.

There is a chance that fresh specimens from Mars will contain actual living things. If that proves to be the case, biologists and microbiologists will be ecstatic. For the first time, we would know for certain that life exists beyond the Earth. And if life exists on two planets in our own little solar system, then the chances must be excellent that it thrives all around the universe.

But suppose the best that future Mars expeditions can find are Martian fossils—less ambiguous than the meteorite traces we already have, but just as dead. Would any genetic material remain, and if it did, could we somehow clone it and bring an ancient Martian organism back to life?

The odds are against it. There is no DNA—the molecule that carries the genetic code—left in comparably ancient bacterial fossils on Earth. Besides, we don't know that Martian life even used DNA. It may have evolved an entirely different means by which to pass on its genetic heritage. And even if we did manage to extract DNA from the fossils, it would be difficult to clone.

If we do find definite Martian fossils, though, it would be good evidence that life was once widespread on the planet. Rather than wasting our effort in futile attempts to revive the dead, it might be better to spend more time and money on exploring Mars. Somewhere—deep beneath the surface, perhaps—the descendants of those ancient fossils might still be thriving. And if we look hard enough, we might find them.

LIFE ON MARS

1698	DUTCH ASTRONOMER CHRISTIAAN HUYGENS' FIRST SPECULATION THAT LIFE MIGHT EXIST ON MARS IS PUBLISHED UNDER THE TITLE KOSMOTHEOROS.
1719	MARS COMES SO CLOSE TO EARTH THAT ITS BRIGHTNESS IN THE SKY CAUSES PANIC.
1854	ENGLISH OBSERVER WILLIAM WHEWELL REPORTS GREEN SEAS AND RED LAND ON MARS AND DISCUSSES POSSIBLE LIFE FORMS.
1877	THE NEW YORK TIMES RUNS THE EDITORIAL "IS MARS INHABITED?"
1907	THE WALL STREET JOURNAL CLAIMS "PROOF BY ASTRONOMICAL OBSERVATIONS...THAT CONSCIOUS, INTELLIGENT HUMAN LIFE EXISTS UPON MARS."
1908	AMERICAN ASTRONOMER PERCIVAL LOWELL PUBLISHES "MARS AS THE ABODE OF LIFE" IN MAGAZINE FORM. HE DESCRIBES INTELLIGENT MARTIAN LIFE.
1911	THE NEW YORK TIMES RUNS A STORY HEADED "MARTIANS BUILD TWO IMMENSE CANALS IN TWO YEARS".
1947	WERNHER VON BRAUN'S PROPOSED MARS EXPEDITION—A FLEET OF 10 SHIPS AND 70 CREW—IS PUBLISHED AS THE MARS PROJECT.
1965	MARINER 4 MAKES THE FIRST SUCCESSFUL FLY-BYOF MARS ON JULY 15. EXPLORATION OF THE PLANET FROM SPACE BEGINS.

RED NOT DEAD

Until the first planetary probes of the space age relayed their observations, many astronomers were sure that life thrived on Mars. The planet's seasonal color changes were put down to the summer growth of vegetation, and there were more arguments about the nature of Martian plants than about their existence.

But the Mariner fly-bys in 1971 and the Viking landers in 1976 changed all that. Mariner 9 revealed Mars to be cold and arid, scorched by ultraviolet radiation—and sterile. Experiments carried out by the Vikings appeared to confirm the lifelessness of the Martian soil: The Red Planet was almost as desolate as the Moon.

But as the data was assessed, scientists came to believe that Mars was once a very different place. All over the planet were gullies and eroded channels—signs that water had once flowed. Strewn boulders on the Martian plains resembled flood debris on Earth. And the shape and structure of some of the Martian craters suggested that incoming meteorites may have struck wet ground—a Martian swamp, or perhaps even a shallow ocean.

For liquid water to have existed on Mars' surface, the planet must have been much warmer than it is today, and its atmosphere much thicker. Four billion years ago, just 500 million years after the formation of the solar system, the Martian environment could have been just right for the evolution of life. Conditions on Mars would certainly have been better than they were on Earth at the same time: Mars is much smaller than the Earth, and almost 50 million miles (80 million km) farther from the Sun, so it would have cooled more quickly from its original molten state.

IF LIFE EVOLVED

Early Martian life, if it existed, may have resembled the living things that emerged on Earth a few hundred million years later. In warm, shallow water, organisms akin to blue-green algae on Earth would have appeared.

DID LIFE FIND A NICHE?

Once, water surged across the surface of Mars, and life may have flourished. Then the water dried up, the atmosphere thinned and the environment became inhospitable. But could life have adapted to these harsh conditions—and still be lurking somewhere on the planet?

DIGGING UP PROOF
Viking 1's sampling arm dug out soil from the Martian surface for analysis (right). No signs of life were found. But scientists then were looking for chemical reactions with CO_2, such as those that appear in life forms on Earth. We now know that life can be sustained by other chemicals, such as hydrogen and iron.

HOT HOUSE
Life may have thrived under the surface of Mars in the boiling water of hot springs. These are found near volcanoes on Earth, driven by the heat from the underground magma. A volcanic cleft in La Fournaise volcano (right) on the island of Réunion has hot water seeping through a crack in the rock.

CAVE DWELLERS
Shalbatana Vallis channel (running top to bottom in the image at right) is thought to be the result of an underground river that carved a long cave before breaking through the surface to form the channel. Some scientists believe that such caves were the last havens for Martian life.

IRRIGATION SCHEME

U.S. ASTRONOMER PERCIVAL LOWELL CAUSED A SENSATION IN 1908 WHEN HE PUBLISHED HIS BOOK *MARS AS THE ABODE OF LIFE*. AFTER 15 YEARS OF CAREFUL OBSERVATION, HE WAS CONVINCED THE PLANET WAS COVERED BY A NETWORK OF CANALS (RIGHT) DUG BY AN INTELLIGENT MARTIAN RACE TO BRING WATER FROM THE POLES TO THE EQUATOR. AT THE INTERSECTIONS OF THE CANALS, LOWELL EVEN NOTED "OASES." SADLY, BOTH OASES AND CANALS TURNED OUT TO BE OPTICAL ILLUSIONS.

Gradually they would have spread across Mars' surface, finding a foothold wherever it was moist and sunlit.

Again, as on Earth, these organisms would have begun the slow process of altering the atmosphere from a blanket of carbon dioxide to the oxygen-rich mixture that would fuel more advanced life forms. But there the two planets' biological histories diverged. On Earth, the algae dominated for eons before more advanced types of life evolved: It took more than 3 billion years from the dawn of life to the arrival of the first complex animals.

On Mars, though, complexity never had the time to evolve. Faster than life could compensate, the planet began to chill. Much of its atmosphere was lost to space, and as the core cooled, geological activity came to an end. One by one, the great volcanoes whose outgassing might have replenished the thinning air sputtered into extinction. The age of liquid water came to an end.

Probably, any Martian life ended with it. But it could not all have died at once. Forever seeking warmth and wetness, Martian organisms would have retreated into the deep valleys where a little water still remained, or into hot springs around

the dying volcanoes. The battle for survival would have been fought on a microscopic scale, but it was still an epic—and life may not have lost.

Against the relentless transformation of the planet's environment, life had only one weapon. But it was a powerful one: the ability to adapt. The volcanoes died slowly—some were still active less than a billion years ago—and Martian organisms may have had a chance to carve themselves a new niche. Certainly, their descendants would have been bred for toughness. Deep inside the planet, in a few favored locations, they might still be hanging on.

PHOBOS AND DEIMOS

Dark and dusty, Phobos and Deimos orbit close to their parent planet. The surfaces of these tiny Martian moons are pitted with large craters and covered by a loose layer of broken rock, absorbing almost all the sunlight that reaches them. The moons are probably small asteroids—cosmic debris left over from the formation of the solar system—that were later snared by Mars' gravitational pull. The trap is a lethal one for Phobos: Its orbit is decaying and it is on a collision course with the Red Planet.

WHAT IF...

...PHOBOS CRASHES INTO MARS?

While Deimos orbits at a safe distance from Mars, Phobos is spiraling slowly toward eventual destruction. The planet's gravitational pull is reeling in the moon at a rate of 60 feet (18 m)per century. But a collision between Phobos and Mars may never occur—the moon may first suffer the fate of being broken up by the planet's tidal forces.

As Phobos orbits around Mars, the side of the moon that faces the planet is subject to a greater gravitational force than the opposite side. The difference between the forces puts the little moon under a constant strain that could be enough to tear it apart.

Phobos orbits close to the *Roche limit*—the point where a planet's tidal force is powerful enough to destroy a moon. Whether Phobos will be demolished in this way depends upon its cohesive strength. The moon's density suggests that, like many asteroids, it probably consists of carbonaceous chondrite and pea-size nodules of various minerals. This composition would make the moon brittle and likely to crack and fracture.

The largest crater on Phobos, 6-mile (10-km) wide Stickney Crater, may offer clues to the moon's cohesive strength. If the moon can withstand such a large impact, it may be able to resist Mars' tidal forces—at least for a long time to come. But Phobos' orbit will continue to decay under the influence of Mars' gravitational field. If Phobos remains in one piece, it may be pulled through the thin Martian atmosphere before finally crashing onto the planet's surface.

Alternately, the enormous force of the impact that produced Stickney may also have weakened Phobos, and the large grooves that run outward from the crater may be fractures. With so many splits, the moon may fall victim to Mars' tidal forces, ending up as a rocky ring of debris around the planet. Either way, astronomers have plenty of time to prepare for the dramatic event: Phobos is expected to stay in one piece for at least another 40 million years.

Phobos is so close to Mars that it will eventually plummet to a violent death on the planet's surface—unless tidal forces first tear the moon apart and reduce it to a ring of orbiting debris.

PHOBOS AND DEIMOS

PHOBOS		DEIMOS
16 MILES BY 12 MILES (26 KM BY 19 KM)	DIMENSIONS	MILES BY 6 MILES (16 KM BY 10 KM)
5,830 MILES (9,380 KM)	DISTANCE FROM MARS	. . . .14,580 MILES (23,460 KM)
12 TRILLION TONS (11 TRILLION TONNES)	MASS	TRILLION TONS (1.8 TRILLION TONNES)
7 HOURS 39 MINUTES	LENGTH OF DAY	. . . . .30 HOURS 18 MINUTES
0.32 DAYS	ORBITAL PERIOD	. . . . .1.26 DAYS
1.08°	ORBITAL INCLINATION	. . . . .1.79°
0.0151°	ORBITAL ECCENTRICITY	. . . . .0.00033°
11.3	VISUAL MAGNITUDE	. . . . .12.40

TINIEST MOONS

Small, dark and fast-moving, the two Martian moons went unnoticed in the blackness of space until 1877, when U.S. astronomer Asaph Hall picked them out as Mars made a particularly close approach to the Earth. For the time, it was quite an achievement. Phobos and Deimos are among the smallest known satellites in the solar system. Too small to have formed in a regular, spherical shape, they measure only about 16 by 12 miles (25 by 19 km) and 10 by 6 miles (16 by 10 km) respectively.

From Earth, even the most powerful telescopes can make out no more of the moons than dim spots. For the first detailed views of Phobos and Deimos, astronomers had to wait for the U.S. Mariner 9 probe in 1971. Mariner's pictures were followed by more images taken by the Viking 1 Orbiter as it neared Mars in 1976—the centennial of the moons' discovery.

Unlike Earth's moon—a chunk of metallic rock that was probably ripped from its parent planet early in its creation—the Martian moons seem to be adopted companions, stray asteroids that passed so close to Mars that they have been trapped by its gravity.

Although their matching dark exteriors might suggest they should possess similar surface features, there is a sharp contrast between the two moons' appearance. Images from the Viking missions showed Deimos to have a comparatively smooth exterior. Few craters are present, and many have been partially filled up by the bits of broken rock and dust that cover the moon.

A TEXTURED HISTORY

The surface of Phobos is more dramatic, testimony to a turbulent impact history. It is covered in deep ridges, steep hills and numerous craters, one of which—the 6-mile-wide (10-km-wide) Stickney—stretches across almost half the moon's diameter. Stickney would be visible from the surface of Mars.

DARK AND DUSTY WORLDS

DEIMOS

BATTLE SCARS
This Viking 1 image (left) shows crater chains and deep chiseled grooves known as striations—probably the result of the impact that created Stickney, the largest crater on Phobos.

HALL'S FAME
Hall Crater (left) is named after Asaph Hall, the astronomer who discovered Phobos in 1877. Almost 4 miles (6.5 km) across, the crater was photographed by Viking at the moon's south pole.

TINY AND DIM
Viking 2 took these closeups of Deimos. The image above shows the moon's surface from 19 miles (30 km) away. We can see craters, covered by rocks and dust about 55 yards (50 m) thick, and large blocks of debris (above center) 10–35 yards (10–30 m) across, probably ejected by the impacts. The image at right shows the moon's elongated shape at its northern tip.

Stickney Crater

PHOBOS

Phobos orbits at a mean distance of only 5,830 miles (9,380 km) from the planet's center, and appears in the sky as almost half the size of a full Moon seen from Earth—although it is much darker. The smaller Deimos, orbiting at 14,580 miles (23,460 km), shows up in the Martian sky as a tiny disk, only just discernible to the naked eye.

Phobos and Deimos both orbit east to west, but as seen from the surface of Mars, they seem to travel in opposite directions. This apparent paradox is caused by the orbital periods of the two moons. Phobos whizzes around Mars in just 7 hours 39 minutes, so it circles the planet more than four times in the course of the Martian day of 24 hours 37 minutes. But Deimos travels slower than Mars rotates. Left behind in the sky, it appears to move in an easterly direction.

JUPITER

The king of the planets, giant Jupiter orbits the Sun far beyond Mars and the asteroid belt. Yet despite its great distance, it shines brilliantly in Earth's skies, and is frequently the brightest planet apart from Venus. Jupiter's brightness is largely due to its size—its diameter is nearly twelve times that of Earth. Unlike the inner planets, however, this giant world is composed almost entirely of gas—largely the same light hydrogen gas that makes up the Sun itself, but with an upper layer of more complex chemical compounds that create chaotic and colorful weather systems such as the famous Great Red Spot, a storm large enough to swallow Earth whole. Deeper inside the planet, the hydrogen gas is compressed to liquid form, while at the centre there may be an Earth-sized planet, crushed under the weight of the huge atmospheric envelope. Like all the outer planets, Jupiter has a large family of satellites, dominated by Io, Europa, Ganymede, and Callisto, each of which is about the size of our own Moon or slightly larger. These four giant satellites, discovered by Italian astronomer Galileo in the 1600s and often called the Galilean moons, are complex worlds in their own right.

The first true-color photograph of Jupiter taken with the Wide Field Planetary Camera on the Hubble Space Telescope. Cloud formations in the atmosphere of Jupiter, containing small crystals of frozen ammonia and compounds of carbon, sulfur, and phosphorus, create the colorful belts and whorls. The temperatures of Jupiter's clouds are extremely cold, about −280°F (−173°C).

JUPITER

T he fifth planet from the Sun, Jupiter is by far the largest in the solar system. Over 1,300 Earths could fit inside it, and it is more than twice as massive as all the other planets put together. Jupiter has a complex weather system, which generates the bands of clouds that swirl across its surface and also includes the planet's best-known feature, the Great Red Spot. Like its neighbors in the outer solar system, Jupiter has rings. It also has at least 63 moons, which have been likened to a miniature solar system.

WHAT IF...

...JUPITER BECAME A STAR?

J upiter radiates more heat into space than it receives from the Sun. Up until the 1940s, scientists believed that nuclear reactions like those that power stars were taking place within the planet. But now we know that Jupiter gives out heat because it is slowly shrinking under the pressure of its own gravity. This causes the gas giant to compress and heat up, much like air does in a bicycle tire when the tire is pumped up.

Scientists believe that to generate the kind of pressure needed to trigger the nuclear reactions that make stars shine, Jupiter would have to be 80 to 100 times its present size. If the planet had been big enough to become a star, the solar system would be very different today.

Assuming that the planets had been able to form at all, they would by now be in complex and possibly unstable orbits around two suns. And on Earth, assuming that we, too, had evolved, life would be interesting, to say the least. Each day would bring two dawns and two sunsets, with two periods of daylight and two of darkness, all varying in length according to the season. There would also be two sets of seasons. Sometimes these would cancel each other out and sometimes they would reinforce each other, resulting in scorching summers and ferocious winters.

In the less dramatic scenario envisaged by science-fiction writer Arthur C. Clarke, Jupiter might one day begin to "shine" like a very small, weak star. This would not have such a great effect on the solar system at large, but it could warm Jupiter's larger moons to the point where we might be able to colonize them.

Life on Earth would be affected, too. For much of the time there would be two suns visible, our regular Sun and the Jupiter star, or Sol 2 as Clarke calls it. Far fainter than the Sun, Sol 2 would still be 50 times as bright as the full Moon—enough to light up the night sky for much of the year. But the Earth would also receive extra heat as well as light, resulting in dramatic, and probably unpleasant, changes to our planet's climate.

Arthur C. Clarke's novel 2061: Odyssey Three *visualizes a future in which Jupiter has been turned into a star and its moon Ganymede has become warm enough to be colonized. Ganymede's atmosphere is seeded with bacteria that generate oxygen, and within about 100 years, the air on the satellite is breathable and plants brought from Earth are growing there.*

JUPITER PROFILE

JUPITER		EARTH
88,846 MILES (142,983 KM)	DIAMETER	7,973 MILES (12,831 KM)
3° 10'	AXIS TILT	23° 27'
4,329 DAYS (11.86 EARTH YEARS)	LENGTH OF YEAR	365 DAYS
9 HOURS 55 MINUTES 29 SECONDS	LENGTH OF DAY	24 HOURS
483.7 MILLION MILES (778.4 MILLION KM)	DISTANCE FROM SUN	93.5 MILLION MILES (150 MILLION KM)
−186°F (−85°C)	SURFACE TEMPERATURE	59°F (15°C)
2.53 G	SURFACE GRAVITY	1 G
HYDROGEN (90%), HELIUM (10%), METHANE (0.07%)	ATMOSPHERE	NITROGEN (79%), OXYGEN (21%)
700 MILLIBARS	ATMOSPHERIC PRESSURE	1,000 MILLIBARS
HYDROGEN (90%), HELIUM (10%), METHANE (0.07%)	COMPOSITION	SILICON (60%), ALUMINUM (15%)

GIANT PLANET

Like the other gas giants of the solar system—Saturn, Uranus and Neptune—Jupiter has no solid surface. The planet has a rocky core, but most of it consists of gases that become more and more dense toward the center until they eventually turn to liquid. The striking patterns observed by space probes and telescopes are not surface features but clouds. Their bands, swirls and eddies are the outward signs of the immensely powerful weather engine that drives Jupiter's atmosphere.

The clouds have arranged themselves into 19 clearly defined bands in shades of red, amber and brown. The winds in adjacent bands blow in opposite directions at speeds of 250 mph (400 km/h) or more. The cloud bands are probably the outer surfaces of thick layers of atmospheric material that rotate around the planet and extend deep into its interior. The bands themselves are remarkably stable. Although the cloud patterns within them are constantly changing, there are features in Jupiter's cloudscapes that have been there for many years, or even centuries. The best-known of them is the aptly named Great Red

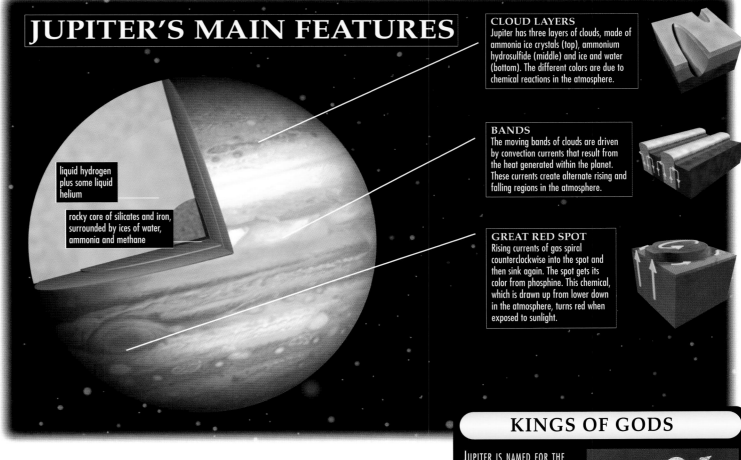

JUPITER'S MAIN FEATURES

liquid hydrogen plus some liquid helium

rocky core of silicates and iron, surrounded by ices of water, ammonia and methane

CLOUD LAYERS
Jupiter has three layers of clouds, made of ammonia ice crystals (top), ammonium hydrosulfide (middle) and ice and water (bottom). The different colors are due to chemical reactions in the atmosphere.

BANDS
The moving bands of clouds are driven by convection currents that result from the heat generated within the planet. These currents create alternate rising and falling regions in the atmosphere.

GREAT RED SPOT
Rising currents of gas spiral counterclockwise into the spot and then sink again. The spot gets its color from phosphine. This chemical, which is drawn up from lower down in the atmosphere, turns red when exposed to sunlight.

BETHLEHEM'S STAR?

THE STAR OF BETHLEHEM MAY HAVE BEEN THE CONJUNCTION (COMING TOGETHER) IN THE SKY OF JUPITER AND ONE OR MORE OF THE OTHER BRIGHT PLANETS. SEVERAL CONJUNCTIONS OCCURRED AROUND THE TIME THE "STAR" IS SAID TO HAVE APPEARED.

Spot, a vast anticyclonic storm that is three times the size of the Earth and has existed for at least 300 years.

Jupiter's atmosphere is about 5,000 miles (8,050 km) thick and consists mostly of hydrogen. There is also some helium and small quantities of methane and ammonia, plus traces of other compounds. Beneath these gases is an ocean of hot liquid hydrogen. Even at more than 3,150°F (1,732°C), the hydrogen does not boil away: It is kept under a pressure some 90,000 times that of the atmosphere on Earth.

Jupiter's liquid hydrogen layer is over 30,000 miles (48,280 km) thick. Far beneath it, under what scientists believe may be a layer of water and ammonia, is a rocky core measuring around 4,200 miles (6,760 km) across.

Deep within the planet, the pressure reaches 45 million Earth atmospheres and temperatures rise to more than 20,000°F (11,000°C). Under these extreme conditions, the liquid hydrogen takes on some of the characteristics of a metal: Electric currents flow through it and generate Jupiter's magnetic

KINGS OF GODS

JUPITER IS NAMED FOR THE KING OF THE ROMAN GODS, WHO IS USUALLY DEPICTED HURLING A THUNDERBOLT. HE WAS ALSO KNOWN AS JOVE AND, TO THE ANCIENT GREEKS, AS ZEUS. THE PLANET'S MOONS ARE NAMED FOR OTHER CHARACTERS IN THE GREEK MYTHS OF ZEUS, MOST OF THEM HIS LOVERS.

field, which, after the Sun's, is the strongest in the solar system.

CHANGING VIEWS OF JUPITER

From king of the gods to king of the planets, Jupiter has maintained a regal pedigree throughout history. As scientists resolved the planet through more powerful telescopes, they discovered the most intriguing family of moons in the solar system—and the longest-lived storm. The wildest view of Jupiter formed in the 19th century, when it was seen as a star that was slowly burning away to become a planet. And although we now know that it had no such stellar past, our picture of Jupiter is by no means complete.

...WE LEARNED MORE ABOUT JUPITER?

Centuries of observation and theoretical studies have given astronomers a good picture of Jupiter's present and previous lives. For instance, we know how Jupiter formed. It took shape as small clumps of rock and ice, called planetesimals, rammed together to build an ever-bigger body. As it grew, the young Jupiter swept up hydrogen and helium gas left over from the birth of the Sun, and swelled to the giant proportions we see today.

But not every episode of the Jovian saga is so well understood. Scientists still do not know precisely where in the solar system Jupiter formed. Several factors suggest that it may have started out farther from the Sun and migrated inward to its present orbit.

One of the strongest pieces of evidence for Jupiter's relocation is its composition. When the Galileo probe plunged into the planet's atmosphere in December 1995, it found two to three times the levels of nitrogen, argon, krypton and xenon gases that scientists had expected. At Jupiter's current distance from the Sun—about five times the Earth's distance (five astronomical units, or AU)—conditions in the early solar system should have been much too warm for these gases to exist.

Jupiter may have inherited the elements from planetesimals that came together more than 40 AU from the Sun, where conditions were much colder. But many astronomers say that is unlikely, because the gases should have boiled away as the newborn planet moved into warmer territory. Alternatively, perhaps it is our models of planetary formation that are wrong: If these gases did indeed survive Jupiter's long journey, then conditions in the early solar system may have been much colder than astronomers imagine.

Jupiter may also have paid the Sun a closer visit in the distant past. A thick disk of gas and dust around the Sun may have acted as a brake that slowed Jupiter down and let it spiral closer to the star. Jupiter's own gravity could have helped to drive the planet closer still, as it batted billions of planetesimals out of the solar system.

This theory is supported by the behavior of Jupiter-size planets in other solar systems. Many of these bodies are only a few million miles from their parent stars—a tiny fraction of Jupiter's distance from the Sun. Astronomers say that there is no way for gas giant planets to form so close to a star. They must have formed farther out, and somehow spiraled inward.

Clues about Jupiter's birth are hidden in the atmospheres of the giant planets. But we need many more probes to uncover the whole truth, and our views of Jupiter will undoubtedly change again.

JUPITER'S BIGGEST CHANGES

1610	GALILEO GALILEI DISCOVERS THE MOONS OF JUPITER
1665	JEAN DOMINIQUE CASSINI DISCOVERS AND STUDIES THE GREAT RED SPOT
1690	CASSINI DISCOVERS THAT JUPITER'S POLES AND EQUATOR ROTATE AT DIFFERENT RATES
1932	RUPERT WILDT DETECTS METHANE AND AMMONIA, WHICH IMPLIES THAT JUPITER IS MAINLY HYDROGEN
1973	PIONEER 10 IS THE FIRST SPACECRAFT TO FLY PAST JUPITER
1979	VOYAGER 1 AND 2 FLY PAST JUPITER AND DISCOVER VOLCANOES ON IO
1995	GALILEO ENTERS ORBIT AND DROPS PROBE INTO JUPITER'S ATMOSPHERE

MIGHTY SIGHTS

When Galileo Galilei (1564–1642) turned his small telescope toward Jupiter in 1610, he changed our concept of the cosmos. Under Galileo's gaze, Jupiter grew into a tiny disk flanked by four tiny "stars" that moved back and forth. He realized that they were moons, orbiting Jupiter as our Moon orbits the Earth.

The discovery shattered the notion of crystal spheres supporting the stars and planets, and demonstrated that the heavens are not fixed and unchanging, as religious dogma of the day maintained. It also supported the then-controversial idea that the Earth is not the center of the universe, but is simply one small body orbiting the Sun.

The importance of Galileo's discovery seems fitting for Jupiter. Our concepts of the planet have always been big, and Jupiter has never disappointed. The planet's brightness led Babylonian, Greek and Roman stargazers to associate it with their supreme deities. It owes its modern name to Jupiter, king of the Roman gods. Only with the invention of the telescope did anyone realize that Jupiter was not a star, but a planet like our own.

But even as they compared Jupiter to the Earth, early 19th-century astronomers invariably found that their observations led them back to the Sun. They claimed that the vast, swirling storms that dominated Jupiter's surface were the result of "vast, unexpended stores of internal heat." Through the new telescopes, Jupiter's visible disk appears to blur toward the edges—evidence for light bent through a deep atmosphere like that on the Sun.

THE SUN THAT NEVER SHONE

From these apparent similarities, scientists concluded that Jupiter was a sun in decay, set on a course that would eventually turn it into a terrestrial planet like the Earth. They believed that Jupiter's intense heat would ultimately subside, that its turbulent clouds would condense into oceans and dry land appear.

We now know that Jupiter has never burned

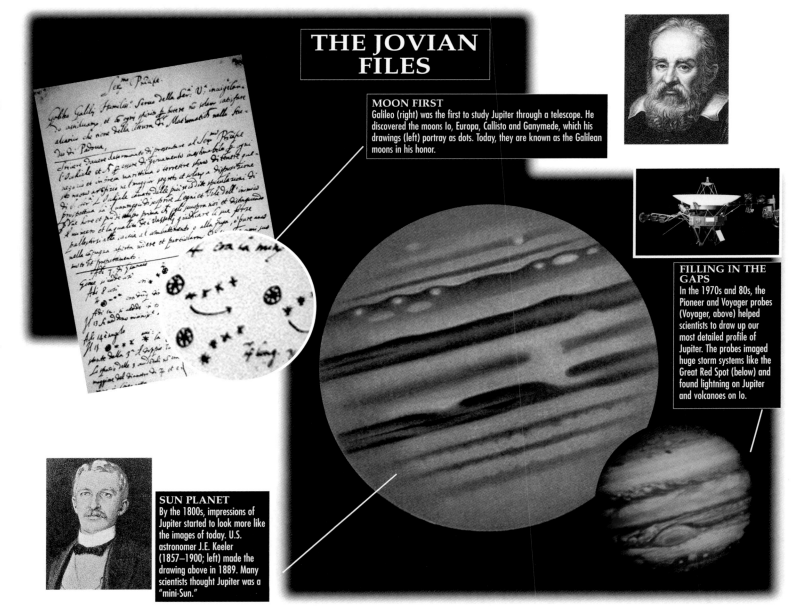

THE JOVIAN FILES

MOON FIRST
Galileo (right) was the first to study Jupiter through a telescope. He discovered the moons Io, Europa, Callisto and Ganymede, which his drawings (left) portray as dots. Today, they are known as the Galilean moons in his honor.

FILLING IN THE GAPS
In the 1970s and 80s, the Pioneer and Voyager probes (Voyager, above) helped scientists to draw up our most detailed profile of Jupiter. The probes imaged huge storm systems like the Great Red Spot (below) and found lightning on Jupiter and volcanoes on Io.

SUN PLANET
By the 1800s, impressions of Jupiter started to look more like the images of today. U.S. astronomer J.E. Keeler (1857–1900; left) made the drawing above in 1889. Many scientists thought Jupiter was a "mini-Sun."

nuclear fuel, and today, the planet has been downgraded from a star in decline to a sun that never shone at all. Because Jupiter is similar to the Sun in terms of hydrogen content, some astronomers describe it as a "failed star"—Jupiter

would have to be about 80 times more massive to trigger fusion reactions. The only energy that comes from within is the result of gravitational contraction and the slow release of heat left over from its formation.

Modern technology has sharpened our view of Jupiter, but the planet has many more secrets to reveal—some of which may make our current ideas sound as far-fetched as the notion of a Jovian sun.

JUPITER'S ATMOSPHERE

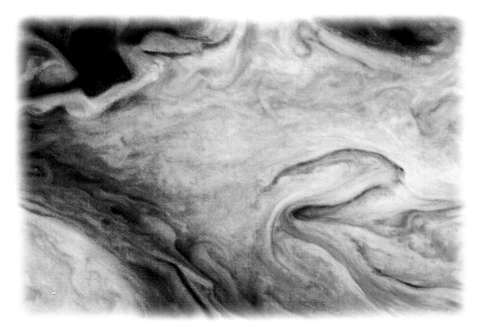

Jupiter is the largest planet in the solar system—bigger than all the others put together—and its weather is on a scale to match. Turbulent winds, fierce lightning and raging storms keep the atmosphere constantly churning. Three multicolored cloud layers wrap the entire globe in ever-changing, swirling patterns, rotating in bands that move either with or against the planet's spin. The Voyager and Galileo probes have sent back vivid pictures and valuable data that reveal the complexity of Jupiter's atmosphere.

WHAT IF...

...WE COULD WATCH THE GALILEO PROBE DESCEND?

July 13, 1995: The Galileo probe is 50,660,000 miles (81,529,370 km) from Jupiter and about a billion miles (1.6 billion km) from Earth. The mission reaches a critical phase as the umbilical cord connecting the orbiter and the probe is broken and the two spacecraft are flung into their carefully planned trajectories. The orbiter will later take up its calculated path around the gas giant and continue to photograph the planet and its moons for several years. But a different fate awaits the probe.

Now it is December 7, 1995, and the Galileo probe has just woken from its quiet cruising mode. Six hours later it begins a kamikaze dive into Jupiter's atmosphere. It falls at a top speed of 106,000 mph (170,590 km/h), causing the plasma in its path to heat to a temperature of 28,000°F (15,537°C). Any ordinary probe would vaporize instantly in these conditions, but NASA engineers spent nearly 20 years laboring to develop this craft, and after about 112 seconds it is still intact despite the fearsome heat.

Parachutes deploy after 170 seconds and the probe begins a more leisurely descent into a haze of smog above Jupiter's clouds. Soon it is falling through the first cloud layer, past swirling layers of ammonia ice. The temperature is a frigid –238°F (–114°C), but it will increase as the probe continues its descent. At the second cloud deck, the temperature is –90°F (–32°C). Twenty-four minutes after atmospheric entry, the probe reaches the final cloud deck—a layer of water ice where the temperature is 32°F (0°C)—and the probe picks up radio bursts of distant lightning. As it drops deeper, the winds become fiercer, and the probe is buffeted mercilessly. Plumes of gas from the interior are rising all around it. Controllers back on Earth will be disappointed to learn that the probe has apparently plunged into a Jovian desert.

Sinking deeper still, the probe is now getting very hot. The hostile conditions are starting to affect the probe's instruments, and eventually, 115 miles (185 km) below the cloud tops, it ceases to transmit, having traveled 400 miles (645 km) through the atmosphere.

A few miles farther down, the probe starts to melt. The pressure at this level is 24 times that of sea level on the Earth, and the severe conditions finally destroy the probe. But far above, the waiting orbiter has received its signals—the first ever sent from inside a gas giant and prepares to beam them back to the Earth.

ATMOSPHERIC FACTS

TYPICAL TEMPERATURES

EDGE OF SPACE	–236°F (–113°C)
32 MILES DOWN	–136°F (–58°C)
62 MILES DOWN	62°F (17°C)
93 MILES DOWN	260°F (127°C)

MAXIMUM STORM SIZE........8,000 BY 16,000 MILES (12,870 BY 25,750 KM) (GREAT RED SPOT)

MAXIMUM WIND SPEED ......250 MPH (400 KM/H)

ATMOSPHERIC COMPOSITION

HYDROGEN	86.4%
HELIUM	13.6%
WATER	0.1%
METHANE	0.21%
AMMONIA	0.07%
HYDROGEN SULFIDE	0.008%

LIQUID SKY

Like all the gas-giant planets, Jupiter is a spinning sphere of liquid. There is no "surface" at the planet's center—the atmosphere simply gets thicker the deeper it goes, until the pressure is so great that it causes gases to turn into liquid, becoming unlike anything we would choose to call an atmosphere.

Jupiter's atmosphere is made up mainly of hydrogen with lesser amounts of helium, making it very similar to the Sun: If it were a lot bigger, nuclear reactions could start in Jupiter's center and cause it to burn like a star. It is the smaller amounts of heavier elements that cause cloud layers to form high in the atmosphere. The diverse palette of colors here gives us clues to the chemical composition, with reds and browns suggesting the presence of sulfur and phosphorus.

The winds are much stronger on Jupiter than on Earth. Without geographical features to get in the way, the winds whip around the huge globe in distinct weather bands. There are two equatorial bands, four bands in the northern hemisphere and three in the southern hemisphere. These bands are either *belts* or *zones*. The belts are dark-colored and move in one direction; the zones are light-colored and hurtle in the opposite direction. All of them stay in the same latitudes and have done so for at least 90 years—as long as astronomers have been peering at Jupiter through modern telescopes. Some scientists suspect that the bands are not merely atmospheric, but extend deep into the planet's interior.

STORMY WEATHER

Jupiter is famous for the storms that move relentlessly within the different bands. The larger storms can go on for years or even decades, although we still don't know how they manage to last so long. Storms at the same latitude can race at different speeds, sometimes overtaking each other or merging. In 1998, three great white ovals that had chased each other around the globe for decades became just two, as one of them caught

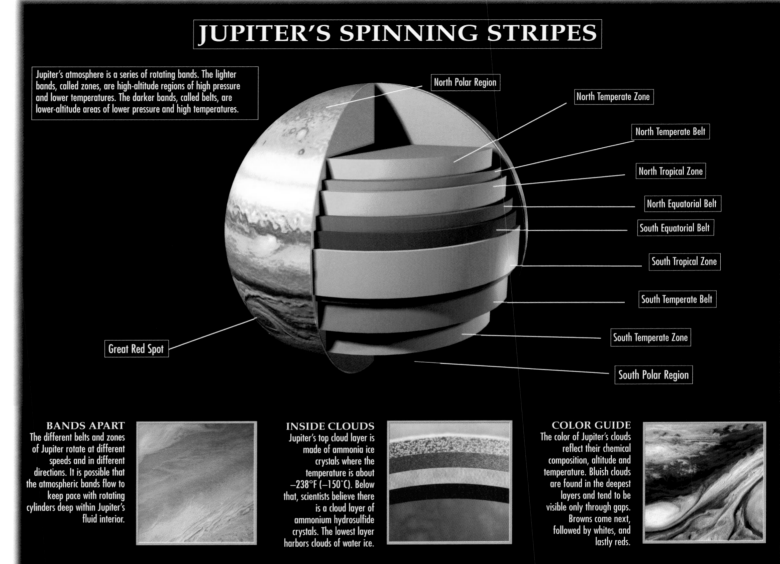

JUPITER'S SPINNING STRIPES

Jupiter's atmosphere is a series of rotating bands. The lighter bands, called zones, are high-altitude regions of high pressure and lower temperatures. The darker bands, called belts, are lower-altitude areas of lower pressure and high temperatures.

North Polar Region

North Temperate Zone

North Temperate Belt

North Tropical Zone

North Equatorial Belt

South Equatorial Belt

South Tropical Zone

South Temperate Belt

South Temperate Zone

Great Red Spot

South Polar Region

BANDS APART
The different belts and zones of Jupiter rotate at different speeds and in different directions. It is possible that the atmospheric bands flow to keep pace with rotating cylinders deep within Jupiter's fluid interior.

INSIDE CLOUDS
Jupiter's top cloud layer is made of ammonia ice crystals where the temperature is about −238°F (−150°C). Below that, scientists believe there is a cloud layer of ammonium hydrosulfide crystals. The lowest layer harbors clouds of water ice.

COLOR GUIDE
The color of Jupiter's clouds reflect their chemical composition, altitude and temperature. Bluish clouds are found in the deepest layers and tend to be visible only through gaps. Browns come next, followed by whites, and lastly reds.

and swallowed another. But the most famous storm of all, the Great Red Spot, is big enough to hold two whole Earths and has been twirling around Jupiter for at least 150 years—and shows no signs of disappearing. Just as on Earth, Jupiter's violent weather is powered by heat. As

gas warms, it expands and rises, creating eddies and swirls. The Sun's light is weak, so scientists think the storms are caused by warm plumes rising from Jupiter's boiling interior. In turn, the storms create turbulence that powers the banded jet streams in their endless rotation.

Most sunlight on Jupiter is absorbed at the equator, but temperatures are similar at all latitudes. The Sun's heat is somehow redistributed, yet scientists are unsure how. The space age may have revealed many new aspects of Jupiter's atmosphere, but there is still more to learn.

JUPITER'S GREAT RED SPOT

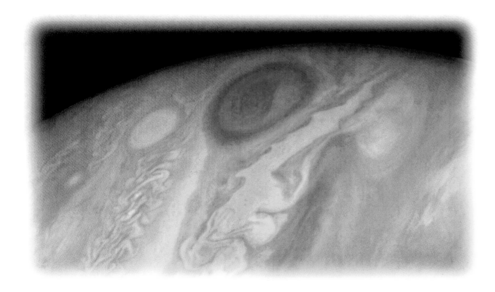

Jupiter, giant of the solar system, is host to an appropriately gigantic—and long-lived—storm. Twice the diameter of the Earth, the colossal hurricane known as the Great Red Spot has raged across the southern hemisphere of the planet since 1665 for certain, and perhaps for much longer. But despite almost 350 years of observation—and the close attention of several space probes—no one really knows how the Great Red Spot was born, why it is so durable or how it will come to an end.

WHAT IF...

...WE COULD TRAVEL THROUGH JUPITER'S RED SPOT?

When the Galileo spacecraft reached Jupiter in December 1995, it dropped a small probe into the giant planet's atmosphere. Thanks to its protective shielding, the probe penetrated 373 miles (600 km) down through the clouds before pressure and heat annihilated it.

Given the powerful forces involved, it was an impressive feat. But the Jovian atmosphere extends many thousands of miles beyond. To learn anything of conditions deep inside the planet, we will need more robust machinery—or a shortcut. Could the Great Red Spot act as a tunnel for a future probe to explore Jupiter's lower depths?

Such a future probe would first fall through the layer of murky haze above Jupiter's clouds. As it enters the top of the Red Spot, the temperature is around –207°F (–97°C). Its onboard computers have aimed it squarely at the Spot's center, avoiding the 250 mph (400 km/h) winds that hurtle counterclockwise around the edges. Upwelling gases push against the probe's descent, and help it decelerate as it drops through layers of turbulent, windblown clouds of ammonia crystals, separated by vast, open stretches of clear atmosphere.

There are at least two cloud layers—the probe should be able to give us a definitive number, if it survives long enough—before a final layer of ice fragments. Since the temperature has now risen to around freezing point, there could even be water droplets, too. Atmospheric pressure has reached five times Earth-normal, and is increasing fast.

Already, the probe has sent back streams of data—enough, perhaps, for scientists to finally determine the cause of the Spot's strange colors and the source of its energy. The team at mission control certainly hopes so, because the probe cannot last much longer.

Now it is falling through the hydrogen that makes up the bulk of Jupiter's atmosphere. Around 5,000 miles (8,050 km) below, gaseous hydrogen slowly merges into the hot liquid hydrogen that makes up most of Jupiter's interior. But the probe will never get that far. No matter how strongly its makers have built the machine, the millions of pounds of atmospheric pressure will crush it long before it reaches that unimaginable ocean.

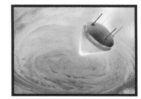

Heatshield glowing, an atmosphere probe plunges into the heart of Jupiter's Great Red Spot. If it survived long enough, the probe might explain a few mysteries.

RED SPOT STATISTICS

DISCOVERY	1665, BY JEAN DOMINIQUE CASSINI (1625–1712)
LENGTH (1999)	14,300 MILES (23,000 KM)
LENGTH (MAXIMUM OBSERVED)	24,250 MILES (39,000 KM)
WIDTH (1999)	7,700 MILES (12,400 KM)
WIDTH (MAXIMUM OBSERVED)	9,000 MILES (14,480 KM)
HIGHEST WIND SPEED	250 MPH (400 KM/H)
TEMPERATURE (CLOUD TOPS)	–207°F (–97°C)
DURATION	330+ YEARS

ETERNAL STORM

When French astronomer Jean Dominique Cassini turned his primitive telescope toward Jupiter in 1665, he noted a distinct elliptical ring on the planet's southern hemisphere. Cassini had made the first sighting of what was to become known as the Great Red Spot. Most early observers thought that it was an island floating on a Jovian ocean. In fact, the Red Spot is a colossal storm that rages within Jupiter's atmosphere and bulges 10 miles above the planet's cloud-tops.

At the edge of the Red Spot, winds blow counter-clockwise at speeds of more than 250 mph (400 km/h). They circle the slow-moving clouds at the center every seven to eight days, in a manner similar to big tropical storms on Earth. Storms, though, are by their nature short-lived. But the Great Red Spot has endured more than three centuries, and perhaps much longer. The secret to its stamina lies in the combination of Jupiter's deep, layered atmosphere and the huge planet's fast rotation.

Even on Earth, planetary spin plus the Sun's heat create wind patterns that allow hurricanes and typhoons to last for days or weeks. On Jupiter, the day is less than 10 hours long. Given the planet's 88,846-mile (142,983-km) diameter, that means that its equator hurtles along at nearly 30,000 mph (48,280 km/h)—almost 30 times faster than on Earth.

MYSTERIOUS VORTEX

At the distance of Jupiter's orbit, almost 500 million miles (800 million km) from the Sun, solar energy is less significant than on Earth. But the planet itself generates vast quantities of heat as it shrinks slowly under the force of its own gravity. Convection from the interior helps fuel the racing bands of cloud that give Jupiter its characteristic appearance. These cloud bands are driven in alternate directions by hurricane-force winds. The Red Spot, squeezed between east- and west-rotating bands, may have begun life when a hot spot below the clouds sent a bubble of warmer gas up to the top of Jupiter's

atmosphere. Once in place, it became a semipermanent, whirling eddy trapped between the spinning cloud bands.

The Red Spot is fed by more gas upwelling from deep inside Jupiter, as well as by lesser, more temporary storms that the Spot absorbs as they pass. Probably, the storm formation is relatively shallow, since pressures deep in the atmosphere would destroy it. But scientists cannot yet explain the mechanisms that keep the Red Spot going.

They are no more certain of the cause of its coloring. The chemical phosphine, driven up from the planet's interior, may be responsible: It turns red in the presence of sunlight. The theory has not been confirmed by recent observations, and in any case the Spot is not always red. The real explanation may well involve another aspect of Jupiter's complex chemistry.

Just about the only agreement among scientists is that the Great Red Spot will one day vanish. Apparently everlasting to short-lived human observers, it is no more than a transitory blip in Jupiter's atmosphere. Perhaps, once it has gone, a replacement storm will arise to perplex the astronomers of the future.

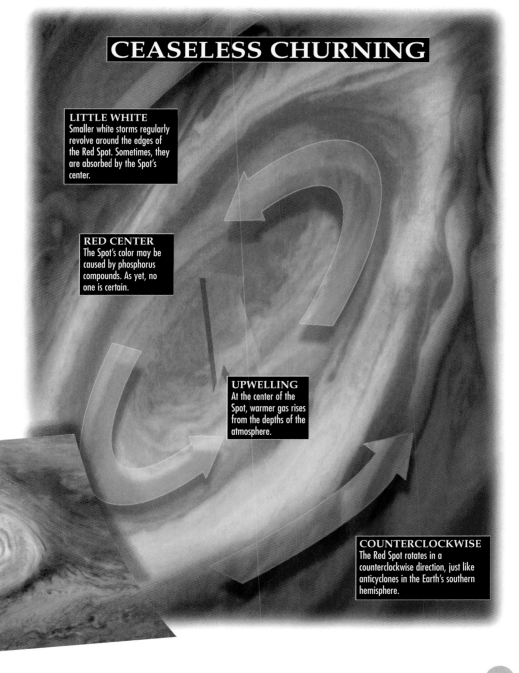

CEASELESS CHURNING

LITTLE WHITE
Smaller white storms regularly revolve around the edges of the Red Spot. Sometimes, they are absorbed by the Spot's center.

RED CENTER
The Spot's color may be caused by phosphorus compounds. As yet, no one is certain.

UPWELLING
At the center of the Spot, warmer gas rises from the depths of the atmosphere.

COUNTERCLOCKWISE
The Red Spot rotates in a counterclockwise direction, just like anticyclones in the Earth's southern hemisphere.

INTRICATE
The Spot has a complex, almost organic form. This infrared image, shown in false colors, reveals details of its shape and structure. Clouds shown as blue are deepest in Jupiter's atmosphere. Pink areas are high and thin, sometimes reaching 20 miles (32 km) above their surroundings. The image is a composite of 18 separate pictures taken over a six-minute period by the Galileo probe in June 1996.

JUPITER'S MAGNETISM

The giant planet Jupiter lies at the center of a mighty web of magnetic force. Created by powerful forces deep in the planet's interior and boosted by the planet's rapid, 10-hour rotation period, the Jovian magnetosphere forms an enormous bubble of energy that seethes with electrically-charged particles. At the center of the magnetosphere is a disk of plasma—torn atoms—through which the volcanic moon Io plows. At its outer extreme—the magnetotail—Jupiter's magnetism extends beyond the orbit of Saturn.

WHAT IF...

...A SPACESHIP TRAVELED INTO JUPITER'S MAGNETOSPHERE?

Radiation levels in some regions within Jupiter's magnetosphere are greater than those in a nuclear reactor. Jupiter's radiation almost overwhelmed the Pioneer probes in the 1970s—each of the two craft was subjected to around 1,000 times the dose that is lethal to humans. When the Galileo spacecraft reached Jupiter in 1995, its orbit was carefully calculated in order to minimize radiation exposure. Only when the spacecraft had completed its main mission did its operators send it plunging closer to Jupiter to measure the magnetism and radiation inward of Io.

In January 2001, astronomers had a unique opportunity to study the magnetosphere when the Saturn-bound Cassini spacecraft flew past Jupiter. Although it stayed almost 9 million miles (14.5 million km) away, it was able to measure the outer regions of the magnetosphere, while Galileo, closer in, carried out similar observations. Together, the probes were able to show how the magnetosphere interacts with the solar wind.

Some day, we may want to send humans to Jupiter or at least to the Jovian moons—just as in Stanley Kubrick's classic movie, *2001: A Space*

Odyssey. Such a journey, if it happens at all, will be a long way behind the movie's optimistic schedule. Present-day space technology is nowhere near capable of sending humans to Jupiter, and is unlikely to catch up with the Hollywood version until the middle of the 21st century. At that point, the development of a 2001-style nuclear spaceship could turn out to be the easiest part of the Jupiter mission. We would still have to deal with the problem of Jupiter's magnetosphere and its radiation belts.

Without extensive shielding—shielding too heavy for any spacecraft to carry—a human crew would soon sicken and die. Complex onboard electronics would be bombarded into uselessness. The disabled ship could end up trapped forever in Jupiter orbit, doomed by the most powerful magnetic field of all the planets.

Perhaps future scientists will dream up a lightweight means of deflecting or absorbing dangerous radiation. If not, we may have to accept that there are places in our solar system where we simply cannot go.

The nuclear spaceship Discovery nears Jupiter in 2001: A Space Odyssey. In the movie, the ship's computer, HAL 9000, murdered most of the crew. In reality, Jupiter's radiation would probably have killed them first—and fried HAL's circuits, too.

MAGNETOSPHERE FACTS

STRENGTH OF THE MAGNETIC FIELD AT THE EQUATOR	4.28 GAUSS
OFFSET BETWEEN MAGNETIC AND ROTATIONAL AXES	9.6°
DISTANCE BETWEEN MAGNETIC CENTER AND PLANET CENTER	0.131 JOVIAN RADII
LENGTH OF MAGNETOSPHERE TAIL	400 MILLION MILES (644 MILLION KM)
DISTANCE BETWEEN JUPITER AND IO	262,200 MILES (421,970 KM)
STRENGTH OF ELECTRICAL CURRENT LINKING JUPITER WITH IO	3 MILLION AMPS

LINES OF FORCE

Jupiter's magnetic field is the largest structure in the solar system—it extends about seven million miles into space on Jupiter's sunward side and 100 times further on its other side. Solar winds prevent the field from stretching further towards the Sun, and give the magnetosphere a tail that points toward Saturn's orbit.

The origins of this enormous magnetic force lie deep within the planet. Most of Jupiter is made up of hydrogen—only a small part of the planet is solid. Yet this solid core of molten rock is 15 times heavier than the Earth. Under immense pressure from the outer layers of gas, the hydrogen surrounding the core is so compressed that it behaves like a liquid metal. Inside this liquid metallic hydrogen are convection currents that spin around very fast, inducing electrical currents. The result is a magnetic field 20,000 times stronger than the Earth's.

The shape of this magnetosphere is manipulated by another force that Jupiter generates—centrifugal force. The planet's short rotation time and rapid spin combine to exert immense pressure on the magnetosphere. Near the planet, the lines of magnetic force arch from the northern to the southern hemispheres—in much the same way as iron filings arrange themselves around a bar magnet. But farther away from the planet, the centrifugal force throws the plasma trapped in the magnetic field outward. This pushes out and distends the magnetic field lines around Jupiter's equator.

Jupiter's electrical force is amplified by its moon Io. Volcanic geysers on Io throw out around 30 billion tons (27.2 billion tonnes) of matter every year. Most of this falls back onto the moon's surface, but some is lost into space. Under the influence of Jupiter's magnetic field, a mixture of gas and other material spreads out along Io's orbit to form a "doughnut" around Jupiter. Known as the Io torus, the doughnut is a concentrated mix of electrons and ions.

The magnetic field and Io's torus co-rotate at a faster speed than Io orbits, causing plasma to flow around the back of the moon. As Io is brought through the magnetic field, an electrical path opens up between Io and Jupiter's atmosphere. The electrical flux reaches millions of amps and causes surges of radiation that generate huge aurorae as they encounter the outermost layers of Jupiter's atmosphere.

Space probes and the Hubble Space Telescope have yielded images of these astonishing displays. But no human is ever going to watch the show close-up in person. Most of the action takes place in ultraviolet light. And even if a crewed spaceship could make the trip, radiation exposure would kill all aboard.

JUPITER'S MAGNETOSPHERE

The size of Jupiter's magnetosphere varies with the changing pressure of the solar wind. Jupiter's fast rotational spin produces a strong centrifugal force, which causes a concentration of particles around the magnetic equatorial plane. Here, the particles form a large disk of thin, ionized gas—plasma—crossed by strong electrical currents that interact with the magnetic field.

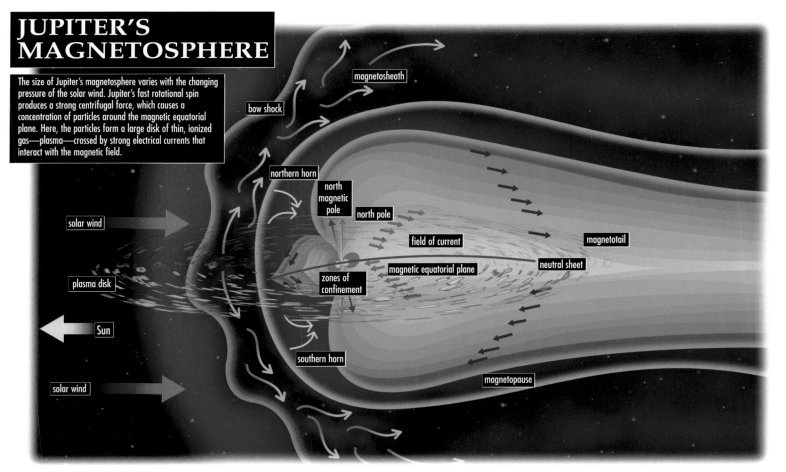

magnetosheath

bow shock

northern horn

north magnetic pole

north pole

field of current

magnetotail

solar wind

magnetic equatorial plane

neutral sheet

plasma disk

zones of confinement

Sun

southern horn

solar wind

magnetopause

JOVIAN RADIO

CHARGED PARTICLES CIRCLING JUPITER'S MAGNETIC LINES PRODUCE RADIO WAVES. THESE EMISSIONS, FIRST DETECTED IN 1955, PROVIDED EVIDENCE OF JUPITER'S ENORMOUS MAGNETIC FIELD. UNTIL THEN, ONLY EARTH WAS KNOWN TO HAVE A MAGNETIC FIELD. AMATEUR RADIO ENTHUSIASTS, INCLUDING A GROUP AT TAUNTON SCHOOL IN ENGLAND (RIGHT), PICKED UP SIGNALS FROM JUPITER USING LARGE BUT SIMPLE YAGI ANTENNAS.

JUPITER'S MOON SYSTEM

A grand total of 63 moons are known to orbit the giant planet Jupiter. The four largest were discovered by Galileo in the 17th century, but nearly 300 years passed before the fifth was found. Amalthea—and Jupiter's other small moons—proved to be nothing like the larger Galilean satellites. The lumpy, irregularly shaped objects are more like asteroids than moons. Their tiny size and great distance from Earth make them hard to study: Only visits by the Voyager 1 and Galileo spacecraft have revealed any of their features.

WHAT IF...

...WE KEEP FINDING MOONS?

All the giant planets—Jupiter, Saturn, Uranus and Neptune—have extensive families of small satellites. More are regularly discovered by spacecraft and Earth-based telescopes—Uranus has 27, and Saturn at least 47, but Jupiter's family is the largest of all.

However, Jupiter's moons also fall into an unusual pattern. Most moons orbit within a few degrees of the plane of their parent planet's equator, because they have formed from the same spinning gas cloud as the planet itself. The sheer size of the Galilean moons may have disrupted the formation of any other regular satellites further out, or such satellites might have been flung out of the Jovian system in their youth. The outer satellites orbit at radical angles to Jupiter's equator. Many of their orbits are quite elliptical, and a larger number even orbit Jupiter in the "wrong" direction.

Astronomers are at a loss to explain why Jupiter has so few regular satellites. But the probable history of the outer moons is becoming clearer. The skewed orbital angles mean that the outer moons were almost certainly asteroids captured from interplanetary space by Jupiter's gravity. According to one theory, just two original objects may have been responsible for the two major families of outer satellites. Each one then suffered collisions with other bodies. Massachusetts Institute of Technology researcher Schelte Bus has supported this theory with his work on asteroid collisions. "Sometimes these collisions are powerful enough to result in a catastrophic disruption," he says, "which leaves families—fragments of the original parent asteroid—traveling in similar orbits."

Two other irregular satellites orbit Jupiter at highly inclined angles, and cannot be linked to any other known object. It is likely that these are small captured asteroids in their own right.

MINOR MOON ROUNDUP

Name	Diameter	Mean Distance from Center of Jupiter	Orbital Inclination	Discovery Date
Metis	37 x 21 miles (60 x 34 km)	79,510 miles (127,960 km)	0°*	1979
Adrastea	16 x 12 x 9 miles (26 x 19 x 14 km)	80,140 miles	0°*	1979
Amalthea	155 x 80 miles (249 x 129 km)	112,650 miles	0.4°	1892
Thebe	72 x 52 miles (116 x 84 km)	137,880 miles	0.8°*	1979
Leda	68 miles (109 km)	6,893,000 miles	27°	1974
Himalia	106 miles (171 km)*	7,133,000 miles	28°	1904
Lysithea	15 miles (24 km) *	7,282,000 miles	29°	1938
Elara	50 miles (80 km)*	7,293,000 miles	28°	1905
Ananke	12 miles (19 km)*	13,200,000 miles	147°	1951
Carme	19 miles (31 km)*	14,000,000 miles	163°	1938
Pasiphae	22 miles (35 km)*	14,600,000 miles	148°	1908
Sinope	17 miles (27 km)*	14,700,000 miles	153°	1914

*UNCERTAIN

KING'S CONSORTS

Jupiter has at least 59 small satellites—four inner moons that orbit in near-perfect circles, and a blizzard of outer ones in orbits of varying eccentricity. We still have much to learn about these tiny siblings of the four large moons discovered by the astronomer Galileo. But by observing their orbits, analyzing their color and gleaning data from the few images we have, scientists are piecing together an ever-better understanding of them.

The largest and most photographed of the small satellites is Amalthea. The most surprising discovery has been the moon's color—Amalthea is the reddest object in the solar system. Astronomers believe the color comes from a layer of sulfur ejected from Io's violent volcanoes. The extremely active Io hurls vast quantities of material into space, and its volcanic substances fall in a stream toward Jupiter. Caught in its path, Amalthea is splattered red. But even odder than its red coating are the mysterious green patches that appear on the major slopes. At present, these are unexplained.

Thebe, Metis and Adrastea, the remaining inner moons, were caught on camera by Voyager 1 during its 1979 flyby of Jupiter, but were not photographed properly until the Galileo probe arrived in the 1990s.

Around the same time, improvements to Earth-based telescopes led to a boom in the numbers of irregular outer satellites known to orbit all of the outer planets. We know even less about Jupiter's outer moons, orbiting beyond Callisto, than we do about the inner moons. One group, clustered at about 7 million miles (11.2 million km) from Jupiter, circle in the normal direction—that is, the same way the planet spins. An outer group of moons lying at about 14 million miles (22 million km) have wild, elliptical orbits and *retrograde* rotation—in other words, they orbit backward. This backward motion adds weight to the theory that Jupiter's small outer moons are captured asteroids. An asteroid could have come from any direction, and would have had a 50% chance of ending up in a backward orbit. Moons that formed along with their parent planet, on the

other hand, orbit in the same direction as the planet rotates. It could be that a large asteroid hurtled toward Jupiter and broke into four pieces before being thrown backward into orbit.

Scientists also think the outer moons are former asteroids because they appear to be rich in carbon, or *carbonaceous*. A group of carbonaceous asteroids known as the Trojans travel in the same orbit as Jupiter, but always ahead of, or behind, the planet by 60° of the orbit. So the moons could well be escaped Trojans. However, we may have to wait many years before another robot visitor to the Jupiter system can help us shed more light on the problem of the moons' origins. Galileo is scheduled to fly past the outer moons Elara and Himalia, so perhaps it will shed some light on the problem of the moons' origins. But we might have to wait quite a long time before another robotic visitor arrives on the scene to take the first pictures of the other outer worlds.

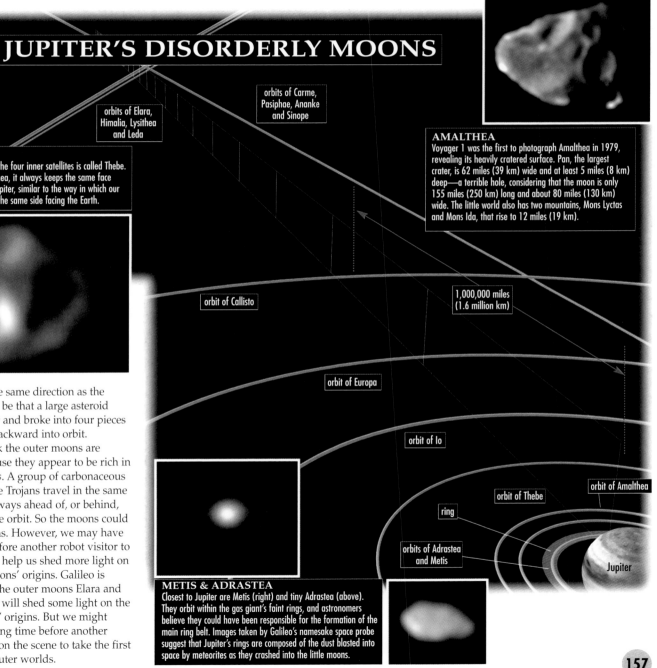

JUPITER'S DISORDERLY MOONS

orbits of Carme, Pasiphae, Ananke and Sinope

orbits of Elara, Himalia, Lysithea and Leda

THEBE
The last moon of the four inner satellites is called Thebe. Along with Amalthea, it always keeps the same face pointed toward Jupiter, similar to the way in which our own Moon keeps the same side facing the Earth.

AMALTHEA
Voyager 1 was the first to photograph Amalthea in 1979, revealing its heavily cratered surface. Pan, the largest crater, is 62 miles (39 km) wide and at least 5 miles (8 km) deep—a terrible hole, considering that the moon is only 155 miles (250 km) long and about 80 miles (130 km) wide. The little world also has two mountains, Mons Lyctas and Mons Ida, that rise to 12 miles (19 km).

orbit of Callisto

1,000,000 miles (1.6 million km)

orbit of Europa

orbit of Io

orbit of Amalthea

orbit of Thebe

ring

orbits of Adrastea and Metis

Jupiter

METIS & ADRASTEA
Closest to Jupiter are Metis (right) and tiny Adrastea (above). They orbit within the gas giant's faint rings, and astronomers believe they could have been responsible for the formation of the main ring belt. Images taken by Galileo's namesake space probe suggest that Jupiter's rings are composed of the dust blasted into space by meteorites as they crashed into the little moons.

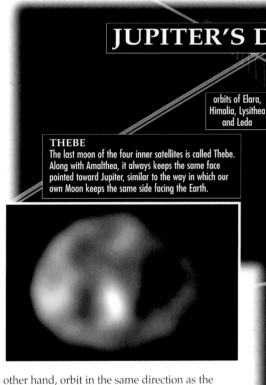

IO

Similar in size to our own Moon, Io is the third largest of the four big moons of Jupiter. It is also the most volcanically active body in the solar system. On Io, massive volcanoes launch plumes of sulfur over 200 miles (320 km) into the sky, coating the surface in sulfur and frozen sulfur dioxide gas. The volcanoes get their power from "tides" in Io's molten crust that are created by the gravitational pull of the giant planet Jupiter. These tides squeeze and stretch Io, heating the interior to the point where it explodes.

WHAT IF...

...THE VOLCANOES ON IO STOP ERUPTING?

Were it not for tidal heating, volcanic activity on Io would have died out 2 to 3 billion years ago. In the early days of Io's existence, it was the moon's rotational speed that raised tides in the crust and generated heat at the core. Now, Io's rotation has slowed, and it always keeps the same face toward Jupiter as it orbits the planet.

Io's orbit around Jupiter is not circular, but elliptical, which is why the planet's gravity continues to raise tides on its moon. If Io were Jupiter's only moon, the tidal heating would have absorbed energy from Io's orbit, causing it to become circular. Then the tidal heating would have stopped. But Io's orbit remains elliptical, because of the gravitational pull of two of Jupiter's other large moons, Europa and Ganymede. For every circuit Ganymede makes around Jupiter, Europa makes two and Io four. This gives Io regular "tugs," keeping the moon's orbit elliptical, and ensuring that its volcanic activity continues.

But scientists have calculated that Io's volcanoes release heat faster than tidal heating could generate it. This suggests that the current volcanic activity is driven by heat generated in the past, when the tidal effect was stronger—which it could have been because the pattern of gravitational tugs on the moon is so complex. It seems likely that the volcanic activity comes in bursts that peak every few hundred million years. If so, we are probably witnessing one now.

...WE COULD GO THERE?

If Io has just passed a peak of volcanic activity, it is conceivable that in a few thousand years this activity will dwindle away to nothing—in which case, humans may one day decide to go there. But even without its sulfur plumes and lava flows, Io is a hostile place. There is no oxygen or water, and the thin, choking atmosphere would be poisonous. The fuming lakes of liquid sulfur are unlikely to make a pleasant vacation spot, while the ever-present danger of volcanic eruptions will deter even the most intrepid visitor. Probably the best way for tourists of the future to view Io's spectacular scenery will be from a safe distance!

JUPITER I: IO

ASTRONOMICAL NAME	Io (JUPITER 1)
YEAR OF DISCOVERY	1610
DISCOVERER	GALILEO GALILEI
ORBITAL PERIOD	1.77 EARTH DAYS
AVERAGE DISTANCE FROM JUPITER	262,000 MILES (421,650 KM)
DIAMETER	2,260 MILES (3,640 KM)
MASS	1.21 TIMES THE MASS OF EARTH'S MOON
AVERAGE SURFACE TEMPERATURE	–215°F (–102˚C)
ESCAPE VELOCITY	1.59 MILES/SEC (2.6 KM/SEC)

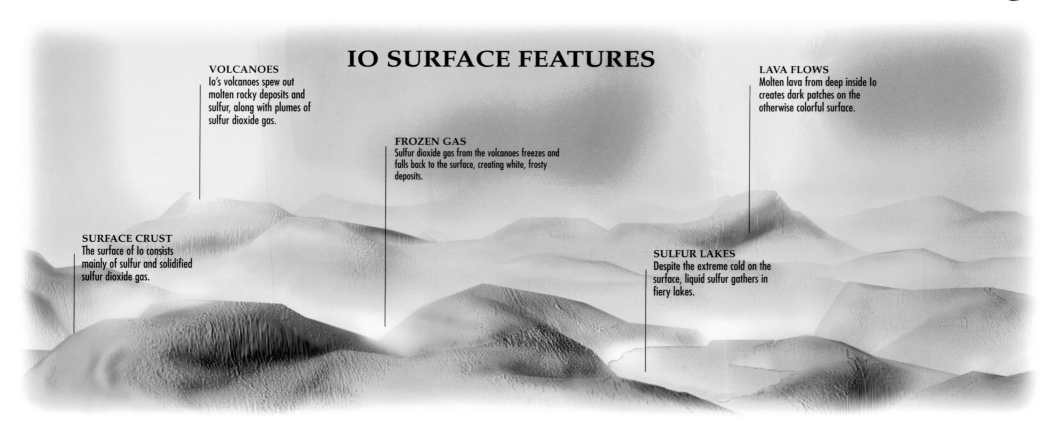

IO SURFACE FEATURES

VOLCANOES
Io's volcanoes spew out molten rocky deposits and sulfur, along with plumes of sulfur dioxide gas.

FROZEN GAS
Sulfur dioxide gas from the volcanoes freezes and falls back to the surface, creating white, frosty deposits.

LAVA FLOWS
Molten lava from deep inside Io creates dark patches on the otherwise colorful surface.

SURFACE CRUST
The surface of Io consists mainly of sulfur and solidified sulfur dioxide gas.

SULFUR LAKES
Despite the extreme cold on the surface, liquid sulfur gathers in fiery lakes.

COSMIC PIZZA

When the first Voyager space probe flew past Io in March 1979, scientists could hardly believe their eyes. Though they had been forewarned that Io was volcanic—not the dull, frozen world they had once thought—not one of them was prepared for the sight of volcanoes spewing plumes hundreds of miles into space, or for the colorful, orange-yellow, cratered surface resembling a giant cosmic pizza.

The images returned by Voyager revealed hundreds of volcanoes covering 5% of the surface, 11 of them active. The largest, Pele, was gushing material 200 miles (320 km) into space at speeds of up to 2,000 miles per hour (3,220 km/h).

ACTIVE WORLD

Four months after Voyager 1, when Voyager 2 flew past Io, there were many changes. This proved just how volcanically active Io is. At first, scientists thought that the volcanoes were erupting molten sulfur. But sulfur alone is not stable enough to create the steep cliffs and volcanic craters that are visible on the surface. Also, the temperatures inside the volcanoes must be far too high for sulfur to remain in a liquid state. It now seems that the lava flows contain rocky deposits similar to those on Earth, and the sulfur is sprayed out in the plumes.

The eruptions are driven by the expansion of sulfur dioxide gas that is superheated below Io's crust. After erupting through the surface, the gas freezes in the coldness of space and settles on the surface as a frosty white layer. But where does the heat come from?

On Earth, it is the radioactive decay of elements such as uranium that keep its core hot. But Io is so small that any heat generated in this way would quickly be lost to space. Instead, it appears that Io is heated by friction due to the strong gravitational field of Jupiter. Just as the Sun and Moon raise tides in the oceans of the Earth, so Jupiter raises tides in the body of Io. This constant flexing keeps the interior molten. When NASA's Galileo spacecraft reached Jupiter in the late 1990s, Io was one of its major targets. It found that more dramatic changes had occurred on the surface of Io since 1979, including many new active volcanoes, sulfur deposits and lava flows.

ERUPTION

WHEN VOYAGER 1 FLEW PAST IO IN MARCH 1979, IT CAUGHT THE VOLCANO LOKI ERUPTING. THE MASSIVE PLUME WAS PLAINLY VISIBLE, BLASTING A JET OF FINE PARTICLES AND GAS AN ESTIMATED 200 MILES (320 KM) OUT INTO SPACE.

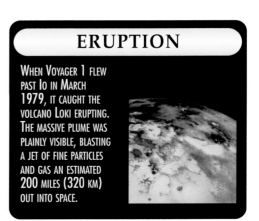

EUROPA

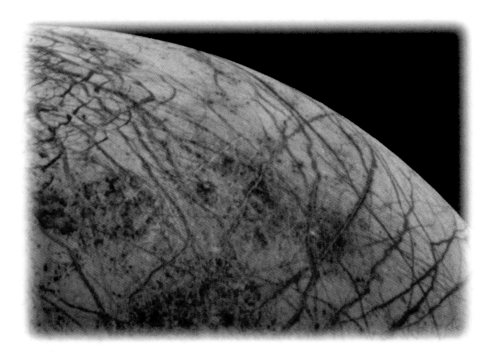

Europa is the smallest and arguably the strangest of the four Galilean moons of Jupiter. The others—Ganymede, Io and Callisto—are peppered with ancient impact craters, yet the surface of Europa is almost completely smooth, with hardly any features more than a few hundred feet high. Europa's high reflectivity—five times that of the Moon—offers a clue: The surface consists of a single sheet of ice that is cracked and fragmented in places by heat from the moon's interior. Below the ice lies a vast ocean of water—an ocean that some scientists have speculated may harbor life.

WHAT IF...

...THERE WERE SOME FORM OF LIFE ON EUROPA?

Many astronomers would gamble that if we ever find life elsewhere in our solar system, it will be beneath the ice that covers Europa. Titan, Saturn's largest moon, is also a candidate, but it is very cold. And Mars, which is virtually without water, is probably sterile now, even though it may have been otherwise in the past. Two key ingredients make Europa so promising: Heat and water.

Europa is almost certainly subject to strong tidal stresses caused by the gravitational pull of its parent planet, Jupiter. Although the stresses are unlikely to be as strong as those on Io, which is closer to the giant planet, they are probably powerful enough to flex the moon's interior and induce heat, in the same way that a spoon gets warm if you repeatedly bend and straighten it.

Substantial amounts of heat generated over a long enough time could have melted Europa's icy shroud below a depth of as little as 70 miles (113 km). This would have allowed any stray hydrocarbons to combine with water in an organic "primordial soup" similar to the one that existed on Earth billions of years ago. On Earth, this soup spawned the amino acids that constitute the building blocks of life. Perhaps a similar process has taken place in the depths of the Europan ocean.

Even today, life on Earth still flourishes in a variety of seemingly hostile environments. Primitive bacteria line the mouths of deep-ocean geothermal vents, where the temperature can reach 450°F (232°C). And bacterial life has even taken hold deep within the ice of the south polar caps.

If life is as resilient on Europa as it has been on Earth, it may well have evolved beyond the stages of primitive, single-celled bacteria clustered around underwater volcanoes. Strange, fish-like creatures might swim blindly through the dark waters, or there may be alien beings of types that we cannot even begin to imagine. We are unlikely to know for certain unless we actually go there. Unfortunately, such an exciting prospect still lies a great many years in the future.

Any higher life-forms that may have evolved on Europa will be aquatic creatures, living in the pitch-dark ocean that lies beneath the moon's icy crust.

JUPITER II: EUROPA

ASTRONOMICAL NAME	Europa (Jupiter II)
YEAR OF DISCOVERY	1610
DISCOVERER	Galileo Galilei
ORBITAL PERIOD	3.55 Earth days
DISTANCE FROM JUPITER	416,880 miles (670,903 km)
DIAMETER	1,951 miles (3,140 km)
MASS	5.37 x 10¹⁹ tons (53.7 million trillion tons/48.7 million trillion tonnes)
SURFACE TEMPERATURE	−260°F (−127°C)
ESCAPE VELOCITY	1.26 miles per second (2 km/sec)

ANATOMY OF EUROPA

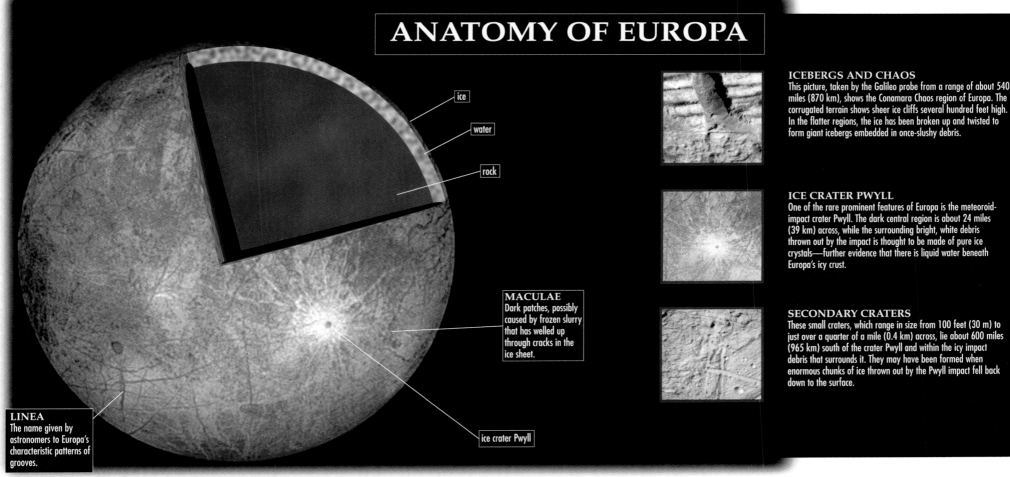

ice

water

rock

ICEBERGS AND CHAOS
This picture, taken by the Galileo probe from a range of about 540 miles (870 km), shows the Conamara Chaos region of Europa. The corrugated terrain shows sheer ice cliffs several hundred feet high. In the flatter regions, the ice has been broken up and twisted to form giant icebergs embedded in once-slushy debris.

ICE CRATER PWYLL
One of the rare prominent features of Europa is the meteoroid-impact crater Pwyll. The dark central region is about 24 miles (39 km) across, while the surrounding bright, white debris thrown out by the impact is thought to be made of pure ice crystals—further evidence that there is liquid water beneath Europa's icy crust.

SECONDARY CRATERS
These small craters, which range in size from 100 feet (30 m) to just over a quarter of a mile (0.4 km) across, lie about 600 miles (965 km) south of the crater Pwyll and within the icy impact debris that surrounds it. They may have been formed when enormous chunks of ice thrown out by the Pwyll impact fell back down to the surface.

MACULAE
Dark patches, possibly caused by frozen slurry that has welled up through cracks in the ice sheet.

LINEA
The name given by astronomers to Europa's characteristic patterns of grooves.

ice crater Pwyll

CRACKED ICEBALL

When astronomers studied the first close-up pictures of Europa, taken by the Voyager 1 and 2 spacecraft in 1979, they were astonished by the virtual absence of surface features. There were few craters, no valleys, no canyons, no volcanoes or mountains and no flood plains or impact basins. Almost the only distinguishing feature of Jupiter's fourth-largest moon was a network of dark grooves, some stretching for thousands of miles, that criss-crossed their way across the flat and highly reflective surface.

It quickly became clear that Europa's bright surface was a vast ocean of frozen water. But the moon's comparatively high density—roughly twice that of neighboring Callisto—implied that not very far beneath the ice lay a rocky interior.

Further calculations suggested that the ice sheet extends no farther than about 70 miles (113 km) below the surface and that underneath lies an ocean of water. Scientists suspect that the ocean is kept in a liquid state by heat generated by the gravitational effects of nearby Jupiter. They think that beneath the ocean, Europa is composed of solid rock—perhaps with a molten inner core.

The grooves on the surface are almost certainly fractures in the ice caused by Europa's internal heating mechanism. It seems that where new cracks appear, a dark slurry of water and rocky debris rises from below the surface and freezes instantly in temperatures that plunge as low as −260°F (−127°C). As the slurry freezes, it expands and forces the broken ice apart in giant wedges the size of whole cities or even states.

For this reason, Europa's terrain is constantly changing. Any impact craters have long since been obliterated, and the ice is too thick for any other features to show through.

EUROPA'S OCEAN

The detailed images of Europa from the Galileo space probe in the late 1990s revealed objects that are only tens of feet across. The discovery of what appear to be areas of relatively pure ice around the moon's few impact craters are further evidence that beneath the ice lies water. If this is so, Europa's dark ocean is the most substantial body of water yet discovered in the solar system outside of the Earth. With its internal heat source, Europa may have all the ingredients necessary for primitive life to form and flourish.

161

GANYMEDE

...GANYMEDE BECAME WARMER?

The moons of Jupiter represent a potential paradise for planetary scientists, and one day people from Earth will have the chance to explore them. At first sight Ganymede looks like the ideal base from which to explore the rest of the Jupiter system. Even so, it would still involve living for long periods in an extremely hostile environment.

In some ways, life on Ganymede would be similar to living on the Moon: There is no atmosphere and, as on the Moon, astronauts would weigh about one-seventh of their weight on Earth. Unlike the Moon, however, the surface of Ganymede is covered in ice and remains bitterly cold even at high noon.

On Ganymede, the Sun shines at less than 4% of its brightness on Earth, which is why spacecraft that venture this far, such as the Galileo probe, use nuclear generators instead of solar panels to produce their electrical power. If the Sun became brighter, Ganymede would be more hospitable; and if Jupiter were a star, rather than a planet, life there would be totally transformed.

In his novel *2061: Odyssey Three*, science fiction author Arthur C. Clarke writes about a future in

which Jupiter has been turned into a star by a mysterious alien intelligence. The moons of Jupiter become a miniature solar system, warmed by this new sun. Colonists from Earth establish a base on Ganymede to explore the nearby moons and set up an observatory to monitor the Jupiter-star.

In the story, the parts of Ganymede that are turned toward Jupiter are gradually transformed by the warming rays of the new sun. The melting surface wipes out most of the features recorded by the Voyager missions and creates shallow seas as large as the Mediterranean. What was once an icy wasteland gradually turns into a habitable region with a temperate climate.

Fifty years after Jupiter becomes a star, the steadily rising atmospheric pressure reaches a point where spacesuits are no longer needed and people can walk freely on the surface. The colonists still need oxygen masks. But because biologists have seeded Ganymede with oxygen-generating bacteria, they expect there to be a breathable atmosphere within a few more decades.

Fantasy? Yes. But Clarke's story reminds us just how much our comfortable home on Earth depends on the life-giving rays of the Sun.

Ganymede, the biggest of Jupiter's four big moons, is the largest satellite in the solar system. Images from the Voyager and Galileo probes reveal an icy, pock-marked world, patterned with intertwined dark and light patches. The dark areas are ancient and heavily cratered; the light areas are younger and are inscribed with mysterious parallel grooves and ridges. Ganymede may once have had oceans of water that gradually froze as the moon cooled. If it were to warm up again, it might be possible for people to live there.

JUPITER III: GANYMEDE

	GANYMEDE	MERCURY	THE MOON
DIAMETER	3,270 MILES (5,262 KM)	3,032 MILES (4,870 KM)	2,160 MILES (3,476 KM)
MASS (EARTH MASSES)	.0247	.055	.0123
SURFACE GRAVITY (EARTH'S = 1G)	.145G	.38G	.165G
DENSITY (EARTH'S = 1)	0.35	.98	0.61
DAYTIME TEMPERATURE	−180°F (−82°C)	660°F (349°C)	265°F (129°C)

WORLD OF FROZEN OCEANS

At first glance, Ganymede resembles the Moon, with dark patches much like the lunar "seas" surrounded by lighter uplands. But whereas the craters and mountains of the Moon are rocky, the features of the surface of Ganymede are sculpted from ice.

The dark areas cover 40% of Ganymede and appear to be the oldest parts of the surface. Judging by the amount of cratering, they are well over 3 billion years old—similar in age to the cratered regions of the Moon.

The lighter areas are less cratered and so are believed to be younger. They are distinguished by a remarkable system of parallel ridges up to 2,000 feet (610 m) high, which run for thousands of miles across the surface.

Scientists believe that the darker areas are parts of the original icy crust of Ganymede—possibly an ocean of water that later froze solid as the moon cooled. If this is so, then the liquid water that remained below may have welled up through cracks in the ice sheet to create the lighter, ridged terrain. The action of water freezing in the cracks could have forced the dark "continents" to spread apart, much like continental drift on Earth. Alternatively, the ridges could be folds caused by those same continents crashing into one another. Whatever the reason for the formation of the ridges, the motion stopped long ago, and Ganymede's surface froze solid.

Ganymede's craters are much softer than those on the Moon, with no central peaks or jagged ramparts. The younger craters have fresh deposits of icy debris around them, but the walls of the oldest craters have almost sunk back into the ice.

Observations with the Hubble Space Telescope have revealed an extremely thin atmosphere of oxygen around Ganymede. This is believed to be the result of bombardment of the surface ice by radiation and charged particles from the Sun.

MAGNETIC FIELD

A big surprise from the Galileo mission was the discovery that Ganymede has its own magnetic field. This implies that the moon has a small metallic core (or, less likely, a liquid metallic layer), but there is another element to the field that appears to be created by interaction with Jupiter's own magnetosphere. This induced field seems to be created by a layer of conductive liquid below Ganymede's surface. Excitingly, the most likely candidate would be a saltwater ocean beneath the crust.

GHOST

THE DARK AREAS OF GANYMEDE SHOW THE GHOSTLY REMAINS OF CRATERS, UP TO 250 MILES (400 KM) ACROSS, THAT HAVE SINCE BEEN ERODED UNTIL THEY ARE BARELY VISIBLE. THESE "GHOST CRATERS" ARE ALSO CALLED "PALIMPSESTS," THE TERM USED FOR ANCIENT MANUSCRIPTS THAT HAVE BEEN ERASED AND USED AGAIN.

SURFACE FEATURES

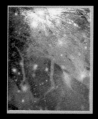

THIS PICTURE OF GANYMEDE WAS TAKEN BY THE VOYAGER 2 PROBE. IT SHOWS WHITE TRAILS OF DEBRIS SPREADING OUT FROM AN IMPACT CRATER ACROSS LIGHT-COLORED REGIONS OF "YOUNG" ICE. THE DARK AREAS ARE ANCIENT ICE, WHICH MAY BE OVER 3 BILLION YEARS OLD.

THE SURFACE OF GANYMEDE

ATMOSPHERE
Ganymede has a very thin atmosphere of oxygen.

ANCIENT ICE
The dark areas of ice are thought to be the oldest parts of Ganymede's surface.

YOUNG ICE
The lighter colored areas of the surface are regions of relatively young ice.

IMPACT CRATERS
Ganymede's surface is pockmarked with the craters caused by the impact of interplanetary rocks.

JUPITER
Ganymede is about 660,000 miles (1.06 million km) away from its parent planet, Jupiter.

CALLISTO

Callisto, the outermost of the Galilean moons of Jupiter, is a scarred and battered world. Its surface is deeply pitted with impact craters that date back some 4 billion years to the early days of the solar system. Since then, it has been almost impact-free and has hardly changed at all. If Callisto possessed a source of internal heat energy, its icy outer shell would have continuously cracked and reformed, obliterating all signs of early cratering. Instead, this cold, dead world remains forever frozen in time.

WHAT IF...

...THE ICE WERE THINNER?

Because the icy outer shell of Callisto is frozen solid and up to 200 miles (320 km) thick, its surface has remained virtually unchanged for billions of years. But had the ice been as thin as it is on Europa—only about 50 miles (80 km) thick—then the surface would have been very different.

Fifty miles (80 km) is still a substantial depth, far deeper than a meteoroid could penetrate. But impacts on relatively thin ice generate shock waves that are capable of cracking the surface and releasing fresh ice or water from below.

Flooding would almost certainly have been more common on Callisto if the ice were thinner, and today there would be more plains and fewer surviving craters. Even so, the surface could never become as smooth as that of Europa: Devoid of an internal energy source, Callisto lacks the heat to produce a regular flow of material from beneath the surface.

The ice on Callisto is also extremely dirty, because of the pulverized fragments of billions of years of meteoroid impacts that are embedded within it. If the ice were to thaw, the surface would turn to slushy soup containing everything from dust particles to boulders the size of small mountains.

No natural process that is likely to occur in the foreseeable future could even begin to melt such a large quantity of ice—save the unlikely prospect of Jupiter becoming a star. Instead, we will have to wait a few more billion years, until the Sun becomes a red giant and increases in brightness, before the age-old Callistian winter begins to thaw.

If Jupiter became a star, its heat would slowly start to melt the 200-mile-thick (320-km-thick) ice on Callisto. The craters would gradually fill with murky water and their walls would start to collapse. Eventually, if the temperature rose high enough, the surface of Callisto would become an ocean.

JUPITER IV: CALLISTO

ASTRONOMICAL NAME	CALLISTO (JUPITER IV)
YEAR OF DISCOVERY	1610
DISCOVERER	GALILEO GALILEI
ORBITAL PERIOD	16.69 EARTH DAYS
AVERAGE DISTANCE FROM JUPITER	1.17 MILLION MILES (1.9 MILLION KM)
DIAMETER	2,983 MILES (4,800 KM)
MASS	117 MILLION TRILLION TONS (106 MILLION TRILLION TONNES)
AVERAGE SURFACE TEMPERATURE	−180°F TO −320°F (−82°C TO −160°C)
ESCAPE VELOCITY	1.51 MILES PER SECOND (2.4 KM/SEC)

JUPITER'S BATTERED MOON

Bigger than the Moon and slightly smaller than the planet Mercury, Callisto is one of the four large satellites of Jupiter discovered in 1610 by Italian astronomer Galileo Galilei (1564–1642). It is the second-largest of these "Galilean moons;" Ganymede is the largest. Yet its surface could not be more different from that of its big brother, or those of the other Galileans.

As seen from the Earth, the most obvious feature of Callisto is that its surface is comparatively dark. It is hard to make out much detail with Earth-based telescopes, and much of what is known about Callisto has been deduced relatively recently—from photographs taken by the Voyager 1 and 2 spacecraft that flew past Jupiter in 1979.

During these encounters, Callisto was revealed to be a battered world that is peppered with craters at a density unseen elsewhere. Most of the craters are less than 60 miles (97 km) across, but two very large impact sites, hundreds of miles in diameter, serve as landmarks on a world that would otherwise look the same from almost any angle.

One reason why Callisto is so pockmarked could be that it lacks a substantial internal heat source. Ganymede has (or had) a hot interior that in the relatively recent past has modified the moon's surface by cracking it and allowing liquid water to well up from below. But because Callisto is farther from Jupiter than Ganymede and is less affected by the planet's gravitational pull, there are no internal stresses to generate an equivalent heat source.

THICK SHELL

Despite the lack of heat, Callisto's structure still seems to have played a part in shaping its surface. Astronomers speculate that the moon is 40% ice and 60% rock and iron. These materials seems to be distributed in layers, with the ice forming an outer shell up to 200 miles (320 km) thick—far thicker than on the similarly structured Ganymede and Europa.

Meteoric impacts on Ganymede and Europa have cracked their fragile icy shells and modified the surface accordingly. On Callisto, the

equivalent impacts have left permanent scars: The surface may have slumped in places, but is otherwise much as it was more than 4 billion years ago, when it was bombarded by the last major pieces of debris left over from the formation of the solar system. This also explains why Callisto's surface is dark—debris from the impacts has pitted and blackened the ice.

Callisto has one final secret. The Galileo probe discovered that it has a magnetic field, most likely explained by a layer of liquid saltwater below the surface. How such a layer could survive in this deep-frozen world, though, is a mystery.

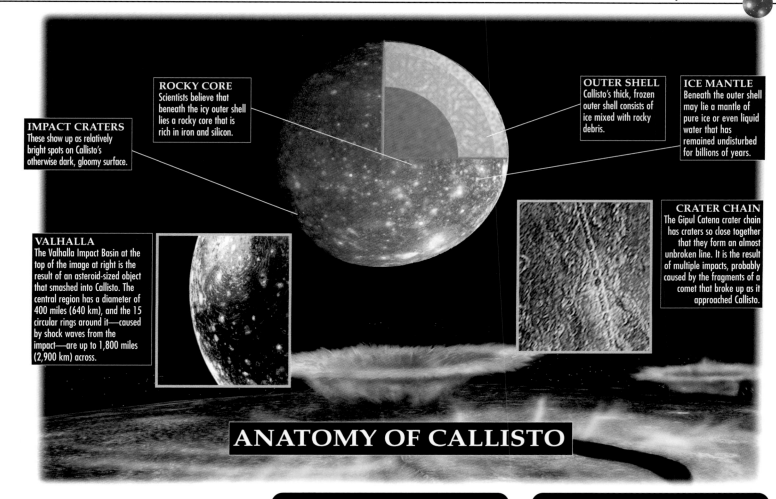

ROCKY CORE
Scientists believe that beneath the icy outer shell lies a rocky core that is rich in iron and silicon.

IMPACT CRATERS
These show up as relatively bright spots on Callisto's otherwise dark, gloomy surface.

OUTER SHELL
Callisto's thick, frozen outer shell consists of ice mixed with rocky debris.

ICE MANTLE
Beneath the outer shell may lie a mantle of pure ice or even liquid water that has remained undisturbed for billions of years.

VALHALLA
The Valhalla Impact Basin at the top of the image at right is the result of an asteroid-sized object that smashed into Callisto. The central region has a diameter of 400 miles (640 km), and the 15 circular rings around it—caused by shock waves from the impact—are up to 1,800 miles (2,900 km) across.

CRATER CHAIN
The Gipul Catena crater chain has craters so close together that they form an almost unbroken line. It is the result of multiple impacts, probably caused by the fragments of a comet that broke up as it approached Callisto.

ANATOMY OF CALLISTO

IMPACT SCARS

THE CRATERS ON CALLISTO WERE FORMED BY THE IMPACTS OF METEOROIDS, COMETS AND ASTEROIDS. WHEN SUCH AN OBJECT CRASHES INTO A PLANET OR MOON, IT VAPORIZES BOTH ITSELF AND SOME OF THE SURFACE THAT IT STRIKES. MATERIAL SPRAYS OUT OF THE IMPACT REGION, LEAVING A CRATER WITH A DIAMETER TYPICALLY 100 TIMES THAT OF THE INCOMING OBJECT. IF THE IMPACTOR IS SUFFICIENTLY LARGE—MORE THAN A FEW MILES ACROSS—THE GROUND BELOW THE CENTER OF THE CRATER MELTS THEN "REBOUNDS" AFTER THE IMPACT. THE RESULT IS A CHARACTERISTIC CENTRAL PEAK IN THE CRATER.

DOOMED NYMPH

IN GREEK MYTHOLOGY, CALLISTO WAS A NYMPH, A SPIRIT WHO LIVED IN MOUNTAINS, TREES AND LAKES. THE SUPREME GOD, ZEUS—JUPITER TO THE ROMANS—WAS SO FOND OF HER THAT HIS JEALOUS WIFE, HERA, TURNED CALLISTO INTO A BEAR. ZEUS SET THE BEAR IN THE SKY, IN THE FORM OF THE CONSTELLATION URSA MAJOR—THE GREAT BEAR.

SATURN

S aturn was the outermost planet known before the invention of
the telescope, and is famous for its spectacular system of rings,
discovered by Galileo in the early 1600s but not properly
described until they were observed by Dutch astronomer Christiaan
Huygens in 1656. The rings are now known to consist of billions of
icy fragments, each in an independent orbit around the planet. They
may be fragments of a comet broken up by Saturn's enormous
gravity, or by a collision with one of the planet's huge family of
moons. Saturn itself is a smaller version of Jupiter, similar in
composition, though with a layer of haze in its upper atmosphere that
masks the activity of its weather systems. The satellite system
contains a variety of very different icy satellites, dominated by Titan,
a moon larger than the planets Mercury and Pluto, with a complex
atmospheric system driven by the chemical compound methane.

*This image of Saturn's south pole and the southern side of the
planet's rings was produced by Hubble's Wide Field Planetary
Camera using 30 different color filters. Filtering and combining
different wavelengths of light allows researchers to better
interpret data to unveil the secrets of Saturn.*

SATURN

...WE COULD GO TO SATURN?

Saturn and its satellites are bleakly inhospitable to any kind of earthly life. Yet more than one science-fiction writer has suggested building a honeymoon hotel on a convenient moon. It would be much harder to reach than Niagara Falls—with present-day rocket technology, the journey would take many years—and honeymooners would have to stay under a protective dome or wear spacesuits. But the view is probably the most awe-inspiring in the solar system.

Beneath the intricately patterned rings, themselves one of the solar system's greatest tourist attractions, guests at the honeymoon hotel would see Saturn's swirling, ever-changing banded clouds, and could marvel at the assortment of moons that complete the giant planet's retinue.

A visit to Saturn's surface is almost certainly impossible, even with far more advanced technology than our own. For a start, the planet does not have a solid surface. And even if it had, the temperatures and pressures down below the atmosphere would simultaneously scorch and crush even the best-protected visitor. At such densities, the atmosphere would be almost as opaque as solid rock.

But if some kind of super-technology should ever overcome these difficulties, tourists would discover an extraordinary fact: Starting off from a landing site near the planet's pole, they would steadily lose weight as they traveled toward the equator.

There are two reasons for this weight loss. First, the planet spins so fast that centrifugal force pushes noticeably against gravity, especially at the equator, where every point on Saturn's surface is whirling along at 7,350 miles (11,830 km) an hour. Second, billions of years of spinning have already made the equator bulge far outward. Because the bulge lifts everything on it an extra few thousand miles from the planet's center of mass, the force of gravity is reduced at the surface.

The difference amounts to about 26%. A chubby couch potato, weighing in at 180 pounds (82 kg) near Saturn's north pole, would shed 46 pounds (21 kg) simply by moving southward to the equator.

From a vantage point in orbit near Saturn, tourists in a future luxury space hotel enjoy a uniquely breathtaking view. But they could never land on the giant planet itself: It has no real surface.

Saturn is an enormous globe of whirling gas—it is made almost entirely of hydrogen and helium—that sits at the center of a complex system of rings and at least 18 moons. The planet and its companions are almost a solar system in miniature. Saturn's rings, which are composed of billions of separate particles and are usually visible in even a small telescope, long ago earned the planet the title "Jewel of the Solar System." But it was not until the Voyager probes reached the planet in the 1970s that astronomers (and everyone else who marveled at the glorious images) were able to take a closer look at the ringed planet.

SATURN PROFILE

SATURN		EARTH
74,898 MILES (120,537 KM)	DIAMETER	7,973 MILES (12,831 KM)
26° 42'	AXIS TILT	23° 27'
10,760 DAYS (29.46 EARTH YEARS)	LENGTH OF YEAR	365 DAYS
10 HOURS 39.4 MINUTES	LENGTH OF DAY	24 HOURS
888 MILLION MILES (14.3 MILLION KM)	DISTANCE FROM SUN	93.5 MILLION MILES (150 MILLION KM)
−292°F (−144°C)	SURFACE TEMPERATURE	59°F (15°C)
0.93 G	SURFACE GRAVITY	1 G
HYDROGEN (97%) HELIUM (3%) METHANE (0.05%)	ATMOSPHERE	NITROGEN (80%), OXYGEN (19%)
1,400 MILLIBARS	ATMOSPHERIC PRESSURE	1,000 MILLIBARS
HYDROGEN (97%) HELIUM (3%) METHANE (0.05%)	COMPOSITION	SILICON (60%), ALUMINUM (15%)

LORD OF THE RINGS

Ever since Galileo pointed his crude telescope at the giant planet back in 1610, Saturn's extraordinary rings have been recognized as a marvel of the solar system. But the planet itself, although less spectacular, is almost as extraordinary. Second only to Jupiter in scale, it is 750 times the size of the Earth.

From space, we can observe only the cloud tops of the giant planet, and even these are often obscured by a yellow haze. The entire atmosphere is divided into distinct bands, similar to Jupiter but not so clearly marked. These cloud bands whirl round the planet in jet streams blowing at up to 1,100 miles (1,770 km) an hour—10 times the speed of an earthly hurricane. Saturn's clouds are a bitterly cold –218°F (–103°C). Those we can see are no more than a frosting of ammonia ice on a huge mass of hydrogen and helium below.

There is nowhere on Saturn that could be described as a planetary surface. If we could send an indestructible space probe down through the clouds, it would record a steady increase in temperature and pressure. Thousands of miles down, the craft's barometer would register atmospheric pressure levels a million times higher than those on Earth. The temperature would rise to match, and would soon reach thousands of degrees.

In such extreme conditions, the hydrogen that makes up much of Saturn's atmosphere behaves very strangely. No longer a gas, it turns into something resembling a liquid metal, capable of conducting electricity. But there is no distinct threshold where atmospheric gas ends and an ocean of liquid hydrogen begins. Beneath the liquid hydrogen is probably a small core of rock, itself in a liquid state.

RINGS AND MOONS

Everything beneath Saturn's cloud tops is hidden from sight. Most of what we know about its interior has been deduced from what we can see of the cloud tops and what we have learned about how the planet behaves.

There is nothing hidden about the rings. Yet these present puzzles of their own. They are composed of billions of little fragments, mainly of ice, ranging from sand-grain size to lumps as big as a house. All of them orbit independently, engaged in an almost fantastically complex dance that is brought to some kind of order by the gravitational pull of Saturn's moons.

These, too, hold unsolved mysteries. Why is Iapetus half-blackened, and how did little Mimas acquire a crater a third the size of the entire moon? The Cassini spacecraft, which arrived at Saturn in 2004, is already helping to solve some of these mysteries through a series of flybys of the Saturnian moons expected to last four years or more.

FLOATER

SATURN IS THE ONLY PLANET IN THE SOLAR SYSTEM LESS DENSE THAN WATER. IF YOU COULD FIND A TUB BIG ENOUGH TO DUNK IT IN, THE GIANT WOULD FLOAT ON THE TOP LIKE AN APPLE IN A WATER BARREL. IN FACT, AT ONLY 69% THAT OF FRESH WATER, SATURN'S AVERAGE DENSITY IS MUCH THE SAME AS MANY APPLES. BY COMPARISON, THE EARTH IS ALMOST EIGHT TIMES AS DENSE AS ITS MUCH BIGGER NEIGHBOR.

SQUASHED

SATURN IS THE MOST FLATTENED PLANET IN THE SOLAR SYSTEM. ITS DIAMETER AT THE EQUATOR IS 74,898 MILES (120,540 KM), BUT ONLY 67,560 MILES (108,730 KM) MEASURED POLE TO POLE. THE EQUATORIAL BULGE IS CAUSED BY THE CENTRIFUGAL FORCE OF THE PLANET'S RAPID ROTATION: SATURN MAKES A COMPLETE REVOLUTION IN JUST 10.5 HOURS. EVEN OUR SLOWER-TURNING EARTH HAS A SLIGHT BULGE AT THE EQUATOR, AND EARTH IS A RIGID, ROCKY WORLD, NOT A GIANT BALL OF GAS.

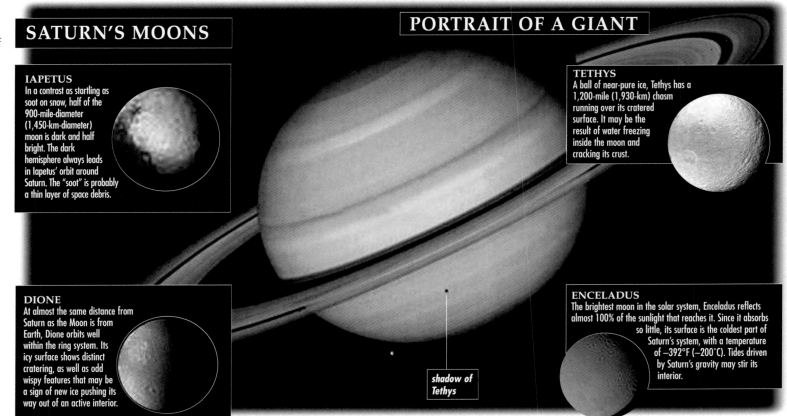

SATURN'S MOONS

PORTRAIT OF A GIANT

IAPETUS
In a contrast as startling as soot on snow, half of the 900-mile-diameter (1,450-km-diameter) moon is dark and half bright. The dark hemisphere always leads in Iapetus' orbit around Saturn. The "soot" is probably a thin layer of space debris.

DIONE
At almost the same distance from Saturn as the Moon is from Earth, Dione orbits well within the ring system. Its icy surface shows distinct cratering, as well as odd wispy features that may be a sign of new ice pushing its way out of an active interior.

TETHYS
A ball of near-pure ice, Tethys has a 1,200-mile (1,930-km) chasm running over its cratered surface. It may be the result of water freezing inside the moon and cracking its crust.

ENCELADUS
The brightest moon in the solar system, Enceladus reflects almost 100% of the sunlight that reaches it. Since it absorbs so little, its surface is the coldest part of Saturn's system, with a temperature of –392°F (–200°C). Tides driven by Saturn's gravity may stir its interior.

shadow of Tethys

CHANGING VIEWS OF SATURN

Once simply a bright point in the night sky, Saturn has been revealed as the most visually interesting planet in the solar system. The planet's ring system is so complex that its appearance baffled the first astronomers to glimpse it through a telescope. It was not until the mid-19th century that the famous rings were explained. Earth-based telescopes and space probes have revealed details of the amazing rings and diverse family of moons—but they have also posed further questions.

WHAT IF...

...WE KNEW MORE ABOUT SATURN?

The progress of technology, from Galileo's telescopes to the Hubble Space Telescope, constantly increases our ability to see the intricacies of the solar system. As far as we on Earth are concerned, Saturn and its rings continue to grow in complexity.

When the Voyager probes made their close flybys of Saturn in the early 1980s, the astronomical community was stunned by the images they captured. The probes revealed that the rings were far more complex than anyone had predicted. As Voyager 1 approached the planet in November 1980, it became clear that there were at least 1,000 rings and ringlets, probably more. Outside the main system, the Voyagers observed the dim F ring. Here, scientists caught a glimpse of what appeared to be braiding—the ring seemed to be made up of various intertwined strands. It was also found that the F ring was kept in place by two shepherd moons—Pandora and Prometheus. Orbiting on either side of the F ring, these two small objects, just 100 miles (160 km) across, keep the ring particles in place with the subtle force of their gravitational fields. It seems likely that this action is a major contributor to the braided structure of the ring.

The Voyagers went on to detect a powerful magnetic field and 1,000 mph (1,600 km/h) winds on Saturn. They also made a close inspection of the planet's largest satellite, Titan.

Nine years after Voyager 2's encounter with Saturn, Mark Showalter, a Stanford University research associate working at NASA's Ames Research Center, spotted a new moon. Showalter had carefully studied over 30,000 Voyager images in the hunt for the object thought to have caused "scalloping" along the edges of the Encke gap in Saturn's outer A ring. Finally, he spotted the tiny culprit. Just 12 miles (19 km) across, the new moon was named Pan.

The Cassini spacecraft has already produced a wealth of new discoveries since its arrival in mid-2004. One of its first tasks was to release the European Space Agency's Huygens lander, which made a successful descent through Titan's cloudy atmosphere in January 2005, and sent back the first pictures from the surface, revealing a world on which methane apparently plays a similar geological role to that of water on Earth. Through 74 orbits of Saturn, Cassini will continue to explore the system, photographing satellites, rings, and the planet itself in a series of daring close encounters, and doubtless revealing new mysteries.

SATURN IN HISTORY

1612	The bodies that Galileo observed on either side of Saturn disappear for the first time
1659	Christiaan Huygens publishes his argument that Saturn's appendages are actually a solid ring
1675	Giovanni Cassini discovers a gap between the rings
1852	Astronomers see Saturn through one of its rings, throwing doubt on the idea that the rings are solid
1857	James Clerk Maxwell argues that Saturn's rings are made of billions of solid particles
1970	Water ice is found in the rings
1979	Pioneer 11 passes within 13,000 miles (20,920 km) of Saturn and photographs a new ring
1980–1	Voyager 1 and 2 discover the so-called shepherd moons, a new ring and the complex structure of the rings
2004	Cassini-Huygens probe enters Saturn orbit

RING SAGA

Galileo Galilei was baffled. "What is to be said concerning so strange a metamorphosis?" the Italian astronomer wrote after a night of observing the planet Saturn with his 30-power telescope. To Galileo (1564–1642), Saturn's strange "appendages" looked like secondary bodies on either side of the planet. The problem was that, over a long period of observation, they had gradually disappeared. Could his telescope be deceiving him?

Almost half a century after Galileo's death, Dutch physicist Christiaan Huygens (1629–95) came up with the answer. Huygens was able to discern that the two mysterious appendages more closely resembled rings than moons. In 1659, he published the results of his observations, stating that the appendages were part of a solid ring girdling the planet's equator. Because Saturn's axis is tilted in relation to the Earth's movement around the Sun, Huygens noted, we see the ring from different inclinations—from about 30° to 0°, when the ring seems to disappear completely. The apparent "metamorphosis" was an optical illusion.

Saturn's were the first planetary rings ever discovered. They were completely novel in our view of the solar system. Huygens argued that a ring that rotated—as every body in the solar system seemed to—would be the natural shape that matter would take in response to the planet's gravity. He assumed the ring was solid.

SATURN THROUGH HISTORY

FIND THE GAP
In 1675, Giovanni Cassini was the first to observe a gap in Saturn's rings (a later sketch is shown here). Using his famous 150-foot (45-m) telescopes, Cassini also discovered four of the planet's moons.

IN FOCUS
William Lassell (1799–1880) was a pioneering telescope builder and astronomer. He made this detailed sketch of Saturn in 1852 from the Mediterranean island of Malta using his own 24-inch (60-cm) telescope.

Sixteenth-century astronomers found it difficult to identify the strange "appendages" of Saturn. This was largely because they had no concept of planetary rings, and so were looking at something completely new to human perception.

A TRILLION MOONS

In 1675, Italian astronomer Giovanni Cassini (1625—1712) caught sight of a black line in the middle of the ring. This gap became known as the Cassini division.

How could the rings be solid and stable? The question, unsolved after 200 years, was the Cambridge University Adams Prize essay question in 1857. The winning entry came from young British physicist James Clerk Maxwell (1831–79). Maxwell correctly argued that solid rings would break up over time. The only way the rings could be stable was if they were composed of billions of unconnected particles, each with their own unique orbit around the planet.

RING RIDDLE

IN 1656, CHRISTIAAN HUYGENS (RIGHT, SEATED) PUBLISHED A TRACT THAT INCLUDED THIS SEQUENCE OF LETTERS: AAAAAAA CCCCC D EEEEE G H IIIIIII LLLL MM NNNNNNNNN OOOO PP Q RR S TTTTT UUUUU. THE LIST CONSISTED OF ALL THE LETTERS USED IN THE LATIN SENTENCE ANNULO CINGITUR, TENUI, PLANO, NUSQUAM COHOERENTE, AD ECLIPTICAM INCLINATO ("THE PLANET IS SURROUNDED BY A SLENDER FLAT RING EVERYWHERE DISTINCT FROM ITS SURFACE, AND INCLINED TO THE ECLIPTIC"). ASSUMING THE MESSAGE COULD NOT BE UNSCRAMBLED BY CLEVER CODEBREAKERS, THE SENTENCE WOULD PUBLICLY RECORD HUYGENS' DISCOVERY WHILE STILL KEEPING IT A SECRET UNTIL HE WAS READY TO PUBLISH A COMPLETE ARTICLE ON THE SUBJECT—AS HE DID THREE YEARS LATER.

VOYAGE OF DISCOVERY
The Voyager probes' visits to Saturn were a milestone in our understanding of the planet. Their images (right) showed us a ring system of breathtaking complexity.

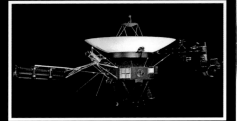

SATURN'S ATMOSPHERE

Saturn's atmosphere is in some ways a low-contrast version of Jupiter's. The two gas giants show marked banding and oval-shape storms, but the colors on Saturn are much more muted. Both planets' atmospheres have a high hydrogen content, but since Saturn's has not been investigated by a probe, astronomers do not know exactly what it contains and at what altitude. The biggest mystery on Saturn, though, is its winds. They are the fastest in the solar system and have a unique circulation pattern.

WHAT IF...

...WE COULD EXPLORE SATURN'S ATMOSPHERE?

On December 7, 1995, a probe from the Galileo spacecraft plunged into Jupiter's atmosphere on a suicide mission. The probe recorded invaluable atmospheric data before contact was lost an hour later. But Saturn's depths remain uncharted, and no such dedicated sortie is currently scheduled, although the Cassini probe has studied the atmosphere from orbit. Maybe in the future it will be possible for astronauts to descend into the atmosphere. What would they see?

Overhead, Saturn's rings arch across the sky as we encounter the first traces of atmosphere, still some 38,000 miles (61,150 km) from the giant planet's center. An enormous rippled sea of yellowish clouds lies below, and as the probe descends through the outermost 125 miles (200 km) of transparent hydrogen, the outside temperature drops steadily from around –200°F (–93°C) to some –320°F (–160°C).

By now the whistling of the atmosphere is evident against the hull: The pressure is about 100 millibars—equivalent to the air pressure on Earth at about 6 miles (10 km) in altitude. This is regarded as altitude zero on Saturn. From now on, the outside temperature starts to climb. After another 70 miles (113 km), the view outside is quite Earth-like as we meet the first cirrus clouds. But these clouds are ammonia, not water, floating in air as dense as the air at sea level on Earth.

Explorers on Saturn may expect to see a pale blue sky created by the outer atmosphere of hydrogen.

The probe falls 50 miles (80 km) deeper, and with the cirrus above, we plunge into a denser cloud deck of thick, yellow, choking ammonia hydrosulfide. For a while the view outside is lost and we fly by instruments only. Then the cloud clears. After this, it remains uncertain what the future explorers would see next. But assuming that Saturn's atmosphere is something like Jupiter's, the next marked transition would occur at a depth of about 160 miles (260 km) below altitude zero, with a layer of Earth-like ice clouds, a temperature just below the freezing point of water, and a pressure of more than five Earth atmospheres.

A few dozen miles after that, the atmosphere outside the probe would begin to blend almost imperceptibly into a sea of liquid hydrogen at a pressure greater than 10 atmospheres. The craft would have cleared the visible atmosphere, just a few hundred miles after entering. From this point downward, Saturn consists of thousands of miles of increasingly dense liquid hydrogen, perhaps with a rocky core at its heart. Any explorers venturing farther would probably not return.

ATMOSPHERIC MAKEUP

SYMBOL	NAME	PERCENTAGE (BY NUMBER OF MOLECULES)
H2	MOLECULAR HYDROGEN	97
HE	HELIUM	3
H20	WATER	NOT YET MEASURED
CH4	METHANE	0.2
NH3	AMMONIA	0.03
H2S	HYDROGEN SULFIDE	NOT YET MEASURED

COLD SOUP

Saturn's atmosphere remains one of the mysteries of the solar system. But by comparing Saturn with Jupiter—whose atmosphere is better understood—and using data collected by Voyagers 1 and 2 and Pioneer 11, astronomers put the puzzle together.

The atmospheres of Jupiter and Saturn seem to be quite similar, each having clouds that encircle the planet in wide belts, parallel to the equator. Part of the reason for the banded structure is that these planets have a very rapid rotation—Saturn spins once every 10 hours 39.5 minutes—that stretches the clouds out. In addition to the banding, the clouds also exhibit an altitude dependence, existing in two or three separate and chemically distinct layers. On both planets the topmost layer is probably a cirrus-type cloud of ammonia (NH_3) ice crystals. On Saturn this may be thicker and relatively lower down than on Jupiter, giving Saturn's underlying cloud decks a washed-out appearance quite unlike the vivid bands of its brightly colored neighbor. Below the ammonia there is probably a layer of frozen ammonium hydrosulfide (NH_4HS) crystals, as there is on Jupiter, or perhaps a combination of water and ammonia.

Like Jupiter with its famous Great Red Spot, Saturn's atmosphere also plays host to some vast storm systems. The more frequent and smaller of these hurricane-like storms are thousands of miles in diameter, and vary in duration from days to years. They tend to form at the boundary between two bands of clouds.

ATMOSPHERIC MYSTERIES

Saturn (below) as seen by the Hubble Space Telescope in October, 1990. The great white storm, which recurs about every 30 years, is clearly visible. The bright clouds, visible from Earth in amateur telescopes, are thought to consist of ammonia ice crystals.

MAGNETIC LIGHT
The Hubble Space Telescope captured this image (above) of Saturn's ultraviolet auroral rings in 1998. The planet has been revealed to have an enormous and powerful magnetic field.

TOO WINDY FOR STORMS

Some of Saturn's atmospheric phenomena are on an altogether vaster scale. At intervals of about 30 years, the planet suffers an outbreak of bright clouds just above the equator. The last outbreak began in September 1990. It rapidly spread around a third of the planet—some 75,000 miles (120,700 km). These events seem to coincide with summer in Saturn's northern hemisphere—but no one knows why.

Such storms are eventually ripped apart by the terrific winds that tear around Saturn. Astronomers believe that the winds are partly a byproduct of the planet's convection system. The theory is that Saturn's interior cools enough for helium to condense out of the upper atmosphere and rain into the interior. As this rain falls, its gravitational energy is converted into heat, setting up giant convection currents that warm the outer portions of the planet in rising cells. These rising currents, whipped up by the planet's rapid rotation, contribute to the vicious winds.

SPIN CYCLE

SATURN OFTEN SHOWS GREAT OUTBREAKS OF WHITE STORM CLOUDS GENERATED WHEN AMMONIA ICE CRYSTALS CONDENSE IN UPDRAFTS OF WARMER AIR. ONE SUCH OUTBREAK IN THE 1790s ENABLED GERMAN-BORN ENGLISH ASTRONOMER SIR WILLIAM HERSCHEL (1738–1822; RIGHT) TO MAKE THE FIRST ESTIMATE OF SATURN'S ROTATION PERIOD. BUT BECAUSE SATURN'S CLOUDS ROTATE FASTER THAN THE PLANET ITSELF, HERSCHEL'S ESTIMATE WAS SLIGHTLY SHORT.

SPOT GAP
A brown spot was photographed by Voyager 2 in 1981. The feature is thought to have been caused by a downdraft creating a gap in the upper clouds and revealing the darker clouds below.

SATURN'S RINGS

Saturn's system of rings is one of the most beautiful sights in the solar system. The planet itself can be seen with the naked eye and the rings viewed through a small telescope. But to see the rings at their spectacular best, we on Earth must turn to the images returned by the Hubble Space Telescope and the two Voyager space probes. These show that instead of just the two big rings visible through a small telescope, there are thousands of narrow ringlets whose paths are guided by the gravity of nearby moons.

A spacecraft could fly safely above or below Saturn's rings, but it would almost certainly be destroyed if it tried to fly through them.

...SATURN LOST ITS RINGS?

Saturn's rings might, on an astronomical timescale at least, be a temporary phenomenon, and we happen to be lucky enough to be around during the few hundred million years when they exist in all their glory. Some calculations show that the big A Ring, for instance, should only last another 100 million years before it collapses and gravity drags it into the planet.

But some astronomers doubt that this will happen and believe that the rings are permanent. They suspect that there are processes at work that

...WE COULD FLY THROUGH SATURN'S RINGS?

The chances are that a spacecraft flying through Saturn's main rings would have a very bumpy ride. In fact, it would almost certainly be destroyed by the impact of ring material. There is a great deal of ice in the rings, most of it small particles but much of it up to snowball size, and you might think that this frozen water would be fairly harmless. But the impact of a snowball moving at several tens of thousands of miles an hour—the speed that a spacecraft would be traveling relative to the ring— would have a more devastating effect than a high-velocity artillery shell.

keep the rings stable, such as the interaction of the "shepherding" moons' gravitational fields. These moons help each other to maintain their own orbits, and also act as a locking mechanism to keep the rings intact.

Saturn would be a far less interesting planet without its rings, and a much less impressive sight. The tiny, insignificant-seeming moons that "shepherd" the rings might be the key to their survival.

RING SIZE

DISTANCE OF INNERMOST (D) RING FROM SATURN	4,080 MILES (6,570 KM)
DISTANCE OF OUTERMOST (E) RING FROM SATURN	74,000 MILES (120,000 KM) (APPROX)
TOTAL SPAN OF RINGS	170,000 MILES (273,590 KM)
THICKNESS OF MAIN RINGS	100–3,000 FT (30–915 M)
NUMBER OF RINGS	7
NUMBER OF RINGLETS	10,000
WIDTH OF NARROWEST RINGLETS	TENS OF YD
TYPICAL SIZE RANGE OF RING PARTICLES	0.5 INCH TO 20 FT (1.3 CM TO 6 M) ACROSS
ESTIMATED DISTANCE BETWEEN LARGER RING PARTICLES	30 FT (9 M)

SATURN'S GLORY

When the Italian astronomer Galileo Galilei studied Saturn in 1610, he saw what he thought were two large moons, one on each side of the planet. But in 1655, the Dutch astronomer Christiaan Huygens concluded that the "moons" Galileo had seen were actually a single, encircling ring structure. Huygens thought that this ring was solid and quite thick.

The first indication that Saturn had more than one ring came in 1676 when the Franco-Italian astronomer Giovanni Cassini observed two rings separated by a gap, rather than a single, solid ring. This gap, between what are now called the A and B rings, is named the Cassini Division. Another gap—in the outer part of the A Ring and now called the Encke Division—was discovered in 1837 by the German astronomer Johann Encke. Although he was the first to notice the gap, he thought it was a dark band.

RING STRUCTURE

Eventually, five rings—the A, B, C, D and E rings—were detected using Earth-based telescopes. A further ring, the F Ring, was discovered by the Pioneer 11 probe in 1979, and yet another, the G Ring, by the Voyager probes in 1980 and 1981.

Astronomers now believe that Saturn's rings are made up of belts of countless particles. Most of these particles range in size from grains of dust to snowflakes, but include some larger chunks of material a few yards across. It is likely that many of the particles are ice, and studies by the Cassini spacecraft suggest they have an almost fluffy consistency.

Close-up pictures have revealed that Saturn's seven main rings are made up of thousands of thin ringlets. The "gaps" between the ringlets still contain particles, but because there are less of them, they reflect less light and so appear darker.

This profusion of ringlets and other mysterious structures suggests that the process that causes the rings to keep their shape is extremely complex. Astronomers believe that it not only involves interactions between the ring particles themselves, but also the gravitational effects of Saturn's moons, some of which "shepherd" the rings to keep them in position.

How Saturn came to possess rings is also a mystery. Some experts believe that they are the remnants of a moon (or moons) that was unable to form because of the powerful gravitational field of Saturn itself. Another suggestion is that the rings are the result of some catastrophic impact, perhaps by an asteroid or a large comet that smashed into a moon orbiting Saturn and left broad bands of debris circling the planet. Yet another theory proposes that the rings are the icy remains of a giant comet that passed close enough to Saturn to be torn apart by the planet's gravity. We may never know for sure.

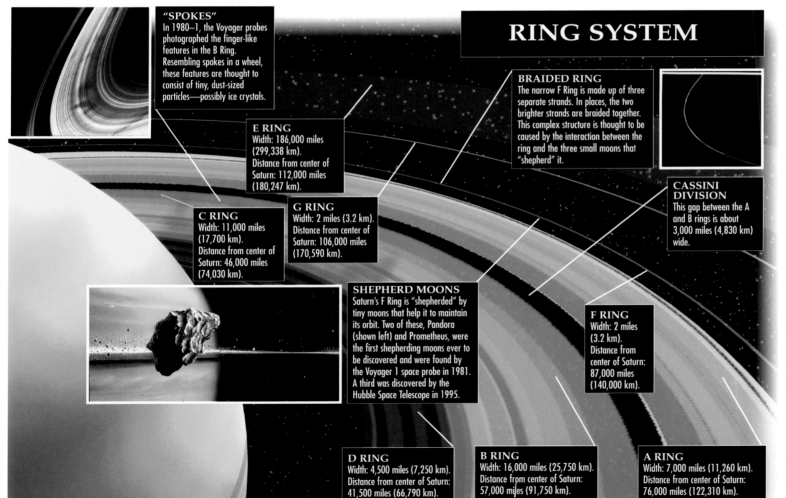

RING SYSTEM

"SPOKES"
In 1980–1, the Voyager probes photographed the finger-like features in the B Ring. Resembling spokes in a wheel, these features are thought to consist of tiny, dust-sized particles—possibly ice crystals.

E RING
Width: 186,000 miles (299,338 km). Distance from center of Saturn: 112,000 miles (180,247 km).

C RING
Width: 11,000 miles (17,700 km). Distance from center of Saturn: 46,000 miles (74,030 km).

G RING
Width: 2 miles (3.2 km). Distance from center of Saturn: 106,000 miles (170,590 km).

BRAIDED RING
The narrow F Ring is made up of three separate strands. In places, the two brighter strands are braided together. This complex structure is thought to be caused by the interaction between the ring and the three small moons that "shepherd" it.

CASSINI DIVISION
This gap between the A and B rings is about 3,000 miles (4,830 km) wide.

SHEPHERD MOONS
Saturn's F Ring is "shepherded" by tiny moons that help it to maintain its orbit. Two of these, Pandora (shown left) and Prometheus, were the first shepherding moons ever to be discovered and were found by the Voyager 1 space probe in 1981. A third was discovered by the Hubble Space Telescope in 1995.

F RING
Width: 2 miles (3.2 km). Distance from center of Saturn: 87,000 miles (140,000 km).

D RING
Width: 4,500 miles (7,250 km). Distance from center of Saturn: 41,500 miles (66,790 km).

B RING
Width: 16,000 miles (25,750 km). Distance from center of Saturn: 57,000 miles (91,750 km).

A RING
Width: 7,000 miles (11,260 km). Distance from center of Saturn: 76,000 miles (122,310 km).

RING SHADOW

THE RINGS OF SATURN DO NOT EMIT LIGHT OF THEIR OWN, THEY MERELY REFLECT LIGHT FROM THE SUN. FOR THIS REASON, WHEN THEY ARE BACKLIT BY THE SUN, THEY APPEAR DARK. THE RINGS CONTAIN ENOUGH MATERIAL TO CAST A CLEARLY VISIBLE SHADOW ON SATURN. IF IT WERE POSSIBLE FOR YOU TO STAND ON SATURN, YOU WOULD SEE TWO TYPES OF SUNLIGHT: UNOBSTRUCTED SUNLIGHT WHEN THE RINGS WERE NOT IN THE WAY, AND DIFFUSE, RING-FILTERED LIGHT WHEN THEY WERE.

SATURN'S MOONS

The moons of Saturn are the most complex and fascinating group of satellites in the solar system. Dominated by mighty Titan, the family also includes the "classical" moons Mimas, Enceladus, Tethys, Dione, Rhea and Iapetus, plus at least 40 smaller bodies. Pictures returned by the Voyager missions in the early 1980s showed Saturn's moons to be cold, icy places that bear the scars of countless ancient impacts. But that's not to say that they are dull worlds: In fact, most of them have yet to reveal their inner secrets.

WHAT IF...

...SATURN CAPTURED ANOTHER MOON?

As the second-largest planet in the solar system, Saturn exerts a gravitational pull that is easily capable of capturing stray objects with unstable orbits and imprisoning them in its own satellite system. In fact, it may have done so already: There is evidence that the gas giant's most distant moon, Phoebe, was once an asteroid that became entrapped by Saturn's steely grip.

Photographs taken by the Cassini probe as it flew into the Saturn system appear to confirm its asteroid-like appearance. In fact, the dark little moon may resemble the Centaurs—a group of asteroid-like bodies whose irregular orbits are confined mainly to the region between Saturn and Neptune. Phoebe also orbits in the opposite direction of Saturn's other moons, which again suggests that it arrived from elsewhere.

Astronomers think that there was a window of about 2 million years during the formation of the solar system when conditions around Saturn were ideal for capturing moons. Prior to this, while the haze of gas in the planet's outer layers was still consolidating, any passing bodies would quickly have been plunged into orbital decay by the resulting atmospheric drag. After the window closed, the relatively stable conditions that by now existed in the solar system made satellite capture far less likely—but not impossible.

The largest of the Centaurs, an icy body known as Chiron (not be confused with Pluto's moon Charon), passes very close to Saturn during the course of its unpredictable orbit. Each time it does so, the gas giant's gravitational pull causes Chiron to make a course correction that propels it on a slightly different orbital path. Astronomers think that it is unlikely that the Centaur will be "ambushed" during its next pass of Saturn, though they cannot tell for sure. Another possible scenario is that Saturn's gravity will nudge Chiron into the inner solar system, causing its thick, icy outer layers to burn off in what could well be the most spectacular comet ever seen.

There are other Centaurs, of course, plus hosts of other icy and rocky bodies farther out in the Kuiper Belt, which lies beyond Neptune. If some unforeseen or little- understood cosmological event nudged one of these pieces of debris into a new orbit, it could easily find itself trapped within the Saturnian system. What would happen to the object then is less easy to imagine. With so many moons in the family, it might find itself with a serious fight for space on its hands.

SATURN'S LARGER MOONS

Name	Diameter (miles/km)	Discovered by	Distance from Saturn (miles/km)	Name	Diameter (miles/km)	Discovered by	Distance from Saturn (miles/km)
Pan	12/19	Voyager 1, 1980	83,033/133,629	Telesto	15/24	Voyager 1, 1980	183,132/294,722
Atlas	21/32	Voyager 1, 1980	85,544/137,670	Calypso	16/26	Voyager 1, 1980	183,132/294,722
Prometheus	68/109	Voyager 1, 1980	86,607/139,380	Dione	695/1,118	J.-D. Cassini, 1684	234,516/377,400
Pandora	55/89	Voyager 1, 1980	88,067/128,855	Helene	20/32	Voyager 1, 1980	234,555/377,480
Epimetheus	75/120	Voyager 1, 1980	94,109/151,454	Rhea	951/1,530	J.-D. Cassini, 1672	327,503/527,065
Janus	118/190	Voyager 1, 1980	94,140/151,504	Titan	3,201/5,151	C. Huygens, 1655	759,385/1,222,112
Mimas	249/400	W. Herschel, 1789	115,282/185,528	Hyperion	174/280	Voyager 1, 1980	920,447/1,481,316
Enceladus	320/515	W. Herschel, 1789	147,906/238,032	Iapetus	910/1,465	J.-D. Cassini, 1671	2,212,992/3,561,353
Tethys	660/1,062	J.-D. Cassini, 1684	183,102/294,674	Phoebe	137/220	Voyager 1, 1980	8,049,720/12,954,769

MOONS OF ICE

Aside from Titan, one of the largest moons in the solar system, Saturn's most interesting satellites are its "classical" moons—those discovered by the astronomers Jean-Dominique Cassini and Sir William Herschel in the 17th and 18th centuries. These six moons form three pairs each of roughly equal size, and all of them bear the scars of ancient impacts that hint at the Saturn family's violent past. Consisting mostly of water and other ices, they are dark, frozen worlds where surface temperatures plummet below –300°F (–149°C).

Yet in spite of their uninviting and lifeless appearance, there are signs that some of these classical moons were once geologically active—and perhaps still are today. Pictures returned by the Voyager probes reveal evidence of fault lines that appear to have been plugged by fluid material from the satellites' interiors. Some of the moons also exhibit lava-like plains, similar to the mare on our own Moon, that appear to have flowed across the surface and covered up the more ancient impact craters.

The puzzle is how bodies that consist mostly of ice could display such apparently volcanic characteristics. In their search for an answer, astronomers have looked to the process that stops a car's radiator from freezing in winter. Just as antifreeze keeps water flowing at well below its normal freezing point, so the large amounts of ammonia thought to be locked up in Saturn's moons may have caused their ices to form a kind of frozen slurry.

VOLCANIC SLUSH

At the incredibly low temperatures that exist on Saturn's moons, an icy, ammonia-based slurry might behave very much like basalt lava. Requiring only minimal heat to get it moving, it could well up through cracks in the crust or creep across the surface like a giant glacier.

But where would the heat come from? Saturn's moons—like our own Moon—must have long since cooled down from the superhot maelstrom that formed them. On Earth,

volcanism is driven by the heat of radioactive decay, which helps to keep the interior molten. But we have yet to find evidence of such decay among Saturn's moons—indeed, as far as the less dense bodies are concerned, it is questionable whether there is any rocky material under the ice at all.

A more likely explanation is that on some moons, it was the heat generated by massive impacts that triggered the process now popularly known as icy volcanism. Another possibility, especially on the inner moons, is that they were wrenched apart and heated by the tidal effects of Saturn's massive gravity, like Jupiter's moon Io. The latest findings from the Cassini probe seem to confirm this, showing that tiny Enceladus has a remarkably fresh surface.

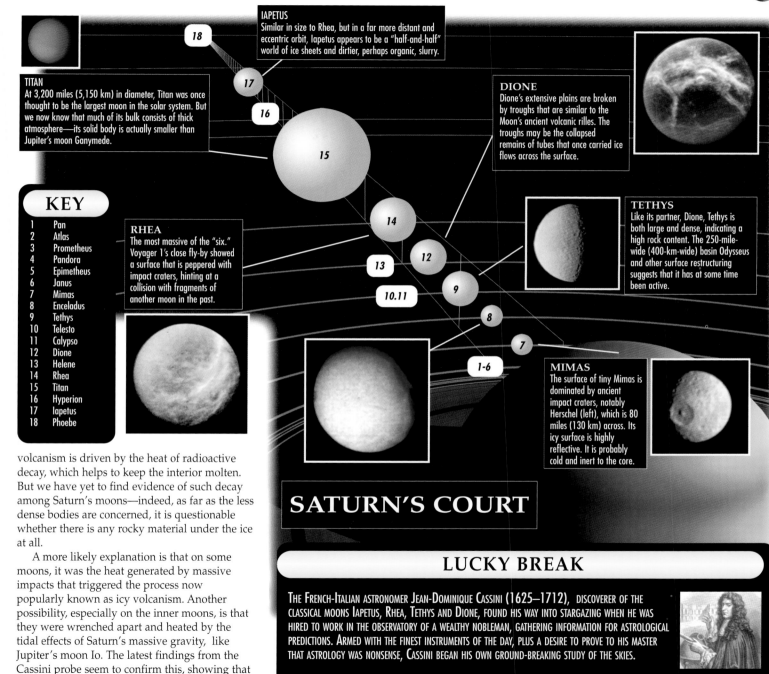

IAPETUS
Similar in size to Rhea, but in a far more distant and eccentric orbit, Iapetus appears to be a "half-and-half" world of ice sheets and dirtier, perhaps organic, slurry.

TITAN
At 3,200 miles (5,150 km) in diameter, Titan was once thought to be the largest moon in the solar system. But we now know that much of its bulk consists of thick atmosphere—its solid body is actually smaller than Jupiter's moon Ganymede.

DIONE
Dione's extensive plains are broken by troughs that are similar to the Moon's ancient volcanic rilles. The troughs may be the collapsed remains of tubes that once carried ice flows across the surface.

TETHYS
Like its partner, Dione, Tethys is both large and dense, indicating a high rock content. The 250-mile-wide (400-km-wide) basin Odysseus and other surface restructuring suggests that it has at some time been active.

KEY

1	Pan
2	Atlas
3	Prometheus
4	Pandora
5	Epimetheus
6	Janus
7	Mimas
8	Enceladus
9	Tethys
10	Telesto
11	Calypso
12	Dione
13	Helene
14	Rhea
15	Titan
16	Hyperion
17	Iapetus
18	Phoebe

RHEA
The most massive of the "six." Voyager 1's close fly-by showed a surface that is peppered with impact craters, hinting at a collision with fragments of another moon in the past.

MIMAS
The surface of tiny Mimas is dominated by ancient impact craters, notably Herschel (left), which is 80 miles (130 km) across. Its icy surface is highly reflective. It is probably cold and inert to the core.

SATURN'S COURT

LUCKY BREAK

THE FRENCH-ITALIAN ASTRONOMER JEAN-DOMINIQUE CASSINI (1625–1712), DISCOVERER OF THE CLASSICAL MOONS IAPETUS, RHEA, TETHYS AND DIONE, FOUND HIS WAY INTO STARGAZING WHEN HE WAS HIRED TO WORK IN THE OBSERVATORY OF A WEALTHY NOBLEMAN, GATHERING INFORMATION FOR ASTROLOGICAL PREDICTIONS. ARMED WITH THE FINEST INSTRUMENTS OF THE DAY, PLUS A DESIRE TO PROVE TO HIS MASTER THAT ASTROLOGY WAS NONSENSE, CASSINI BEGAN HIS OWN GROUND-BREAKING STUDY OF THE SKIES.

TITAN

Titan is the largest moon of Saturn and the second largest in the solar system. Bigger than either Mercury or Pluto, it is a world in its own right. Titan is shrouded in a haze of orange clouds that hang in the unusually dense nitrogen atmosphere, and complex, carbon-based molecules rain down on its icy surface. It may also have rivers and lakes of liquid methane or ethane, and although it is bitterly cold, there is a chance that some form of life may once have managed to evolve there.

WHAT IF...

...THERE WAS LIFE ON TITAN?

Strange though it may seem, the frozen moon Titan is thought to be one of the best hopes for finding traces of life elsewhere in the solar system. Its dense atmosphere resembles that of the young Earth in some ways, and it contains many kinds of the organic (carbon-based) molecules upon which life is based.

Astronomer Carl Sagan, with his colleague Bishun Khare, simulated the atmosphere of Titan in a laboratory by filling a large glass container with a mixture of nitrogen and methane. When they exposed this mixture to radiation similar to cosmic rays and the radiation from the Sun, a brownish deposit formed.

They called this deposit "tholin," from the Greek word for "muddy." The tholin resembled the orange

As the Sun nears the end of its life, in a few billion years' time, it will expand and give off more light and heat. This could warm Titan enough for it to become more Earth-like, and life could emerge from the rich mixture of complex organic molecules covering the surface.

If Titan should become warmer in the far distant future, its thick clouds and the orange haze that hangs below them might clear. This would allow even more of the Sun's light and heat to reach its surface, encouraging the development of any life that evolves there.

haze seen in Titan's atmosphere by the Voyager spacecraft missions. Sagan calculated that if tholin had been forming in Titan's atmosphere, its surface could now be hundreds of feet deep in organic slush.

When Sagan and Khare dissolved tholin in water, it produced a rich mixture of amino acids and many other kinds of complex molecules needed for life. Could life have emerged on Titan from a similar cocktail of chemicals? It seems unlikely—the extremely low temperatures appear to rule out the presence of liquid water and the kind of evolution that has happened on Earth.

However, Titan could have been warmed in the past by volcanic activity or by the impacts of large meteorites, which would have heated and melted the icy surface. As Sagan said, "Could life have started on Titan during the centuries following an impact, when lakes of water or water-slurries briefly formed?"

TITAN STATISTICS

ORBITAL PERIOD	15.94 DAYS
ROTATION PERIOD	15.94 DAYS
DISTANCE FROM SATURN	759,000 MILES (1.2 MILLION KM)
MASS	0.0259 EARTH MASSES
DIAMETER	3,200 MILES (5,150 KM)
MEAN SURFACE TEMPERATURE	−290°F (−143°C)
ATMOSPHERE	97% NITROGEN, 3% METHANE
SURFACE PRESSURE	1.5 EARTH ATMOSPHERES
ESCAPE VELOCITY	1.65 MILES/SEC (2.6 KM/SEC)
(ALL FIGURES ARE APPROXIMATE)	

DEEP FREEZE

One of the largest moons in the solar system, Titan is similar in size to the four large satellites of Jupiter. But while none of these has more than a trace of atmosphere, Titan's is very thick—more than 50% denser than that of the Earth. Titan can only hold on to such an atmosphere because it lies in the cold outer reaches of the solar system.

Voyager 1 revealed that Titan's atmosphere is largely nitrogen—perhaps tapped in ice as the moon formed, and later released. There is also a small proportion of methane, which gives the atmosphere its distinctive orange hue, and Voyager found traces of a dozen or so other organic (carbon-based compounds). The surface, however, was obscured by clouds, and was to remain a mystery for more than two decades.

But astronomers have speculated that the giant satellite probably consists of a mixture of ice and rock, not dissimilar to its neighbors in the Saturnian system. The surface, they suggested, might be covered in slicks of organic chemicals, and there might even be methane rainfall.

HUYGENS ENCOUNTER

Titan's intriguing features made it a prime target for the Cassini Saturn probe. The spacecraft was equipped with imaging cameras capable of seeing through Titan's perpetual methane smog, and carried the European-built Huygens lander, which plunged into Titan's atmosphere early in 2005 and sent back the first detailed images of the surface.

As Huygens descended on its parachutes, high-altitude photographs revealed a startlingly familiar vista—what appeared to be a coastline with river channels, deltas and other erosion features leading down to a darker area— perhaps a lake or sea. Later studies from the Cassini spacecraft itself suggested that the area where Huygens landed is dry at present, but the features were clearly recent, and it seems likely

that methane plays a very similar role on Titan to that played by water on Earth. The latest studies suggest Titan sits at a "sweet spot" in the solar system where methane can exist in solid, liquid and vapour forms—just as water can on Earth.

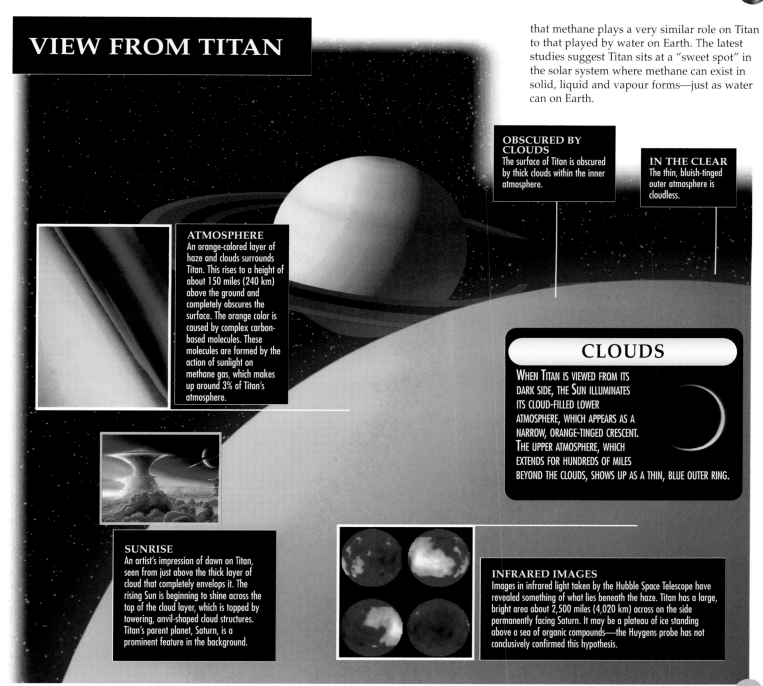

VIEW FROM TITAN

OBSCURED BY CLOUDS
The surface of Titan is obscured by thick clouds within the inner atmosphere.

IN THE CLEAR
The thin, bluish-tinged outer atmosphere is cloudless.

ATMOSPHERE
An orange-colored layer of haze and clouds surrounds Titan. This rises to a height of about 150 miles (240 km) above the ground and completely obscures the surface. The orange color is caused by complex carbon-based molecules. These molecules are formed by the action of sunlight on methane gas, which makes up around 3% of Titan's atmosphere.

CLOUDS

When Titan is viewed from its dark side, the Sun illuminates its cloud-filled lower atmosphere, which appears as a narrow, orange-tinged crescent. The upper atmosphere, which extends for hundreds of miles beyond the clouds, shows up as a thin, blue outer ring.

SUNRISE
An artist's impression of dawn on Titan, seen from just above the thick layer of cloud that completely envelops it. The rising Sun is beginning to shine across the top of the cloud layer, which is topped by towering, anvil-shaped cloud structures. Titan's parent planet, Saturn, is a prominent feature in the background.

INFRARED IMAGES
Images in infrared light taken by the Hubble Space Telescope have revealed something of what lies beneath the haze. Titan has a large, bright area about 2,500 miles (4,020 km) across on the side permanently facing Saturn. It may be a plateau of ice standing above a sea of organic compounds—the Huygens probe has not conclusively confirmed this hypothesis.

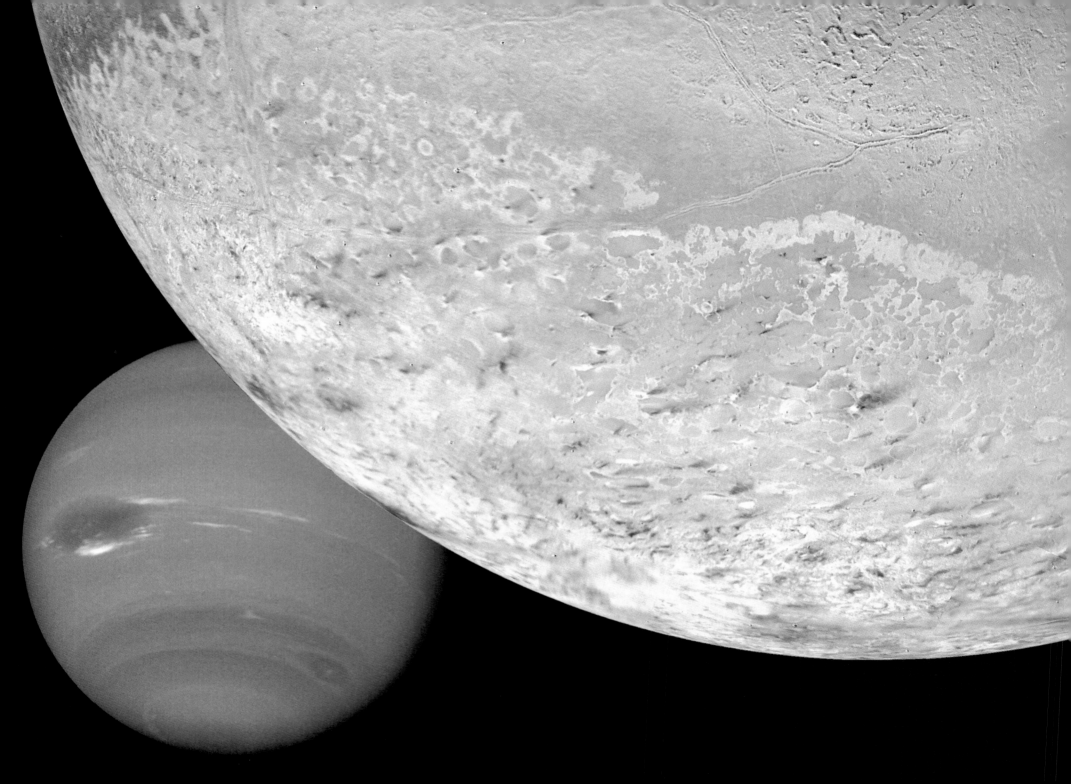

URANUS AND NEPTUNE

The Solar System's two outer giants are near-identical twins, but with some major differences in their orbits. Uranus lies twice as far out as Saturn, and Neptune farther still. Both planets are sometimes called "ice giants"—considerably smaller than Jupiter and Saturn, they consist of a mix of chemical compounds such as methane, and their interiors are thought to take the form of slushy "ices" of these various compounds. Their composition also gives them their distinctive blue-green colour (though Neptune is significantly bluer). Aside from these similarities, the planets are very different from one another. Uranus orbits the Sun "on its side," with its axis tilted so far from the vertical that parts of the planet experience days and nights lasting more than 40 years. Perhaps because of this, Uranus shows far less activity than Neptune, despite receiving far more heat from the Sun. Both planets have ring systems and large families of satellites, the largest of which is Neptune's giant icy moon Triton.

This computer-generated montage shows Neptune as it would appear from a spacecraft approaching Triton, Neptune's largest moon. Triton's surface is mostly covered by nitrogen frost mixed with traces of condensed methane, carbon dioxide, and carbon monoxide.

URANUS

T he giant blue-green globe of Uranus is big enough to swallow the Earth 64 times over. It has a strong magnetic field, a family of at least 27 moons and, after Saturn, the most impressive system of rings in the solar system. Beneath its almost featureless exterior, Uranus is essentially a fluid world. An atmosphere rich in hydrogen thickens imperceptibly, with no clear break between gas and liquid, into a world-wide ocean where water and methane predominate.

WHAT IF...

...WE COULD VISIT THE CENTER OF URANUS?

T o visit Uranus and travel to its core, we would have to build a spacecraft that could journey safely through the planet's atmosphere—and then dive deep into its immense and hostile ocean.

As the ship dropped toward the planet, it would first pass through mists of hydrogen, helium and methane in the upper atmosphere. All the while, the temperature would be falling and by the time the ship reached the highest clouds—composed of methane—it would reach –365°F (–220°C). Beneath them, the temperature would begin to increase with depth and growing pressure; the ship would cross two more cloud layers, one of ammonia and the other of water vapor.

Deeper still, the atmosphere below the lowest cloud layer would steadily become denser; the mists of water vapor, hydrogen and methane would become increasingly opaque. Soon, the ship would be traveling through liquid—an ocean of water, liquid hydrogen and liquid methane. Transition

Traveling down toward the center of Uranus, a spacecraft would first pass through the blue-green atmosphere and its cloud layers, then through the immense ocean and finally reach the rocky core.

from atmosphere to sea would be gradual, but about a third of the way to the planet's center, the visiting spacecraft would have to become a submarine.

The pressure and temperature of the surrounding ocean would increase as the ship neared Uranus' rocky core, which is similar in size to the whole planet Earth. There, at the very center, the ship could experience temperatures as high as 13,000°F (7,204°C), with more than a ton of material crushed into every cubic foot (0.03 cubic m).

URANUS PROFILE

URANUS		EARTH
31,673 MILES (50,973 KM)	DIAMETER	7,973 MILES (12,831 KM)
97° 55'	AXIS TILT	23° 27'
30,685 DAYS	LENGTH OF YEAR	365 DAYS
17 HOURS 14 MINUTES	LENGTH OF DAY	24 HOURS
1.78 BILLION MILES (2.86 BILLION KM)	DISTANCE FROM SUN	93.5 MILLION MILES (150 MILLION KM)
–365°F (–185°C) (CLOUD TOPS)	SURFACE TEMPERATURE	59°F (15°C)
0.92 G	SURFACE GRAVITY	1 G
HYDROGEN (82%), HELIUM (15%), METHANE (2%)	ATMOSPHERE	NITROGEN (80%), OXYGEN (19%)
0.1 MILLIBAR (CLOUD TOPS)	ATMOSPHERIC PRESSURE	1,000 MILLIBARS
WATER, AMMONIA, METHANE, MOLTEN ROCK, HYDROGEN, HELIUM	COMPOSITION	SILICON (60%), ALUMINUM (15%)
27	NUMBER OF MOONS	1

GREEN GIANT

Uranus is a giant planet—four times the diameter of the Earth—but it is so far away that it is all but invisible to the naked eye. So although the planets in the solar system out to Saturn have been known since antiquity, Uranus was not discovered until 1781. Its discovery doubled the dimensions of the known solar system, but astronomers knew little of this huge world until the Voyager 2 probe passed within 51,000 miles (82,077 km) of it on January 24, 1986—almost two centuries after its discovery.

The images sent back by Voyager were somewhat surprising. No surface details of Uranus can be seen from Earth, but even close up, Voyager could see only a featureless blue-green globe. The planet has none of the strikingly colored atmospheric bands typical of Saturn and Jupiter, even though, like these two worlds, its atmosphere is composed almost entirely of hydrogen and helium.

Voyager 2 did show that the atmosphere on Uranus has a band-like structure, but it can only be seen clearly in high-contrast false-color images. Uranus is so cold that its clouds form very low down in its atmosphere, their colors hidden from view by the blue-green atmosphere above them.

LIQUID GAS

The interior of Uranus is also unlike those of Saturn and Jupiter, which consist mainly of hydrogen. Its density is higher, a sign that heavier gases and liquids are present. Although Uranus is often called a gas giant, its actual composition is almost certainly mainly liquid—hydrogen makes up only 15% of the planet's mass, compared with more than 80% of Jupiter's, and almost all of it is in the atmosphere. Uranus probably has a rocky core surrounded by an ocean of water, liquid methane and liquid hydrogen. The ocean—hot, under the prevailing pressure—gradually blends into the atmosphere. Like all the other gas giants, Uranus has no distinct surface to separate "air" from "ground."

The rings of Uranus were first seen in 1977 when the planet passed in front of a background star. The star appeared to flicker as it went behind the planet, a sign that something was blocking its light. Earth-bound telescopes detected a total of nine rings; the Voyager 2 mission found two more. They are all very narrow and dark, typically as black as coal; and like those of Saturn, the material they are made of ranges in size from dust specks to rocky lumps up to 30 feet (9 m) across. The innermost ring, designated 1986U2R by astronomers, is about 24,000 miles (38,620 km) from the center of Uranus. The outermost and brightest, the Epsilon Ring, is around 6,000 miles (9,650 km) farther out.

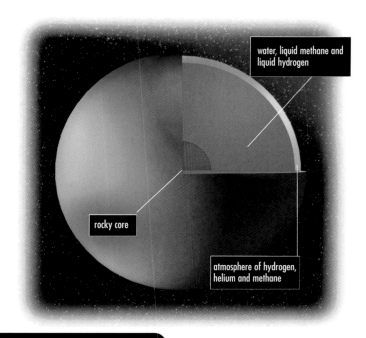

water, liquid methane and liquid hydrogen

rocky core

atmosphere of hydrogen, helium and methane

DISCOVERY

URANUS WAS DISCOVERED IN 1781 BY SIR WILLIAM HERSCHEL, A BRITISH MUSICIAN OF GERMAN ORIGIN. HE NAMED THE NEW FIND GEORGIUM SIDUS ("GEORGE'S STAR") FOR HIS PATRON, KING GEORGE III, BUT THE NAME WAS WIDELY CHALLENGED. GERMAN ASTRONOMER JOHANN BODE SUGGESTED THAT THE NEW PLANET BE CALLED URANUS, AND THIS NAME WAS OFFICIALLY ADOPTED IN 1850.

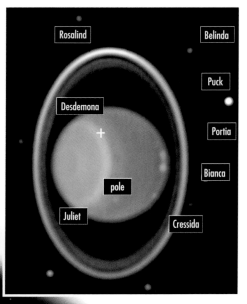

Rosalind

Belinda

Puck

Desdemona

Portia

Bianca

pole

Juliet

Cressida

RINGS
Uranus has 11 rings, nine of which were detected from Earth. The other two were discovered in 1986 by the Voyager 2 probe. This probe also found 10 new moons, two of which act as "shepherd moons" and keep the outer ring, Epsilon, in position.

MOONS
Until the Voyager 2 encounter with Uranus, only five moons of Uranus were known; two were found in 1787 (Titania and Oberon), two in 1850 (Ariel and Umbriel), and the last as recently as 1948 (Miranda). Most are named after characters from the works of William Shakespeare. Voyager 2 found another 10 (of which eight are visible in this false-color picture), and in 1997, Earth-based telescopes found 12 more, bringing the total up to 27.

URANUS: ITS RINGS AND MOONS

URANUS' ATMOSPHERE

A space traveler hurtling by Uranus would see nothing but an inscrutable blue-green billiard ball: Without any surface markings, the planet wouldn't even appear to be rotating. But viewed through color filters at close range, or through the Hubble Space Telescope, atmospheric features such as racing wind bands and lethal clouds of methane gas begin to show themselves. Voyager 2's fly-by revealed that Uranus has certain similarities with its previous targets, Jupiter and Saturn—but also some important differences.

WHAT IF...

...WE COULD FIND OUT MORE ABOUT URANUS' ATMOSPHERE?

When Voyager 2 embarked on its mission of discovery, one of NASA's key aims was to find out exactly how much hydrogen and helium exists in gas giants' atmospheres. The hope was to find a surviving sample of a primordial solar nebula—the cosmic "soup" from which both the Sun and the planets formed—and so learn more about the earliest origins of the solar system.

On Jupiter and Saturn, Voyager found too much hydrogen: In the pressure-cooker conditions of the two planets' interiors, the helium separates out and falls like rain into the core. But when Voyager arrived at Uranus, it found the ratio of hydrogen to helium to be just about what scientists had expected. Because of Uranus' smaller size, atmospheric pressures and temperatures are lower than those on Jupiter and Saturn. Uranus may therefore be an example of primordial material that has been deep-frozen in time—perhaps the best, apart from comets, that we are likely to find.

To analyze this primordial material, and to solve the many mysteries of the Uranian atmosphere, further probes will have to be sent. Voyager 2 is the only spacecraft so far to have passed close to Uranus, and it was not even optimized to study the planet—

its primary goals were Jupiter and Saturn. Neither was Voyager able to see beyond the impenetrable layers of methane in the upper atmosphere. But in 1995, the Galileo probe completed a successful pioneering mission by diving headlong into the hostile atmosphere of Jupiter. This achievement was technically as difficult as any attempt to penetrate a gas-giant planet is ever likely to be—Jupiter is enormous, with extreme atmospheric pressures and an inferno-like interior. Now we know that it would be possible to send a similar probe to Uranus. It is merely a question of having the will and the finances to do it.

In the long term, at least, the chances are good. NASA has various projects on the drawing board to send low-cost but scientifically invaluable unmanned probes into deep space. Encouraged by the success of its "Discovery Program," which has already dispatched missions to the Moon, to Mars, and to rendezvous with comets and asteroids, the space agency has yet more ambitious plans for the future.

Then there is the Pluto-Kuiper Express, a budget-conscious mission that will journey to the edge of the inner solar system to gaze on the only planet yet to be visited by a spacecraft from Earth. Unless the manned space program completely drains NASA of its budget, we may yet get a chance to unravel the mysteries of our distant blue-green neighbor.

URANIAN ATMOSPHERE FACTS

THICKNESS	.4,500 MILES (7,242 KM)
COMPOSITION	.HYDROGEN (83%) HELIUM (15%) METHANE AND OTHERS (2%)
WIND SPEEDS	.UP TO 400 MPH (644 KM/H)
MEAN SURFACE TEMPERATURE	.AROUND −320°F (−160°C)
MEAN SURFACE PRESSURE	.14.5 POUNDS (6.6 KG) PER SQUARE INCH (6.5 SQUARE CM)
PRESSURE AT AMMONIA CLOUD LAYER	.116 POUNDS (32 KG) PER SQUARE INCH (6.5 SQUARE CM)
PRESSURE AT WATER CLOUD LAYER	.1,740 POUNDS (790 KG) PER SQUARE INCH (6.5 SQUARE CM)
TEMPERATURE AT AMMONIA CLOUD LAYER	.AROUND −200°F (−93°C)
TEMPERATURE AT WATER CLOUD LAYER	.AROUND 35°F (1.6°C)
AVERAGE PLANETARY DENSITY	.0.047 POUNDS PER CUBIC INCH (0.0021 KG PER 6.5 CUBIC CM) (1.3 TIMES AS DENSE AS WATER)

COSMIC ICEBOX

Despite being classed as a "gas giant," Uranus bears little resemblance to those mighty gas-balls Jupiter and Saturn. Although the Uranian atmosphere is mainly hydrogen and helium—like all the gas giants— only part of the planet is composed of gas at all. Beneath Uranus' 4,500-mile-thick gassy shroud, scientists believe there is a liquid mantle—a vast, scalding-hot ocean of methane, ammonia and water—and an Earth-sized rocky core. As a result, Uranus is sometimes described as a hybrid planet, partway between a gas giant and a rocky world like Earth. Of all the planets in the solar system, only Neptune closely resembles it.

As well as hydrogen and helium, the relatively thick atmosphere contains smaller amounts of methane and ammonia. The methane absorbs red wavelengths from the sunlight that shines on the outer atmosphere, which is what gives the planet its distinctive blue-green color.

During Voyager 2's encounter with Uranus in 1986, different colored filters were used on each of its eight cameras. This made it possible to photograph Uranus in more detail, and revealed icy clouds and fast winds. Uranus has three cloud decks: an upper one of methane ice, a middle layer of ammonia and a lower one of water ice. The lower two are permanently hidden by the methane layer. There is also a high-altitude haze layer above the methane clouds. Racing clouds showed that the winds blow mostly in the same direction as the planet rotates, but vary with latitude. Faint bands were also indicated, yet Voyager found no evidence of raging storms. Unless maelstroms of the type seen on Jupiter and Saturn are buried deep within the atmosphere where they can't be seen, we must assume that Uranian weather is fairly calm. This is probably due to the temperature.

Uranus is as chilly as Neptune, even though Neptune is 1.6 times farther from the Sun. Both planets have an average temperature of just –353°F (–178°C). The only explanation for Uranus' mysterious lack of warmth is that, unlike the other gas giants, the planet generates little or no internal heat. On Jupiter, storms are

thought to be powered by plumes of hot gas that rise from the interior, but Uranus seems to lack the means to stir up storms in this way.

Uranus also has unusual seasons. Because the planet is tipped on its side, it rolls around the Sun like a beer barrel rather than spinning like a top. During its 84-year-long orbit, the poles point at the Sun for a period of 42 years at a time. But its sunlit regions are actually colder than the

night side—in the upper atmosphere, at least. The reasoning is complex, but scientists suspect that Uranus may behave like a giant refrigerator in which the Sun's energy somehow sets up flows of material within the planet's interior that cause it to lose heat. It is one of many oddities of the Uranian atmosphere that remain unexplained.

NOW VOYAGER

THE VOYAGER 2 FLYBY IN 1986 GAVE US VAST AMOUNTS OF DATA ON URANUS—MORE THAN HAD BEEN GATHERED IN OVER 200 YEARS OF GROUND-BASED OBSERVATION. VOYAGER DETECTED THE FIRST CLOUDS AND WINDS AND STUDIED THE PLANET'S COMPOSITION AND TEMPERATURES. THE PROBE ALSO DISCOVERED AURORAE ON THE NIGHT SIDE AND A MYSTERIOUS "ELECTROGLOW" (RIGHT) ON THE DAY SIDE THAT SCIENTISTS STILL DO NOT FULLY UNDERSTAND.

THE HEART OF THE MATTER

According to data collected by Voyager 2, the gaseous Uranian atmosphere envelopes a planet containing more solid material. Gas flows with the atmosphere driven by the heat of the Sun may account for the surprisingly low temperatures.

CLOUD SPOTTING
In the warmest latitude bands, where the atmospheric layers are penetrated by sunlight, Voyager 2 discovered the first direct evidence of Uranian weather—a top layer of clouds composed of methane ice crystals (white smudges on the images to the left). Voyager also found evidence of clouds in its Radio Science experiment. As Uranus passed between the probe and Earth, the radio signals sent to Earth were weakened by the methane in the planet's atmosphere. The signals dipped twice during the probe's transit, so scientists deduced that the probe had flown over methane clouds on either side of the planet.

GASEOUS ATMOSPHERE
Composed mainly of hydrogen and helium, plus methane, ammonia and water, it is estimated to be around 4,500 miles (7,240 km) thick.

LIQUID MANTLE
A vast ocean of boiling ammonia, methane and water mingled with rocky deposits that is kept in a liquid state by massive gravitational pressures.

PREVAILING WINDS
Prevailing wind directions on Uranus vary according to latitude—as do wind speeds. Some of the fastest winds are found on or around 60°S of the equator, where they race at up to 220 yards (200 m) per second—faster than the planet's rotational speed.

PURPLE HAZE
This image, taken by Voyager 2 with colored filters that were later enhanced by color contrast, shows Uranus' sunlit south pole shrouded in haze or smog. This is believed to be made up of acetylene and other chemicals that are generated by the action of sunlight on the methane in the planet's atmosphere.

ROCKY CORE
Thought to be around 3,000 miles (4,830 km) in diameter and about the same mass as the Earth.

north pole

equator

south pole

RINGS OF URANUS

...URANUS' RINGS ARE THE DUST OF CRUMBLING MOONS?

The puzzling nature of Uranus' rings caused U.S. Naval Observatory astronomer Thomas Van Flandern to propose a radical theory of their origins. In a paper written two years after their discovery, he argued that the rings' narrow, roughly circular form should be unsustainable. Collisions between particles within the rings should make them spread outward over time. And the microscopic particles between the Lambda ring and the planet ought to be dragged down into the planet itself by collisions with gas atoms escaping from Uranus' atmosphere. Van Flandern is not alone—most astronomers accept that the dust particles between the rings must be continually replenished from somewhere.

Van Flandern's explanation was that the rings consist of material shed by a series of small moons orbiting the planet. These moons may be similar to comets, which shed material as they near the Sun. Evidence exists that one of Jupiter's moons, Io, and one of Saturn's moons, Titan, exude gaseous material from their atmospheres into space as they orbit around their planets. But Uranus' moons couldn't do this—they have no atmosphere.

But there could be another explanation. Ring particles could be blasted loose from these small Uranian moons by meteoroid impacts. Data from the Galileo space probe has already linked four tiny Jovian moons to their parent planet's thin rings. Each ring is identical in color to the moon that orbits its outer edge, suggesting that the rings are made up of fragments shed from the moons. But there is no evidence for this in Uranus' satellite system.

In 1985, evidence from Voyager 2 gave rise to another explanation. It found that Uranus' outermost Epsilon ring is made of large fragments several feet wide, kept in place by the gravitational fields of two "shepherd" moons, Cordelia and Ophelia. No shepherd moons were found in the gaps between the inner rings, but so-called "embedded moonlets" might be hidden within the rings themselves. These moonlets could be too small to be detected, but several in each ring could do the shepherding job of one large moon.

The mystery of Uranus' rings may prove impossible to solve from Earth. To locate more than the 27 known moons could take another probe or a fleet of space telescopes working in joint focus. But any moons among the rings would have a limited life span. The Uranian ring system is located inside the planet's Roche limit—the distance within which the planet's tidal forces break up any orbiting bodies.

The rings of Uranus are the darkest objects in the entire solar system—so dark that they went unnoticed for two centuries after the discovery of the planet itself. Even then, the ring system was found by accident, and only some clever detective work revealed the full extent of the narrow, dusky bands. Astronomers are still not sure exactly what Uranus' rings are made of. But one thing is certain: They could not be more different in appearance from the bright rings of Uranus' neighbor, Saturn.

RING STATISTICS

Name	Distance From Planet's Core Miles/km	Width Miles/km	Thickness Miles/km	Albedo
1986U2R	23,560/37,916	1,550/2,494	0.05/0.08	0.03
6	25,941/41,748	0.6–1.8/0.9–2.9	0.05/0.08	0.03
5	26,183/42,137	1.2–1.8/1.9–2.9	0.05/0.08	0.03
4	26,393/42,475	1.2–1.8/1.9–2.9	0.05/0.08	0.03
Alpha	27,726/44,620	.4–7.5/6.4–12	0.05/0.08	0.02
Beta	28,309/45,559	.4–7.5/6.4–12	0.05/0.08	0.02
Eta	29,252/47,077	.0–1.2/0–1.9	0.05/0.08	0.03
Gamma	29,530/47,524	0.6–2.5/0.9–4	0.05/0.08	0.03
Delta	29,946/48,193	1.8–5.6/2.9–9	0.05/0.08	0.03
Lambda (1986U1R)	31,012/49,909	0.6–1.2/2.5–1.9	0.05/0.08	0.03
Epsilon	31,713/19,705	12–62/19–100	0.05/0.08	0.02

BLACK ON BLACK

When 18th-century astronomer William Herschel discovered Uranus, he effectively doubled the size of the solar system. The seventh planet is 1.78 billion miles (2.86 billion km) from the Sun, in an orbit twice as far from the Sun as the sixth planet, Saturn. From Earth, astronomers can discern Uranus as little more than a tiny, indistinct blue-green circle. Because of the planet's great distance from the Earth, its rings were only revealed in 1977, and then merely indirectly. Astronomers took to NASA's Kuiper Airborne Observatory to analyze light from a star as it crossed behind Uranus. Their aim was to find out more about the planet. What they found came as something of a surprise—evidence for a series of narrow rings. These rings formed only the second planetary ring system that has been confirmed since Saturn's rings were found almost four centuries earlier. Yet, after Pioneer 11 flew past Jupiter in 1974, astronomers suspected that Jupiter, too, had rings. This was not confirmed until two years after the disclosure of the Uranian rings.

The revelation was a scientific triumph, but it brought up problems. When Uranus crossed in front of another star in 1978, nine rings showed up. All but the outermost of these were extremely thin and narrow, measuring only a few miles across. How this distinctive state could be maintained was a mystery. Over time, individual circling particles in a ring are bound to collide and, after just a few decades, the rings should start to spread outward.

The outermost and largest of Uranus' rings, dubbed Epsilon, is the closest in appearance to a Saturnian ring, but it too is mysterious. Varying in width between 12 and 62 miles (19 and 100 km) across, the rings are composed of particles that reflect less light than coal dust.

Baffled astronomers wanted to take a closer look at these thin, enigmatic rings and find out what keeps them together. They were in luck. The Voyager 2 spacecraft was launched the same year as the Uranian rings were uncovered. Taking advantage of a rare conjunction of the outer planets, the probe took a gravitational hitch-hike

through the solar system to reach Uranus in late 1985. The rings were a high observational priority.

The experience of photographing Saturn's rings had given NASA mission control some expertise in keeping the spacecraft stable enough to take long exposures. This capability was essential in the low light of the Uranian system, 19 times as far from the Sun as the Earth is. The images Voyager 2 returned showed two additional rings, entirely invisible from Earth, which brought the total of Uranus' rings to at least 11. They also revealed a number of incomplete ring arcs, as narrow as 160 feet (50 m) across. These arcs, made up of faint dust, suggest that the Uranian rings are relatively new and still evolving, perhaps only a few million years old.

Voyager 2 spent only a few months in the vicinity of Uranus and there are no plans for another mission there. But the Hubble Space Telescope's Wide Field and Planetary Camera has managed to resolve the rings of Uranus from Earth orbit. Hubble's new technology provides researchers with the opportunity to check whether the rings undergo any major changes in structure over time and to attempt analysis of their still unknown chemical make-up.

Currently, the leading theory is that the rings are formed out of methane ice that has been cooked into tar-like chains of carbon, hydrocarbons or a mixture of both through long exposure to cosmic radiation. Still, nobody knows for sure.

URANUS' RING SYSTEM

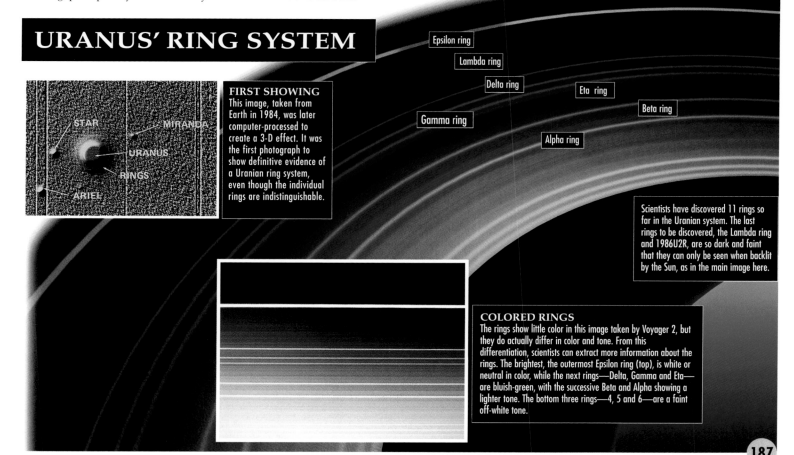

FIRST SHOWING
This image, taken from Earth in 1984, was later computer-processed to create a 3-D effect. It was the first photograph to show definitive evidence of a Uranian ring system, even though the individual rings are indistinguishable.

STAR · MIRANDA · URANUS · RINGS · ARIEL

Epsilon ring
Lambda ring
Delta ring
Eta ring
Beta ring
Gamma ring
Alpha ring

Scientists have discovered 11 rings so far in the Uranian system. The last rings to be discovered, the Lambda ring and 1986U2R, are so dark and faint that they can only be seen when backlit by the Sun, as in the main image here.

COLORED RINGS
The rings show little color in this image taken by Voyager 2, but they do actually differ in color and tone. From this differentiation, scientists can extract more information about the rings. The brightest, the outermost Epsilon ring (top), is white or neutral in color, while the next rings—Delta, Gamma and Eta—are bluish-green, with the successive Beta and Alpha showing a lighter tone. The bottom three rings—4, 5 and 6—are a faint off-white tone.

URANUS' LARGE MOONS

Uranus keeps its many moons well hidden. The five largest satellites are coated in dark material that makes them hard to pick out against the night sky. The last of the big five was found in 1948, but it was four more decades before the Voyager 2 probe resolved their surface features. The images were astounding. As well as the run-of-the-mill impact craters, some of the moons have faults that point to periods of geological activity—and the smallest of the five seems to have had a particularly violent past.

WHAT IF...

...A PROBE VISITED MIRANDA?

Miranda seems to have suffered particularly badly. At first, scientists suggested that the collisions might have shattered the small moon altogether, leaving an orbiting rubble pile that gradually pulled itself back together under its own gravity. Such a catastrophic scenario might have been played out more than once, leaving the scarred world we see today.

More recently, though, most astronomers have changed their minds. It now seems more likely that Miranda was never entirely shattered. Instead, stressed by heat, gravity, and perhaps bombardment, the satellite's interior and exterior were churned up, giving it a patchwork appearance.

More intriguing are several features imaged by Voyager 2 that are unique to this odd little moon. These include a formation that looks like a stack of pancakes, grooves in the ice that make right-angled turns, and steep cliffs several miles high.

Miranda may owe its odd looks to the violent early history of our solar system.

All of the planets and moons suffered intense bombardment by asteroids—many as big as cities, a few the size of small planets—soon after the solar system formed. Asteroid impacts blasted large craters into the surfaces of the solid planets and

moons, and many of these ancient scars are still visible today.

Miranda may have taken this sustained beating even more severely—some scientists think that the collisions shattered the small moon altogether. But most of the remnants of the original satellite stayed close to each other. Eventually, their gravity pulled them back into a moon-shaped pile. In essence, the pieces stitched themselves together, to form a "new" moon from the shattered pieces of the original.

This catastrophic scenario may have been played out more than once. But the suture job was not pretty, and left Miranda with the lumpy face seen by Voyager.

Miranda's scars may be the most interesting places for a future probe to visit. If the moon really did patch itself back together, these sites could reveal details about its reconstruction.

Incidentally, they would also provide some stunning vistas. Ice cliffs are just one of the sights. These stretch many miles long, and tower over three miles (5 km) high before sloping sharply toward the valley floor below. Voyager images indicate that the cliffs are smooth, so they would glisten in the pale sunlight.

Whatever its attractions for the space tourist, to serious astronomers, Miranda's biggest draw will always remain in the moon's mysterious past.

MOON STATISTICS

Name	Distance from center of Uranus (miles/km)	Diameter miles (km)	Orbital Period (days)	Discovery
Miranda	80,700/129,900	298/290 (480/467)	1.4	1948, Gerard Kuiper
Ariel	118,800/191,200	722/718 (1,162/1,156)	2.5	1851, William Lassell
Umbriel	165,300/266,000	727 (1,170)	4.1	1851, William Lassell
Titania	270,800/435,800	982 (1,580)	8.7	1787, William Herschel
Oberon	362,000/582,600	944 (1,519)	13.5	1787, William Herschel

THE BIG FIVE

Uranus, like Jupiter and Saturn, has an impressive array of moons—so far 27 have been discovered. Many of these are so faint that they were only discovered when Voyager 2 paid the planet a close visit in 1986. Others had to wait until ground-based telescopes improved enough to resolve a few tiny moons in distant, elongated orbits.

Before Voyager 2, we only knew of the five largest moons. These are still the most studied, with some of the most whimsical names in astronomy: Ariel, Miranda, Umbriel, Titania and Oberon, all supernatural beings in Shakespeare's plays. All orbit around Uranus' equator.

Unlike the other planets in the solar system, Uranus is tilted on its side, so that its equatorial plane is almost at right angles to the plane of its orbit around the Sun.

The planet may have been knocked over by a collision with an Earth-sized body. Uranus' moons could have formed amid the fall-out of this impact, as debris clouds condensed into new satellites. Alternatively, the five big moons may have circled the original equator, which left them in a pole-to-pole orbit after the big crash.

The planet's gravity slowly dragged the moons into its new equatorial plane. As they moved, the moons were thrown into the path of orbiting debris, as well as into one another. This was a period of intense violence, so scientists were surprised to see only a few impact scars on the frozen worlds of Titania and Ariel. These moons must have been resurfaced by geological activity—so at one point, they must have had hot interiors.

All five moons are made of ices and rock mixed in with carbon and nitrogen compounds. Surface temperatures hover near –350°F (–212°C). Titania and Oberon are the largest members of the quintet—each is just less than half the size of the Earth's moon.

Oberon and the slightly smaller moon Umbriel are the least remarkable of Uranus' big satellites. Impact craters that date to the earliest epoch of the solar system still cover their

surfaces, and there are few cracks and faults to indicate any recent geologic activity.

Ariel and Miranda have had the most active past. The surface of Ariel is scarred by giant canyons, and there is evidence that thick fluids—perhaps a mixture of ices—once flowed through deep channels.

Miranda is the closest moon to Uranus. Its jumbled terrain is the strangest surface in the known solar system. Some scientists think that Miranda's face was modeled by volcanic activity, but others believe that the moon once suffered a giant impact—either way, we may have to wait for another probe to find the truth.

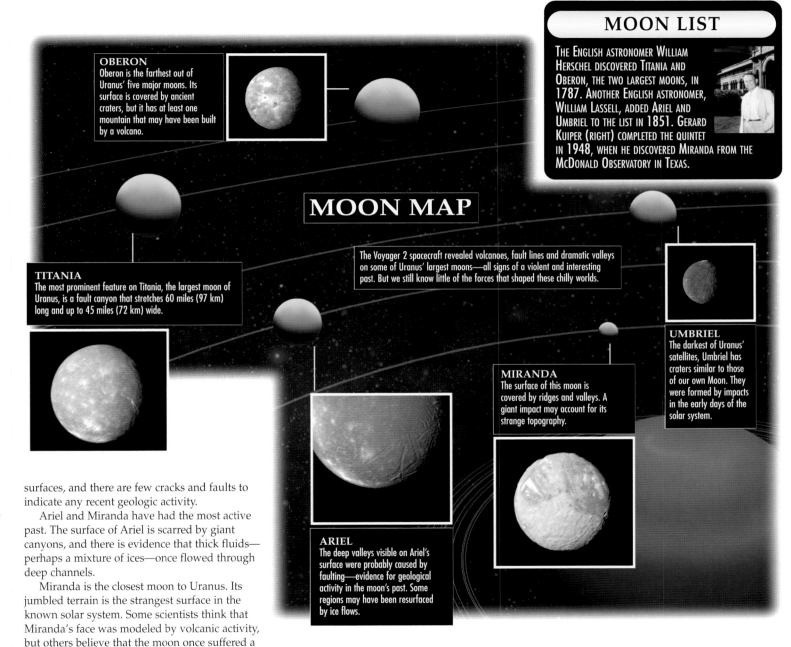

MOON LIST

THE ENGLISH ASTRONOMER WILLIAM HERSCHEL DISCOVERED TITANIA AND OBERON, THE TWO LARGEST MOONS, IN 1787. ANOTHER ENGLISH ASTRONOMER, WILLIAM LASSELL, ADDED ARIEL AND UMBRIEL TO THE LIST IN 1851. GERARD KUIPER (RIGHT) COMPLETED THE QUINTET IN 1948, WHEN HE DISCOVERED MIRANDA FROM THE McDONALD OBSERVATORY IN TEXAS.

MOON MAP

OBERON
Oberon is the farthest out of Uranus' five major moons. Its surface is covered by ancient craters, but it has at least one mountain that may have been built by a volcano.

The Voyager 2 spacecraft revealed volcanoes, fault lines and dramatic valleys on some of Uranus' largest moons—all signs of a violent and interesting past. But we still know little of the forces that shaped these chilly worlds.

TITANIA
The most prominent feature on Titania, the largest moon of Uranus, is a fault canyon that stretches 60 miles (97 km) long and up to 45 miles (72 km) wide.

UMBRIEL
The darkest of Uranus' satellites, Umbriel has craters similar to those of our own Moon. They were formed by impacts in the early days of the solar system.

MIRANDA
The surface of this moon is covered by ridges and valleys. A giant impact may account for its strange topography.

ARIEL
The deep valleys visible on Ariel's surface were probably caused by faulting—evidence for geological activity in the moon's past. Some regions may have been resurfaced by ice flows.

MYSTERIOUS URANUS

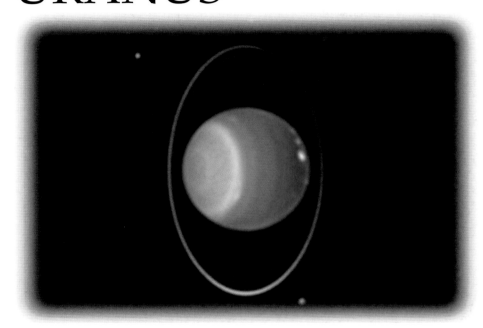

January 24, 1986, was a big day in the largely secret history of Uranus. Voyager 2—the only spacecraft yet to fly by the planet—made its closest approach to the surface and returned data that contained several surprises. It was already known that the planet's axis of rotation was almost sideways to the Sun. But Voyager discovered that Uranus was also unusually cold and that its magnetic field was off-center, suggesting that something very strange was happening at the core. Could these anomalies be linked?

WHAT IF...

...WE COULD FIND A PRACTICAL USE FOR URANUS?

As a cold and distant gas giant, the planet Uranus may seem like an unlikely focus for human attention. But one expert on space resources has suggested a way in which the planet could possibly help to open up a whole new era of space exploitation.

According to John S. Lewis, a professor of planetary science and author of *Mining the Sky,* Uranus's hidden wealth might lie in the plentiful supplies of the gas helium that make up an estimated 15% of its atmosphere. Helium-3—an isotope of the gas that is rare on Earth, but relatively common in all the outer planets—has been can identified by some researchers as an ideal fuel for the high-efficiency, low-waste nuclear fusion reactors that it is hoped could one day supply our energy needs.

In theory, a probe dispatched into the atmosphere of any of the gas giants could extract large quantities of the gas. Several tons could be stored in payload tanks for the return journey to Earth, where the isotope's astronomically high value—calculated by Lewis to be $1 billion per ton at current prices—would more than cover the costs of the trip.

The most difficult part of the project, in Lewis's view, would not be extracting the gas—a fairly straightforward operation—but in getting it back to Earth once it had been gathered. The amount of thrust required to blast a spacecraft on a homeward journey from any of the big planets is enormous, due to the size of their gravitational fields. On Jupiter and Saturn, the largest planets, the pull of gravity would almost certainly defeat any spacecraft that is likely to be built in the foreseeable future. Though Uranus is less than half the size of Jupiter and so exerts a much weaker grip—in fact, its escape velocity is only one-third that of Jupiter—a mining probe would still need a more powerful and efficient propulsion system than any now in use. Even so, the sheer quantity of the potential resources might be a powerful inducement to develop the necessary firepower. By Lewis's calculations, the amount of helium-3 on Uranus alone would be enough to supply the Earth's energy needs for 4 billion years—the time it will take for the Sun to burn out.

UNUSUAL URANUS

EQUATORIAL DIAMETER	15,882 MILES (25,560 KM)
POLAR DIAMETER	15,518 MILES (24,974 KM)
ELLIPTICITY	0.0229
MEAN DENSITY	0.75 OUNCES PER CUBIC INCH (21.3 G PER 6.5 CUBIC CM)
EQUATORIAL ESCAPE VELOCITY	13.2 MILES PER SECOND (21.2 KM/SEC)
EQUATORIAL ROTATION	18.0 HOURS
OBLIQUITY (TILT OF EQUATOR TO ORBIT)	97.86°
MEAN ORBITAL VELOCITY	4.24 MILES PER SECOND (6.82 KM/SEC)
ORBITAL ECCENTRICITY	0.046
ORBITAL INCLINATION TO THE ECLIPTIC	0.77°
SURFACE MAGNETIC FIELD	0.23 GAUSS

TILTED SIDEWAYS

Uranus was already known as a planetary oddball long before Voyager 2 flew by. The planet travels on its side, with its equator at an angle of more than 90° to the direction of its orbit, so its geographic poles take it in turns to point at the Sun. The Uranian year is 84 Earth years long, so that each pole in turn gets 42 years of uninterrupted sunlight. Yet although the planet's poles receive more solar energy than the equator, the equator is hotter—a result that has scientists baffled.

The Voyager 2 data added further mystery. Measurements showed that Uranus' magnetic field is also out of kilter, but not in the direction of its rotation as might be expected. Instead, it lies at an angle of 59° to the rotational axis and, even more bizarrely, does not pass through the center of the planet. The result, one author claimed, is "like finding the Earth's north magnetic pole in Cairo, Egypt, and the south pole in Brisbane, Australia." The source of the magnetic field is also mystery. On other planets, magnetic fields are the result of electric currents generated within their metallic cores. There is no evidence of this on Uranus.

MORE SURPRISES

The planet's weather provided other surprises. It had always been assumed that whichever pole was facing the Sun would be much hotter than the rest of the planet, generating strong polar winds. But Voyager 2 showed that the temperature hardly varies, registering a spine-chilling –417°F (–214°C) at cloud-top level more or less all around the planet. Uranus appears to radiate no more heat than it receives from the Sun, overturning prior assumptions that, like Jupiter and Saturn, it had a hot core.

It was Voyager's close encounter with Miranda, the planet's fifth-largest moon, that provided a possible explanation for these Uranian oddities. Photographs from less than 20,000 miles (32,200 km) up revealed Miranda to be a crazily jumbled world of ridges, faults and clefts in which old and new features seemed to

exist side by side. A possible explanation for this improbable geology is that at some point in the distant past, the moon was shattered by a collision with another body of around Earth size and then "reassembled" by gravity in an almost random way.

Such a cataclysm could also explain many other Uranian anomalies. An impact with a large planetoid might have knocked the entire planet on its side, causing hot material to drain from the core. This in turn could have dissipated any internal heat source and at the same time displaced the magnetic field. It is even possible that Miranda and the other 26 known moons of Uranus are fragments ejected from the collision that have been trapped ever since by the planet's gravitational field.

The evidence needed to prove the theory will be hard to come by, since no additional probes to Uranus are planned. In the absence of unforeseen advances in Earth-based investigation techniques, the oddball is likely to preserve its secrets for decades to come.

FIRST GLIMPSES

URANUS WAS SPOTTED SEVERAL TIMES BEFORE 1781, THE YEAR IT WAS FIRST IDENTIFIED AS A PLANET BY THE GERMAN-BORN ENGLISH ASTRONOMER SIR WILLIAM HERSCHEL. JOHN FLAMSTEED (BELOW)—ASTRONOMER ROYAL AT GREENWICH, ENGLAND—RECORDED IT IN 1690, AS DID GERMAN CARTOGRAPHER TOBIAS MAYER IN 1756, BUT BOTH MISTOOK IT FOR A STAR. SUBSEQUENT INVESTIGATIONS TURNED UP MANY OTHER EARLY SIGHTINGS, HELPING ASTRONOMERS OF THE DAY TO IDENTIFY THE PLANET'S ORBIT AND TO COMPILE TABLES OF ITS FUTURE MOTION.

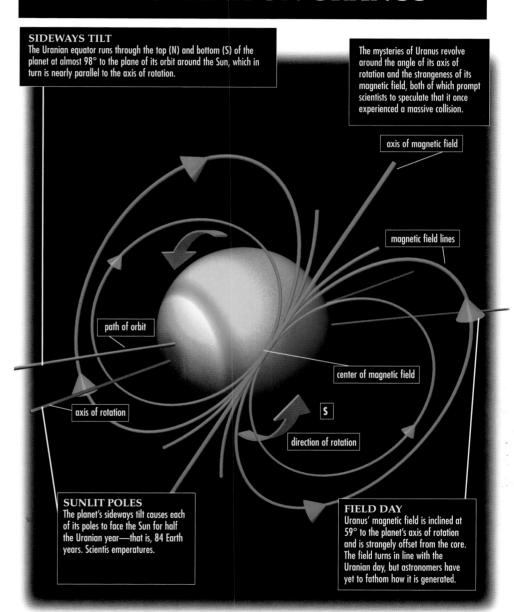

PHENOMENA ON URANUS

SIDEWAYS TILT
The Uranian equator runs through the top (N) and bottom (S) of the planet at almost 98° to the plane of its orbit around the Sun, which in turn is nearly parallel to the axis of rotation.

The mysteries of Uranus revolve around the angle of its axis of rotation and the strangeness of its magnetic field, both of which prompt scientists to speculate that it once experienced a massive collision.

axis of magnetic field

magnetic field lines

path of orbit

center of magnetic field

axis of rotation

S

direction of rotation

SUNLIT POLES
The planet's sideways tilt causes each of its poles to face the Sun for half the Uranian year—that is, 84 Earth years. Scientis emperatures.

FIELD DAY
Uranus' magnetic field is inclined at 59° to the planet's axis of rotation and is strangely offset from the core. The field turns in line with the Uranian day, but astronomers have yet to fathom how it is generated.

NEPTUNE

Remote Neptune, at times the farthest planet in the solar system, was unknown until a century and a half ago. Its year is so long that it still has not completed a full orbit of the Sun since its discovery. Earth-bound observers knew almost nothing about Neptune until the Voyager 2 probe passed by in 1989 and returned spectacular views of this mysterious blue world. But although Voyager established that the planet shares many family characteristics with Jupiter, Saturn and Uranus, Neptune has yet to give up its greatest secret: the source of the heat that rises from the planet's center to drive violent storms in its atmosphere.

WHAT IF...

...YOU WERE ON A MISSION TO NEPTUNE?

Some time in the 21st century, Neptune may become the last staging post on the journey into interstellar space: Tiny, barren Pluto orbits farther out, but Neptune and its moons are far richer in natural resources.

Not that anyone will ever live on Neptune itself—it has no solid surface that we know of. But since the planet is plentifully endowed with hydrogen, it might make an ideal refueling stop for visiting spacecraft powered by nuclear reactors.

Specialized "grazer" probes would dive out of orbit into Neptune's clouds and gulp in raw materials from the atmosphere, then return them to orbiting processing plants. Navigating the high winds would be hazardous, so this task would be left to artificially intelligent robots. A tour of duty on Neptune could see you as the overseer of this robot workforce, with an orbiting space station for a home. But if you longed for solid ground, you might apply for a transfer to the newly established prospecting base on Neptune's largest moon, Triton.

The prospectors who have set up camp here are hardy souls: Triton is the coldest place in the solar system, with an iron-hard surface of frozen nitrogen. Their job is to select sites where robots can mine for water, which is believed to exist on Triton as pockets of ice far below the surface. In the meantime, the base is supplied with regular shipments of water and oxygen from Neptune's orbital processing plants and receives energy in the form of beamed-down microwaves from an orbiting fusion reactor.

Crops have been genetically engineered to grow in the tiny amount of sunlight that reaches this far out. The base is not self-sufficient in food, but the plants help to recycle the carbon dioxide that people exhale. Plant nutrients are synthesized from the frozen nitrogen underfoot and from the methane that is mined on regular expeditions to Triton's polar caps.

No one enjoys these expeditions. The journey is relatively short—Triton is only three-quarters the size of the Earth's Moon. But the polar miners have to contend with the continual hazard of Triton's famous geysers, which erupt without warning and send giant plumes of pressurized liquid nitrogen spiraling over five miles into the empty sky.

NEPTUNE PROFILE

NEPTUNE		EARTH
30,775 miles (49,528 km)	DIAMETER	7,973 miles (12,831 km)
28°19'	AXIS TILT	23°27'
164.79 Earth years	LENGTH OF YEAR	365 days
16.11 hours	LENGTH OF DAY	24 hours
2,794 million miles (4,497 million km)	DISTANCE FROM SUN	93.5 million miles (150.5 million km)
−326°F (−163°C)	SURFACE TEMPERATURE	59°F (15°C)
1.12 G	SURFACE GRAVITY	1 G
Hydrogen (80%) Helium (19%) Methane (1%)	ATMOSPHERE	Nitrogen (78%), Oxygen (21%), Argon (1%)
1,000–3,000 millibars	ATMOSPHERIC PRESSURE	1,000 millibars
Hydrogen (80%) Helium (19%) Methane (1%)	COMPOSITION	Mainly iron and nickel

THE BIG BLUE

The blue-green globe of Neptune keeps a silent watch over the outer reaches of the solar system as it follows its leisurely 168-year orbit around the Sun. Little Pluto, its rocky partner Charon and a host of unknown comets lie beyond, but Neptune is the last of the orderly pattern that comprises the four inner rocky planets—Mercury, Venus, Earth and Mars—and the four outer gas giants—Jupiter, Saturn, Uranus and Neptune itself.

Neptune is so far away from the Sun—about 2.8 billion miles (4.5 billion km) on average—that it is too dim for the naked eye to see. Even the most powerful Earth-based telescopes reveal hardly any details of the surface. To find out more, NASA dispatched the Voyager 2 mission to take a closer look. The spacecraft hurtled past the planet in 1989, closing to within about 3,000 miles (5,000 km) of the cloud-tops of Neptune's thick, hazardous atmosphere.

The Neptunian atmosphere accounts for most of the planet's bulk. It is tormented by winds that blow at up to 1,500 mph (2,400 km/h) and by giant, fast-moving storms like the Great Dark Spot that was identified by Voyager 2.

Atmospheric activity on this scale requires colossal amounts of energy, but the source of this energy is one of the Blue Planet's greatest mysteries. On Earth, our weather is driven by heat from the Sun. Neptune's remote orbit is 30 times farther out from the Sun, which causes solar radiation to dwindle to less than one nine-hundredth of Earth levels. Neptune is also farther from the Sun than Uranus, yet astronomers know that it is warmer. So where does the energy come from?

CRUSHED HEAT

Neptune's heat can only be generated from within the planet itself. The most popular theory is that materials of different densities within the interior have yet to fully separate out—that the planet's gravity is still dragging heavier matter toward the core, creating friction that in turn generates heat.

Scientists have speculated that such enormous heat and pressure would cause Neptune's methane to separate into its component elements—hydrogen and carbon—and that the pressure might compress the carbon into a form familiar to us all: There may be giant diamonds at Neptune's core.

ATMOSPHERE
This makes up most of the planet's bulk and is roughly 80% hydrogen and 19% helium. It is a violently energetic environment, almost certainly due to the heat that emanates from the planet's core.

NEPTUNE'S RING

As the Voyager probes visited Jupiter, Saturn, Uranus and Neptune in turn, they showed that rings are a family characteristic of the gas giant planets. Neptune has at least four narrow, faint rings of dark dust that are believed to be the debris from impacts of tiny meteoroids on its moons. The outer ring is unusual in that it has bright patches where the dust particles appear to be more densely clustered and thus reflect more sunlight. The clusters suggest that the ring is relatively young and has yet to "thin out" around Neptune. The clusters may be the debris of a small moon that has disintegrated within the past few thousand years.

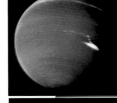

THE SCOOTER
Voyager 2 sent back pictures of a small, bright, eastward-moving cloud that scientists dubbed "the Scooter" (right). True to its name, the cloud scooted around Neptune every 16 hours, blown by the planet's strong winds. Small streaks within the Scooter caused its appearance to constantly change.

NEPTUNE INSIDE AND OUT

DARK SPOT
Voyager 2 revealed a giant storm system (right), dubbed the Great Dark Spot, in the planet's southern hemisphere. The Spot was estimated to be about the same size as the Earth and rotated counter-clockwise. Above the Spot drifted feathery white clouds that resembled the cirrus clouds in the Earth's atmosphere.

MISSED

The Italian scientist Galileo Galilei (1564–1642) almost capped his discovery of Jupiter's moons with the first sighting of Neptune. In 1612, Galileo (right, center) noticed an object near Jupiter, exactly where Neptune would have been at the time. A month later, it had moved—but he never followed up on the clue.

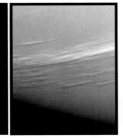

CLOUDS
Neptune's wispy white clouds consist of frozen methane crystals high in the atmosphere. The planet's distinctive blue-green color is due to the presence of methane in the thicker cloud layers below. These layers also contain crystals of ammonia and hydrogen sulfide, and display banding similar to the clouds on Jupiter and Saturn.

GIANT OCEAN
Scientists speculate that beneath the Neptunian atmosphere lies a planet-wide ocean of water, liquid ammonia and methane. Even at "sea level," its temperature is a staggeringly hot 4,000°F (2,204°C). It is the enormous pressure of Neptune's atmosphere that keeps the ocean in a liquid state.

MOLTEN CORE
In common with most of the other gas giants, Neptune may have a small, molten core consisting largely of iron and silicon compounds. The temperature near the core is reckoned to be over 12,000°F (6,649°C).

NEPTUNE'S ATMOSPHERE

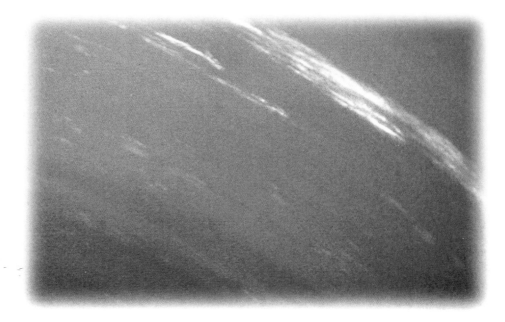

Little was known about conditions on Neptune until the Voyager 2 probe flew by in 1989, on its closest approach to a planet since it was launched 12 years earlier. Voyager discovered that the blue gas giant has some of the wildest weather in the solar system. Jet streams generated by an unknown internal energy source race around the equator and massive storms spin at supersonic speed. Since Voyager, Hubble Space Telescope images have verified that Neptune's atmosphere is as variable as it is turbulent.

WHAT IF...

...YOU COULD FLY THROUGH NEPTUNE'S ATMOSPHERE?

Perhaps someday, a guided tour around Neptune will be possible. If so, it could sound something like this: "All aboard for Flight 703 to Neptune, departing in 15 minutes. On this journey we'll be taking the scenic route. As we pass Neptune's volcanic moon Triton, we'll be slowing down to admire the nitrogen geysers, erupting five miles into the moon's thin atmosphere. Continuing our journey, we'll pass through the orbits of Neptune's five dust rings. From the Earth these rings appear to be disconnected, but closer up you'll see that they go all the way around the planet. You'll also see that the fainter sections of the rings alternate with bright arcs where the ring's particles are more densely packed.

"After passing the inner ring, you'll see a thin red haze, marking Neptune's outermost layer of atmosphere. This is where methane in the atmosphere absorbs the Sun's ultraviolet rays, creating a hot zone. You may also spot faint aurorae generated by solar particles hitting Neptune's magnetic field, which is a lot weaker than the one we have on Earth. Because the axis of Neptune's magnetic field doesn't pass through the center of its globe, it produces aurorae over wide regions of the planet—unlike the Earth, where the Northern and Southern Lights are only visible at the poles.

"As we travel through Neptune's atmosphere, we see very little detail. This far out in the solar system, the Sun's light is very weak. Noon on Neptune looks as dim as Earth half an hour after sunset. But you may be able to discern bright feathery bands of frozen methane clouds and—if the Sun's angle is right—see their shadows on the main cloud layer 50 miles (80 km) below.

"Keep an eye out for any storms that may be brewing. The pilot will slow down for you to observe the clouds whipping around at almost twice the speed of sound. Storms will appear as a clear vortex in the cloud layer. If you peer down, you'll see lightning flickering far below.

"On reaching Neptune's far side, we'll dim the lights in the craft so you can get some sleep. Retracing our steps the next day, you'll find that the storms you saw before will be unrecognizable, since they will have completely changed shape. Orbiting Neptune can be hypnotic—the atmosphere changes constantly, but without any solid land, there are no surface features to give you any bearings. Some people find this relaxing, while others may find it disturbing. After retracing our orbit around Neptune, we will resume our flight back to Earth. Remember to fasten both your seat and floor belts. We hope you enjoy your flight."

ATMOSPHERIC STATISTICS

MEAN CLOUD SURFACE TEMPERATURE	−360°F (−182°C)
ALBEDO (SOLAR RADIATION REFLECTED)	41%
WIND SPEED RANGE	0–1,200 MPH (0–1930 KM/H)
ATMOSPHERIC AEROSOLS	AMMONIA ICE, WATER ICE, AMMONIA HYDROSULFIDE, METHANE ICE
ATMOSPHERIC GASES	PERCENTAGE OR PARTS PER MILLION
HYDROGEN	80%
HELIUM	19%
METHANE	1.5%
HYDROGEN DEUTERIDE	192 PPM
ETHANE	1.5 PPM

RACING WINDS

On August 25, 1989, the Voyager 2 craft passed near Neptune's north pole at an altitude of 3,000 miles (4,828 km). Nearly 3 billion miles (4.8 billion km) away on Earth, observers got their first detailed look at the planet's atmosphere.

Neptune, the outermost of the gas giants, has an atmosphere several thousand miles deep, composed mainly of helium and hydrogen. A small amount of methane absorbs red light—and gives the planet its deep blue color. Because Neptune is so far from the Sun, it receives barely one-thousandth of the solar radiation that drives the weather on Earth. Many scientists assumed that its atmosphere would therefore be featureless.

Instead, Voyager 2 revealed one of the most dynamic weather patterns in the solar system. High in Neptune's frozen atmosphere, streaky bands of methane ice crystals cast shadows on the main cloud layer. This cloud layer displays distinctive zones and features moving at different speeds and in opposite directions.

Along the planet's equator, the clouds blow westward at 900 miles per hour (1,450 km/h)—three times faster than the winds in the strongest tornado on Earth. Farther north or south, at latitudes equivalent to the U.S.–Canadian border or the southern tip of South America, the winds slow down slightly, then pick up again toward the poles to reach 500 miles per hour (805 km/h), this time circling eastward.

About 10° south of the equator, Voyager 2 measured even stronger winds—1,200 miles per hour (1,930 km/h). These winds race past a dark oval area as large as the Earth that rotates counterclockwise. Dubbed the Great Dark Spot, it seems to be a gigantic Neptunian hurricane produced by gases that well up from inside the atmosphere. Voyager's observations revealed that both the Great Dark Spot and the bright cirrus-like clouds fringing it changed size and shape between one planetary rotation and the next.

Where Neptune gets the energy to power such a remarkable weather system is unclear.

Neptune has the same surface temperature as Uranus, which is 1 billion miles (1.6 billion km) closer to the Sun. Scientists have suggested that a vast amount of heat escapes from Neptune's 8,500°F (4,700°C) core, creating *convection currents*, in which warm gas rises and is replaced by cooler gas from beneath. Temperature readings by Voyager indicate that gases rise from the planet's mid-latitudes. Cooling, they drift toward the equator and the poles, where they sink and warm again.

This vertical flow of gases extends to great heights, helping to keep Neptune's upper atmosphere saturated with methane. Turbulent updrafts carry methane to the base of the stratosphere. Here the Sun's ultraviolet radiation converts it to a smog of ethane and acetylene particles, which descends and condenses into ice.

Falling into the warmer troposphere, the ice evaporates into gases and mixes with hydrogen. This reaction regenerates the methane, which is returned to the stratosphere by this ongoing cycle of convection.

THE SCOOTER

DUE SOUTH OF THE GREAT DARK SPOT, VOYAGER FOUND A BRIGHT, IRREGULARLY SHAPED ATMOSPHERIC FEATURE (CIRCLED), WHIZZING EASTWARD AND COMPLETING ONE CIRCUIT EVERY 16 HOURS. NICKNAMED "THE SCOOTER," THE FEATURE WAS PROBABLY A PLUME OF CLOUD RISING FROM A LOWER ATMOSPHERIC LAYER.

LAYERS OF BLUE

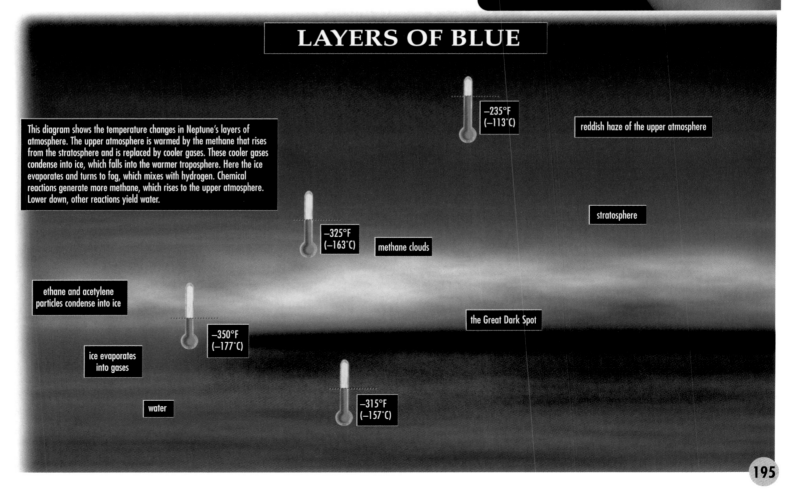

This diagram shows the temperature changes in Neptune's layers of atmosphere. The upper atmosphere is warmed by the methane that rises from the stratosphere and is replaced by cooler gases. These cooler gases condense into ice, which falls into the warmer troposphere. Here the ice evaporates and turns to fog, which mixes with hydrogen. Chemical reactions generate more methane, which rises to the upper atmosphere. Lower down, other reactions yield water.

−235°F (−113°C)

reddish haze of the upper atmosphere

stratosphere

−325°F (−163°C)

methane clouds

ethane and acetylene particles condense into ice

−350°F (−177°C)

the Great Dark Spot

ice evaporates into gases

water

−315°F (−157°C)

NEPTUNE'S RINGS

...NEPTUNE ACQUIRES NEW MOONS OR RINGS?

According to Larry Esposito of the University of Colorado, the Adams ring is a relatively new phenomenon—and it could one day reaccrete to form a new moon. Esposito first made his views known in 1997. He believes that the Adams ring formed just 10,000 years ago, when a comet collided with a 6-mile-wide (9.5-km-wide) moonlet. Comet and moon shattered, forming the Adams ring. But, Esposito claims, the ring is unstable and will reaccrete within another 10,000 years.

Unfortunately, there is no way to verify his ideas. This is because Esposito's theory hinges on the location of the Roche limit—the boundary around a planet within which a body will be ripped apart by tidal forces. In order for the ring to reaccrete, it would have to lie outside the Roche limit.

But scientists still don't know whether the Adams ring lies within the planet's Roche limit or not, mainly because they don't know what Neptune's rings are made of. In particular, they

would need to know the exact density of the material that makes up the ring. This is because the Roche limit is not a fixed boundary—its distance from the planet depends on the density of the orbiting body. Closely packed moons can hold themselves together nearer to the planet than a loosely glued body. The Adams ring lies somewhere within what is effectively a sliding scale of Roche limits, corresponding to the different densities of the material orbiting it. To shed more light on the composition of the rings would probably require another spacecraft to fly by Neptune.

Esposito could be wrong about the Adams ring, but an event scientists are predicting with more confidence is the eventual break up of Neptune's large moon Triton. Dragged relentlessly toward the planet, Triton will pass its Roche limit in about 100 million years. Then, ripped apart by Neptune's powerful gravity, it is quite possible that its remains will form a magnificent new ring circling the planet. Whether or not Neptune's current rings will still be around to accompany the newcomer is a question that only further revelations about Neptune's rings—or time—will tell.

When signs of Neptune's delicate ring system emerged in the 1980s, the existence of rings around all four gas-giant planets was finally confirmed. But the rings looked different from those of the other planets: Neptune's rings seemed to be no more than fragments. Then images from Voyager 2 showed that the rings were complete. Denser segments of material—ring arcs—existed within otherwise extremely thin rings. Now scientists are puzzled about how such curious structures could have formed.

NEPTUNE'S RINGS

Name	Distance*	Radial Width
Galle	26,000 miles (41,800 km)	9 miles (14 km)
LeVerrier	33,000 miles (53,100 km)	68 miles (109 km)
Lassell	33,000–37,000 miles (53,100– 59,500 km)	25,000 miles (40,233 km)
Arago	36,000 miles (57,900 km)	62 miles (100 km)
Adams	39,100 miles (62,900 km)	31 miles (50 km)

Ring Arcs	Length	
Fraternity	6,200 miles (10,000 km)	
Equality	2,500 miles (4,000 km)	
Liberty	2,500 miles (4,000 km)	

* from Neptune's center to ring's inner edge

SPOT THE RINGS

By the early 1980s, the hunt was on for rings around Neptune. This presented astronomers with a real challenge, since the rings are completely invisible from the Earth. Only by observing *stellar occultations*—the eclipsing of a star by another body—could the rings be found. So between 1981 and 1989, various teams of astronomers lined up to observe 50 stellar occultations of Neptune. As many as 100 observations were made. As astronomers watched, a star would pass behind Neptune and become dimmer. But what they were really looking for was a star that dimmed three times— once slightly before, once during and once slightly after passing behind the planet. Only then would a planetary ring have been found. To the amazement of all, what some of the teams in fact found was a star that dimmed just twice. The unexpected result made it look as though Neptune had ring fragments rather than complete rings.

The discovery that there was at least some ring matter circling Neptune was well timed: Voyager 2 was scheduled to fly past Neptune in 1989. The spacecraft was put on a new trajectory because the extent of Neptune's upper atmosphere was unknown and it was feared that the craft might collide with the ring material. As Voyager flew past the planet, it took the first and only photographs of Neptune's rings—rings that proved to be complete after all. In fact, the Voyager images revealed that Neptune has five complete rings.

STRANGE ARCS

Neptune's outermost ring is known as the Adams ring. This ring contains at least three arcs of denser material, which astronomers have named Liberty, Equality and Fraternity. It is thought that the strange arcs are dragged into clumps by the gravity of Galatea, a moon that orbits Neptune just inside the Adams ring. Next comes the thin Arago ring and, after this, the Lassell ring—an immense "plateau" of material about 2,500 miles (4,020 km) wide. Moving closer

toward the planet is the narrow LeVerrier ring and, finally, another wide ring, the innermost Galle ring.

The composition of Neptune's rings is still unknown. They are dark, probably dusty, possibly reddish in color and extremely sparse. Beyond this information, scientists know very little.

The origins of the rings are equally open to speculation, although there are a number of theories. We know that if a large rocky body comes close enough to a gas giant, the body can be torn to shreds by the planet's immense gravity. As a result, many ring theories involve moons that stray too close to the planet and break up, or comets that pass by and suffer the same fate.

Then there are the impact theories, where two small moons collide to create a swarm of debris that spreads around the planet. The debris from the impact is unable to *reaccrete* (clump back together again) if these moons are inside the planet's *Roche limit*—the theoretical boundary around a planet, inside which a body is subject to potentially fatal gravitational forces.

Whichever theory is right, Neptune's enormous gravity can certainly be destructive. It

probably ripped apart some ancient cosmic bodies, leaving the remains scattered in a spinning disk around its equator. Close proximity to a gas giant can be dangerous—turning large bodies into nothing more than dispersed matter, running rings around the planet.

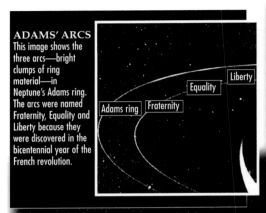

ADAMS' ARCS
This image shows the three arcs—bright clumps of ring material—in Neptune's Adams ring. The arcs were named Fraternity, Equality and Liberty because they were discovered in the bicentennial year of the French revolution.

Adams ring Fraternity Equality Liberty

A total of five rings, traveling in circular orbits around Neptune's equator, have come to light so far. All the rings have been named after astronomers involved in the discovery of Neptune or its moons.

NEPTUNE'S RINGS REVEALED

Galle ring

LeVerrier ring

Lassell ring

Arago ring

Adams ring

NEPTUNE'S MOON SYSTEMS

U ntil 1989, astronomers knew of only two satellites in orbit around Neptune: Triton, a moon the size of a small planet, and Nereid, a distant, mysterious object with a highly distorted orbit. Both were believed to be interloping asteroids captured by Neptune's gravity. But when the Voyager 2 space probe hurtled through the Neptunian system in 1989, it discovered a number of small icy satellites close to the planet. These proved to be the key to unlocking the secret, violent history of Neptune's moons.

WHAT IF...

...NEPTUNE'S MOONS WANDER OFF COURSE?

Neptune's satellite system has had a turbulent past, but it has an even more violent future ahead of it. At present, the inner satellite system has settled down with the moons in roughly circular orbits, while Nereid's path is tilted at a steep angle from the rest of the system, making a collision unlikely. Yet a possible threat could come from the Kuiper belt objects—members of the outer asteroid belt where Triton, Pluto, and Pluto's satellite Charon probably originated. A portion of Pluto's orbit, as it moves closer to the Sun, actually comes within Neptune's orbit, but Pluto's orbital tilt means the two planets never have, and never will, come close.

The real threat to the Neptunian system is Triton. The moon's retrograde orbit is inherently unstable. All satellites experience gravitational tidal forces, which slowly change their orbits over millions of years—our own Moon was once much closer to the Earth, but has slowly spiraled away. In Triton and Neptune's case, the opposite is happening. Because the moon orbits in the wrong direction, the tidal forces act as a drag on its motion, slowing it down a little with every orbit. So far, the result has been Triton's almost perfectly circular orbit. But this orbit is very slowly spiraling closer and closer to Neptune

itself. The tidal forces also create heat inside Triton, which may be responsible for powering the strange geysers of gas and dust photographed by Voyager 2 during its 1989 flyby.

Astronomers believe that Triton's first encounter with Neptune threw an existing moon system into disarray. Billions of years from now, Triton will once again wreak havoc as its decaying orbit takes it through the region inhabited by Proteus and the other inner satellites. Even if there are no actual collisions, the other satellites will be wrenched into different orbits—possibly eccentric ones, like the path that Nereid follows—and may even be ejected from orbit altogether. Meanwhile, the tides on Triton will become stronger still, perhaps even melting the surface. Something similar happened during Triton's initial capture, and was partly responsible for the strangely rippled landscape of the moon today.

Finally, the giant satellite will cross Neptune's Roche limit—the point where the planet's gravity will overcome Triton's structural strength and rip it apart completely. The shattered fragments will spread out to form a new ring around Neptune before they gradually fall to complete destruction in the planet's atmosphere. By the time Triton has fallen to its doom, remote Nereid may be the only moon left in Neptune's satellite system.

SATELLITE STATISTICS

Name	Discovered	Average Radius of Orbit (miles/km)	Eccentricity	Inclination	Diameter (miles/km)
Naiad	1989	29,967/48,227	0.0003	4.7°	36/58*
Thalassa	1989	31,115/50,075	0.0002	0.2°	50/80*
Despina	1989	32,638/52,526	0.0001	0.1°	94/151*
Galatea	1989	38,496/61,953	0.0001	0.1°	100/160*
Larissa	1989	45,701/73,549	0.0014	0.2°	130 x 110/209 x 177
Proteus	1989	73,102/117,646	0.0004	0.6°	270 x 250/435 x 402
Triton	1846	220,438/354,761	0.0000	156.8° (retrograde)	1,681/2,705
Nereid	1949	3,425,868/5,513,400	0.753	07.2°	212/341

*The margin of uncertainty for these figures is more than 10%.

BROKEN FAMILY

Flying by Neptune in 1989, the deep space probe Voyager 2 quadrupled the number of moons known to orbit Neptune. It discovered six new satellites close to the planet: Naiad, Thalassa, Despina, Galatea, Larissa and Proteus, moonlets so small that astronomers have not been able to study them in detail. The largest, Proteus, is 270 miles (435 km) across at most, while tiny Naiad might be only 30 miles (48 km) wide.

In most cases, Voyager's photographs show the inner moons as featureless, irregular blobs. Judging by their albedo—the amount of sunlight they reflect—they are dark and icy worlds. Only the outermost and largest, Proteus, came close enough to Voyager's flightpath to be photographed in detail, revealing the heavily cratered surface of a world too small—and so with too little gravity—to pull itself into the spherical shape of larger moons and planets.

Astronomers think that Proteus might once have been the innermost of a series of medium-size moons like those found around Saturn and Uranus. Theories of satellite formation predict that larger bodies tend to exist farther away from the planet. But Neptune's original satellite system was disrupted violently by a gatecrasher: Triton, the giant moon that now dominates Neptune's system.

DISRUPTIVE VISITOR

Triton may have started life as a planet in its own right, only to be captured by Neptune's gravity. Triton travels in an eccentric or elongated orbit, and its gravity may have disrupted the original moons, smashing them together, tearing them apart, or flinging them out of the system altogether.

The result of this chaos was a ring of debris around Neptune, most of which was swept up by Triton. Neptune's gravity slowed Triton and pulled it into the circular orbit it has today. Only the debris between Neptune and Triton's

orbit remained, gradually clumping together and forming the inner moons.

Recent searches with Earth-based telescopes have again increased the number of Neptunian moons. The new total is 13, and the latest moons are all irregular objects in highly elliptical, tilted orbits. Similar to the recently discovered satellites of the other giant planets, the five new moons are probably captured asteroids or Kuiper Belt objects.

THE UNSEEN MOON

In 1981, astronomers turned their telescopes on Neptune as the planet passed in front of a distant star. They were measuring the telltale variations in the star's light to glean data about Neptune's atmosphere and rings. On this occasion, though, astronomer Harold Reitsema noticed a brief flicker that could not be explained. When Voyager 2 arrived at Neptune, it became clear that this flicker must have been caused by the tiny moon Larissa passing in front of the star.

NAIAD
Neptune's smallest moon is also the nearest to the planet. It completes one orbit every 7 hours 6 minutes, circling about 14,600 miles (23,500 km) above Neptune's cloud-tops. The best photograph we have (above, taken by Voyager 2 in 1989) distorts Naiad—the moon is not really this elongated.

THALASSA
Only 1,150 miles (715 km) farther out than Naiad, Thalassa appears to be an irregularly shaped lump of rock—like all the inner moons. The Voyager photograph above exaggerates Thalassa's length.

DESPINA
The third moon out from Neptune is also the third smallest. All satellite systems tend to fall into this pattern, in which smaller satellites orbit closer to the center than larger ones.

SATELLITE SYSTEM

orbit of Triton (retrograde)

GALATEA
Galatea, the fourth Neptunian moon, was spotted speeding around Neptune at 23,500 mph (37,820 km/h) by Voyager 2 in 1989. Its appearance is distorted in the Voyager photograph at right.

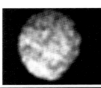

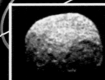

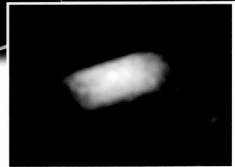

LARISSA
Larissa appears to be very heavily cratered. Like all Neptune's moons, it is named for a water-related character in Greek and Roman mythology.

PROTEUS
If Proteus were any larger, gravity would probably pull it into a sphere. It is one of the darkest objects in the solar system, absorbing 94% of the light that strikes it.

NEREID
Although Nereid was discovered in 1949, the 1989 Voyager photograph above, taken from a distance of 2.9 million miles (4.7 million km), is the best image we have of it.

TRITON

With a surface temperature of –391°F (–199°C), the coldest ever recorded in the solar system, Triton is possibly the last place you would expect to find volcanic activity. Yet this, the largest of Neptune's eight moons, is far from dead. The eerie, pink-colored surface is strewn with sheets of what appears to be frozen liquid that has erupted from deep within the core. Even stranger, Triton's unique terrain boasts giant columns of gas and dust that are driven up to six miles (10 km) into the sky by an unknown source of energy.

WHAT IF...

...YOU COULD TRAVEL TO TRITON?

Triton would be a magnificent world to visit—though perhaps not for an extended vacation and certainly not without adequate protection. Standing on the surface and looking around, you would see a dark, frozen landscape of salmon pink-colored nitrogen "snow." A mauve-tinged atmospheric haze would hug the horizon and the surface would appear noticeably curved and smooth compared with Earth.

Daylight on Triton could easily pass you by. The Sun is 30 times farther from Neptune than it is from Earth and would be little more than a speck in the sky—about the size of Venus as seen from Earth at its closest approach. On Triton, the Sun would be about 900 times fainter than the Sun as we see it, but still brighter than the combined light of more than 550 full moons. Triton's atmosphere is so thin that even with the Sun high in the sky, a great many stars would still be visible.

The most spectacular sight on Triton would be its parent planet Neptune, along with its system of rings. When full, Neptune as seen from the surface of Triton appears some 16 times larger than a full moon as seen from Earth. Because of its size, and because the surface is highly reflective, Neptune supplies more illumination than two of our full moons.

The giant planet would be at its most impressive, however, when Triton passes between it and the Sun. This alignment happens only once every 82 years and probably tints the frozen Triton landscape a chilly shade of blue.

Although Triton is a deeply frozen world, there would be little point in bringing your skates. Ice skating on Earth works because the pressure exerted by the blade on the ice lowers the ice's freezing temperature and allows it to melt, providing a thin layer of water to act as a lubricant.

On Triton, the surface is so cold that even nitrogen freezes solid and water is as hard as rock. Triton's frozen lakes are no more suitable for skating than a regular city sidewalk!

Every 82 years, Triton moves between Neptune and the Sun. It seems likely that reflected sunlight from the giant planet then turns the landscape blue.

TRITON FACTS

COMPOSITION	.75% ROCK; 25% FROZEN WATER, NITROGEN, AMMONIA AND METHANE
DIAMETER	.1,680 MILES (2,704 KM)
MASS	.0.29 MOON MASSES
DISTANCE FROM NEPTUNE	.220,438 MILES (369,245)
TIME TO ORBIT NEPTUNE	.5 DAYS, 21 HOURS
INCLINATION	.157° (TO EQUATORIAL PLANE OF NEPTUNE)
ECCENTRICITY OF ORBIT	.ZERO (IT IS PERFECTLY CIRCULAR)
ATMOSPHERE	.99% NITROGEN, TRACES OF METHANE
SURFACE PRESSURE	.0.0007 EARTH ATMOSPHERES
SURFACE TEMPERATURE	.–391°F (–199°C)
SURFACE GRAVITY	.0.08 G
ESCAPE VELOCITY	.3,240 MPH (5,214 KM/H)

LAND OF PINK SNOW

Triton is the seventh-largest planetary satellite in the solar system and is easily Neptune's largest moon. It was discovered in 1846, only 17 days after Neptune itself, by the British brewer and amateur astronomer William Lassell (1799–1880). Yet although we have been aware of Triton for over 150 years, almost all of what we know about it comes from a single flyby of the Voyager 2 space probe in 1989.

Voyager 2 revealed that Triton is a unique world. Much of the terrain—the coldest in the known solar system—is a striking salmon pink, probably due to a covering of frozen water, methane and nitrogen mixed with dust. There are vast plains that resemble the maria or "seas" on our own Moon but are made of ice, not rock. Other areas contain ridges that resemble the skin of a melon.

The most interesting feature of Triton is that it shows signs of volcanic activity, in the form of giant geysers of gas and dust that erupt miles into the sky. No one is sure what causes these "ice volcanoes." The most likely explanation is that sunlight penetrates the transparent parts of Triton's surface and melts the ice from below, causing local pressure buildups of gas. But since we know that beneath the layers of ice Triton has a rocky core, it is also possible that the pressure buildups are caused by heat generated internally—as on Earth.

UNANSWERED QUESTIONS

Voyager showed that Triton, like Pluto, has an atmosphere consisting mainly of molecular crystals of nitrogen and some methane. This "air" is so thin that the atmospheric pressure at the surface is 70,000 times less than that on Earth. Triton is also about the same size as Pluto, and has a similar composition and density—around 25% ice surrounding a rocky core. This, along with Triton's strange orbit, suggests that the two bodies share a common origin.

Triton not only orbits Neptune in the

SURFACE VIEW
The most striking features on Triton are the "ice volcanoes" that regularly erupt from beneath the surface and the frozen debris that they leave behind. It seems that any impact craters like those found on other moons in the solar system have long since been filled in.

jets blown horizontally by molecular "wind"

Neptune's rings

frozen debris

Neptune

liquid material may well up through cracks

DOOMED MOON

TRITON IS DOOMED, PARTLY BECAUSE OF THE MOON'S STRONG TIDAL INTERACTION WITH NEPTUNE, AND PARTLY BECAUSE IT ORBITS THE "WRONG WAY": ITS DECAYING ORBIT IS TAKING IT EVER CLOSER TO ITS PARENT PLANET. IN LESS THAN 100 MILLION YEARS, TRITON WILL BE ONLY A FEW THOUSAND MILES FROM NEPTUNE AND WILL BE TORN APART BY TIDAL FORCES. SOME OF THE PIECES WILL CRASH INTO THE PLANET; OTHERS MAY FORM A NEW SYSTEM OF RINGS.

SURFACE FEATURES

FROZEN "SEAS"
This Voyager 2 picture shows one of the many plains on Triton that resemble the "seas" on the Moon. The plains seem to be the result of liquid material, probably a mixture of water and ammonia, that has welled to the surface and frozen following meteorite impacts. This could also explain the apparent lack of craters.

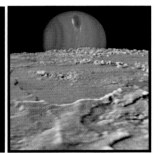

ICE VOLCANOES
Many of Voyager's pictures show geyser-like jets of what appear to be gas, dust and ice particles. These jets erupt several miles into the sky before being blown almost horizontally by Triton's powerful molecular "winds." The eruptions are probably caused by pressure buildups of gas beneath the frozen surface.

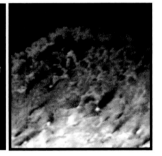

RIDGE FORMATION
Triton has little surface relief and few high mountains. But, in places, the terrain is green-tinged, with depressions and ridges that resemble the skin of a melon. Due to the absence of craters elsewhere, scientists think that the formations are more likely to be the result of subsurface activity than meteorite impacts.

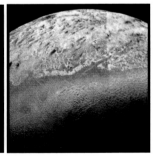

opposite direction to most moons; its orbit is also steeply inclined, at nearly 157° to the equatorial plane of its parent planet. This results in an 80-year "summer" at the poles that lasts for around half of Neptune's 164-year orbit around the Sun. It also suggests that we may have to wait a long time before the many faces of Triton reveal themselves.

MYSTERIOUS NEPTUNE

At a colossal 2.8 billion miles (4.5 billion km) from the Sun, Neptune is sometimes the most distant planet in the solar system—even Pluto strays inside its orbit from time to time. Only one spacecraft, Voyager 2, has ranged as far as the blue planet. Voyager sent back a wealth of data, but generated as many questions as it answered. Neptune's interior is still poorly understood. But most mysterious of all is Neptune's magnetic field: To explain its strange patterns, we could be forced to rethink our theories of planetary magnetism.

WHAT IF...

...WE COULD EXPLAIN NEPTUNE'S MAGNETISM?

Neptune's magnetic field presents a major mystery. Unlike the fields of the Earth, Jupiter or Saturn, it does not line up even roughly with the axis of rotation—so that a compass on Neptune, if one could be put there, would point not north-south but 47° away from that line. Neptune's misaligned magnetism is one of the most puzzling problems in planetary science.

Scientists are not certain how planets get their magnetic fields. The generally accepted theory is that they are caused by some form of internal dynamo. A human-made dynamo generates a current when an electrical conductor moves through a magnetic field. But in the case of a planet, it is the other way around. The dynamo produces a magnetic field by the movement of electrically conducting fluids—such as molten iron-rich rocks or ionized gas—as the planet spins.

This should give a magnetic field that lines up with the planet's axis. In reality, only Saturn's magnetic field is in line with its poles: all the others are out of alignment. Even the Earth's magnetic and rotational axes are offset by 11°. Neptune is just one extreme example—at 59° apart, the poles of Uranus are even more askew.

So why are Neptune's axes so far adrift? An easy answer is that the planet's magnetic field is currently undergoing a reversal of polarity, as the Earth's has done many times in its history. But this theory depends on a huge coincidence. Magnetic reversals are uncommon events upon the Earth, and it seems very unlikely that the Voyager probe would have caught Neptune, let alone Uranus, halfway through this process on its journey to the outer solar system.

The truth probably lies in the dynamo mechanism itself. This dynamo is not created by spin alone, but also by convection currents within the fluid planet. Convection is affected by solar and internal heat, and causes great turbulence within low-density planets such as Neptune and Uranus. These motions need not be symmetrical with the rotation axis, any more than the bubbles of convection in a boiling pan of water are symmetrical with the center of the pan.

In this view, the wildly skewed magnetic fields of Neptune and Uranus seem no more unusual than the slightly off-axis field of Earth. It is even possible that the magnetic field of Neptune is never constant, but moves with relation to the axis all the time—and explorers on Neptune, if they ever exist, would have to rely on more than a compass to find their way around.

GATHERING CLUES

1845–6	ENGLISH ASTRONOMER JOHN COUCH ADAMS AND FRENCH ASTRONOMER URBAIN JEAN JOSEPH LE VERRIER INDEPENDENTLY CALCULATE THE POSITION OF NEPTUNE, BASED ON DEVIATIONS IN THE ORBIT OF URANUS
SEPTEMBER 23, 1846	NEPTUNE IS FIRST OBSERVED BY ASTRONOMERS JOHANN GOTTFRIED GALLE AND HEINRICH LUDWIG D'ARREST
1984	TWO TEAMS OF ASTRONOMERS FROM EUROPE AND THE U.S. FIND EVIDENCE OF RING ARCS AROUND NEPTUNE
AUGUST 1989	FLYBY OF VOYAGER 2 REVEALS FIVE COMPLETE RINGS, SIX NEW SATELLITES AND STUDIES THE BLUE PLANET'S COMPOSITION, ROTATION RATE AND MAGNETIC FIELD

PLANETARY PUZZLES

Neptune has much in common with its neighbor Uranus. These planets are both giants, and the similarities that the Voyager 2 probe found in their colors and atmospheres matched scientists' expectations. But one of the features that Neptune shares with Uranus is much more surprising.

Voyager reported that the rotational axis of Uranus was offset from its magnetic axis by a full 59°—much greater than the angle that separates the magnetic and rotational poles on other planets. Initially, Uranus' oddly aligned magnetic field was attributed to the planet's strange rotation pattern. Other planets spin around more or less upright axes, but Uranus is tipped on its side so that it rolls around the Sun like a barrel. This explanation was not entirely satisfactory, but astronomers accepted Uranus as a unique case—until Neptune threw up a similar dilemma.

When Voyager flew past Neptune, it measured 47° between the planet's magnetic and rotational axes. Neptune, though, does not rotate on its side. This blue planet has an axial tilt of only 30° from the vertical—similar to that of the Earth. In Neptune's case, the strange angle of the magnetic field clearly was not caused by an equally strange angle of planetary rotation.

As if this were not strange enough, the magnetic axes of both Neptune and Uranus pass through the planets off-center—a phenomenon that scientists still have not been able to explain.

PRIMITIVE PLANET

Neptune's atmosphere also raises difficult questions. For a giant planet, it has a high proportion of helium, at 18%, but it is low on hydrogen, at 79%. In fact, Neptune has more helium and less hydrogen than any of the three other giants in our solar system. So where does Neptune's excess helium come from? It seems that Neptune has not gained gas—but that the others have lost it. Around 5 billion years ago, our solar system condensed from a primordial nebula that contained 27% helium: Neptune has simply hung on to more of its original quota than any other planet.

NEPTUNE'S BIG SECRETS

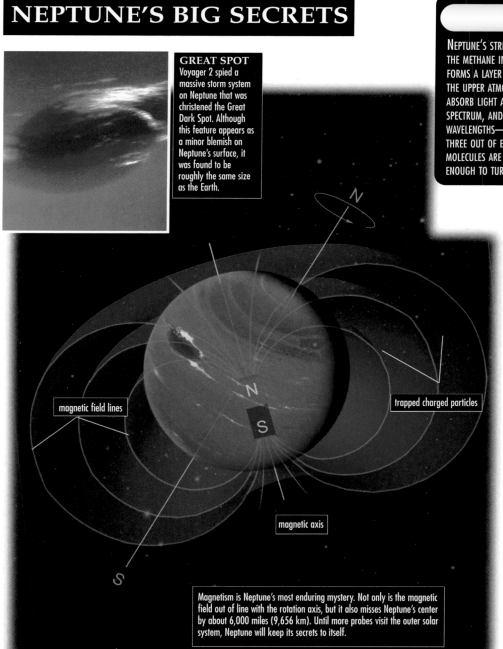

GREAT SPOT
Voyager 2 spied a massive storm system on Neptune that was christened the Great Dark Spot. Although this feature appears as a minor blemish on Neptune's surface, it was found to be roughly the same size as the Earth.

magnetic field lines

trapped charged particles

magnetic axis

Magnetism is Neptune's most enduring mystery. Not only is the magnetic field out of line with the rotation axis, but it also misses Neptune's center by about 6,000 miles (9,656 km). Until more probes visit the outer solar system, Neptune will keep its secrets to itself.

NEPTUNE BLUES

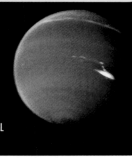

NEPTUNE'S STRIKING BLUE COLOR IS DUE TO THE METHANE IN ITS ATMOSPHERE. METHANE FORMS A LAYER OF ICE CRYSTAL CLOUDS IN THE UPPER ATMOSPHERE. THESE CLOUDS ABSORB LIGHT AT THE RED END OF THE SPECTRUM, AND REFLECT SHORTER WAVELENGTHS—GREEN AND BLUE. ONLY THREE OUT OF EVERY 100 ATMOSPHERIC MOLECULES ARE METHANE, BUT THEY ARE STILL ENOUGH TO TURN THE WHOLE PLANET BLUE.

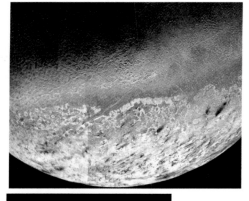

NEW MOON
Neptune's satellite Triton was not always a moon: It once orbited the Sun independently. Triton is as mysterious in the present as the past: We still know little about its orbit, geology and chemistry.

The internal pressures of Saturn and Jupiter are high enough to form metallic hydrogen, a highly compressed gas. Droplets of helium form under enormous pressures and sink through the liquid hydrogen to settle onto the core. Neptune's smaller size results in lower pressures that do not deplete the planet's atmospheric helium in such a way. In its distant orbit, the blue planet has become frozen in time—and if we do solve its mysteries, Neptune will give us a picture of how the giants may have looked when they were young.

PLUTO AND BEYOND

The outer reaches of the Solar System are still shrouded in mystery, but astronomers have begun to learn more about them in recent decades. Tiny Pluto, discovered in 1930 after a deliberate search for a "Planet X" whose gravity affected Neptune, orbits the Sun in an elongated, tilted orbit that sometimes brings it within the orbit of Neptune. However, even with its giant satellite Triton it is too small to affect Neptune, and it now seems that the wobbles in Neptune's orbit can be explained without another giant planet. Pluto remained a "loose end" of the Solar System until the 1990s, when astronomers began to find other objects at the same distance from the Sun. These objects form the "Kuiper Belt," an outer region of icy "minor planets," with Pluto as its largest member. Further out still lies the Oort Cloud, a giant spherical shell of dormant comets, expelled from the region around Neptune early in the Solar System's history.

A computer-generated impression of the view from the Kuiper Belt, a distant zone of icy rocks and minor planets between 4.4 billion miles (7 billion km) and 90 billion miles (145 billion km) from the Sun, which includes Pluto, Charon, and a number of other newly discovered "minor planets."

PLUTO AND CHARON

Pluto, the outermost known planet in the solar system, orbits the Sun once every 248 years at a distance of up to 4.6 billion miles (7.4 billion km). Its moon, Charon, is so similar in size that together they make up the solar system's only double planet, but they are so far from Earth that only the Hubble Space Telescope has been able to make out any surface detail. Despite the lack of data, astronomers have worked out that Pluto is an icy, rocky world and even—at times—has an atmosphere. But there is still much to learn.

WHAT IF...

...WE COULD VISIT PLUTO?

From the barren, icy surface of Pluto, where the average temperature is –364°F (–184°C) and the Sun gives less than a thousandth of the light and heat that it does on Earth, the most impressive sight is Charon, the planet's moon. Charon is much smaller than our own Moon, but it is so close to Pluto that from there it would look about seven times the size of the Moon as seen from Earth. However, Charon is nowhere near as bright. Even though its frozen surface reflects sunlight perhaps up to eight times more efficiently than the Moon does, it appears very dim because it is so far from the Sun.

Charon dominates the sky of Pluto, but you have to be in the right place to see it. Pluto and Charon each rotate once every 6.39 days, which is exactly the time it takes Charon to make one orbit of Pluto. So they swing around their common center of gravity like two balls connected by a piece of string, and the same side of Charon is always facing the same side of Pluto. Charon is only visible from one hemisphere of Pluto, and Pluto can only be seen from one hemisphere of Charon. If you were on the "wrong" side of one of them, you would never see the other. And from Pluto, Charon never appears to move—it is always in the same position, neither rising nor setting; the same is true of Pluto as seen from Charon.

Pluto itself probably resembles Neptune's moon, Triton, and may even have similar features such as ice volcanoes—geyser-like plumes of gas, ice particles and dust that spurt out from its surface. On Triton, these are thought to be caused by sunlight penetrating the surface ice and heating the ice below it. The heated ice melts and vaporizes into gas, which eventually builds up enough pressure to blast out through the surface. A similar process could create ice volcanoes on Pluto.

Fortunately, we may not have much longer to wait before we learn more about Pluto and Charon. NASA's New Horizons probe, destined for Pluto, Charon and the Kuiper Belt beyond them, is scheduled for launch in 2006 or 2007, and if all goes well, it should encounter these tiny outer worlds around the middle of the next decade. Scientists are particularly keen to get a probe to Pluto in a hurry so that they can study its atmosphere, which is expected to freeze to the surface as Pluto retreats from the Sun.

PLUTO AND CHARON

PLUTO	CHARON
DIAMETER . . . . .1,485 MILES (2,390 KM)	DIAMETER . . . . .730 MILES (1,175 KM)
AXIS TILT . . . . .122.5°	DISTANCE FROM PLUTO . . . . . . .12,204 MILES (19,640)
LENGTH OF YEAR . . . . .247.7 EARTH YR	TIME TO ORBIT PLUTO . . . . .6.39 EARTH DAYS
LENGTH OF DAY . . . . .6.39 EARTH DAYS	ROTATION PERIOD . . . . .6.39 EARTH DAYS
DISTANCE FROM SUN . . . . .2.9–4.6 BILLION MILES	SURFACE TEMPERATURE . . . .–364°F (–184°C)
(4.7–7.4 BILLION KM)	SURFACE GRAVITY . . . . .APPROX. 0.02G
SURFACE TEMPERATURE . . . .–364°F (–184°C)	ANGLE OF ORBIT TO PLUTO'S EQUATOR . . . . .98.8°
SURFACE GRAVITY 0.04G	ATMOSPHERE . . . . .POSSIBLY METHANE, NITROGEN AND
ANGLE OF ORBIT TO ECLIPTIC . . . . .17.148°	CARBON MONOXIDE
ATMOSPHERE . . . . .POSSIBLY METHANE, NITROGEN AND	COMPOSITION . . . . .ROCK, ICES
CARBON MONOXIDE	
COMPOSITION .ROCK, ICES	

DISTANT ICEBALLS

Astronomers know very little about Pluto. It is simply too far away, not to mention too small, for even space-based telescopes to reveal much, and it is the only planet yet to be visited by a space probe. But even with the scant information available, it is clear that Pluto is the oddball of the solar system.

Pluto is easily the smallest of the known planets. It is just one-fifth the diameter of Earth and smaller than the solar system's seven largest planetary satellites, including the Earth's own Moon. In fact, Pluto is so tiny that many astronomers originally thought that it was really an escaped moon that had once orbited Neptune. Further evidence for the escaped-moon theory came from Pluto's orbit around the Sun, which is quite unlike that of any other planet and suggests that Pluto entered its present orbit long after the other planets settled into theirs.

Pluto's orbit is the most elongated ellipse of any of the planets. At its most distant from the Sun—4.6 billion miles (7.4 billion km)—Pluto is almost 2 billion miles (3.2 billion km) farther out than when it is at its closest, at which point it passes nearer to the Sun than its neighbor Neptune does. Stranger still, Pluto's orbit is tilted at an angle of some 17° to the orbits of all the other planets.

But in 1978, the discovery that Pluto had a satellite cast doubt on the "escaped moon" theory. As an alternative, some astronomers have suggested that Pluto and its companion, Charon, originated in the Kuiper Belt—the distant ring of rocky and icy debris that extends beyond the orbit of Neptune.

NEVER FAR APART

Pluto and Charon have been labeled the "double planet" because their sizes are closer together than most planet-moon combinations, and the distance between them is far less. Charon is more than half the diameter of Pluto—making it the largest moon in relation to its parent planet—and it orbits its parent at a distance of only about eight Pluto diameters. By comparison,

THE DOUBLE PLANET

DISCOVERY OF CHARON
Charon was discovered on July 2, 1978, by James Christy with the 61-in (1.54-m) telescope at the U.S. Naval Observatory in Flagstaff, Arizona. Officially, Charon is named for the figure in Greek mythology who ferried the souls of the dead across the River Styx into the underworld, which was ruled by Hades—known also as Pluto. Unofficially, Jim Christy also named the moon Charon to honor his wife, Charlene.

THE MOON
Charon is over half the size of Pluto and probably consists of about 70% water ice and 30% rock, with an ice-covered surface.

PLUTO'S ATMOSPHERE
Astronomers first realized that Pluto has an atmosphere in 1980, when the planet passed across a background star. The light from the star diminished as Pluto approached it and increased again as the two separated. From this, astronomers deduced the presence of an atmosphere that is very thin, but which extends a long way. It is at least as deep as Pluto's diameter and may even stretch all the way to Charon, which would give the two bodies a shared atmospheric "envelope" (below). Methane seems to be the principal gas, but nitrogen and carbon monoxide may also be present. The atmosphere is probably created by frozen gas that thaws when Pluto is at its closest to the Sun.

THE PLANET
Pluto is thought to consist of about 70% rock and 30% water ice, with patches of frozen nitrogen, methane and carbon monoxide on its surface.

in the Earth/Moon system, the Moon is just over a quarter the diameter of the Earth and separated from it by 31 Earth diameters.

Pluto's surface is probably covered in frozen nitrogen, methane and carbon monoxide, which vaporize to form a very thin atmosphere when the planet moves closest to the Sun. Charon, by contrast, may have water ice on its surface. But we will not know for sure until the double planet is visited by a space probe—which will hopefully happen within the next decade.

THE SEARCH FOR PLUTO

PLUTO IS THE MOST RECENTLY DISCOVERED PLANET. ITS EXISTENCE WAS FIRST SUGGESTED IN THE 19TH CENTURY TO EXPLAIN SLIGHT WOBBLES IN THE ORBIT OF URANUS, AND EARLY IN THE 20TH CENTURY AMERICAN ASTRONOMERS PERCIVAL LOWELL (1855–1916) AND WILLIAM PICKERING (1858–1938) EACH TRIED TO PREDICT ITS POSITION. IN 1929, CLYDE TOMBAUGH (1906–97, ABOVE) STARTED WORK AT THE LOWELL OBSERVATORY IN FLAGSTAFF, ARIZONA, WITH A TEAM OF ASTRONOMERS HEADED BY VESTO SLIPHER (1875–1969). THEY CARRIED OUT AN EXHAUSTIVE PHOTOGRAPHIC SEARCH FOR THE PLANET, AND ON FEBRUARY 18, 1930, THE 24-YEAR-OLD TOMBAUGH DISCOVERED PLUTO CLOSE TO WHERE LOWELL'S CALCULATIONS HAD SAID IT WOULD BE.

CHANGING VIEWS OF PLUTO

Small and immensely distant, Pluto has yielded few secrets since its 1930 discovery. Astronomers have determined the planet's size and mass, its density, and its surface composition, and they have a fair idea of its surface geology: It has polar caps, for instance, made of methane ice. We also know that Pluto has a moon, Charon, and a very tenuous atmosphere. Aside from these basic facts, though, little else is known about Pluto and Charon. And it could be many years before we find out anything more.

WHAT IF...

...WE COULD LEARN MORE ABOUT PLUTO?

Deep in the uncharted space at the edge of the solar system lies the planet Pluto. This "worldlet" is so far away and so tiny that we know very little about it. But its size and distance are not the only reasons for our ignorance. Easily the most important reason for this lack of knowledge is that Pluto remains the only planet not yet visited by a probe.

In the late 1970s, NASA launched the highly successful Voyager probes to conduct a Grand Tour of the Solar System. Voyagers 1 and 2 got as far as Saturn, after which Voyager 1 was deflected out of the plane of the solar system. But Voyager 2 continued its explorations. It sailed past Uranus in January 1986 and Neptune in August 1989, and since then has been heading for interstellar space. The missions were highly successful, and returned a wealth of information on the outer planets and their moons—but not Pluto. Pluto was poorly placed at the time. After Uranus, it was easier for the probe to continue on to Neptune rather than Pluto, and it could not reach both: The planets were in different directions relative to the probe.

So Pluto remains, 70 years since its discovery, almost as mysterious now as it was back in the 1930s—a sad consequence of the planet's poor celestial positioning. It will be late in the 22nd century before the outer planets will once more line up to make a complete "Grand Tour" possible.

Fortunately, the increasing power of Earth-based telescopes is helping to reveal a little more about Pluto, and there have even been attempts to map the surface, with the help of the Hubble Space Telescope. A new generation of observatories, including the James Webb Space Telescope, Hubble's orbiting replacement, should reveal even more.

However, if we really want to learn more about Pluto, we will have to rely on a dedicated space probe—and fortunately NASA plans to despatch one on an express trip to the outer Solar System in 2006 or 2007. The New Horizons probe will make the journey at breakneck speed, helped by a gravitational slingshot around Jupiter, and should arrive in the middle of the next decade. It will be racing against the oncoming Plutonian winter, which is already thought to be freezing the atmosphere and changing the planet's appearance as it moves further from the Sun.

AN AUDIENCE WITH PLUTO

1905	Percival Lowell begins searching for Pluto
1916	Lowell dies on November 12, at age 61
1919	William H. Pickering captures Pluto on a photographic plate but does not realize it
1930	Clyde Tombaugh finds Pluto on February 18
1976	Infrared spectroscopy reveals methane ice on Pluto
1978	James W. Christy locates Pluto's moon, Charon
1979	Pluto comes inside Neptune's orbit
1979–1985	Pluto's diameter and mass are accurately calculated
1985–1990	Pluto and Charon occult each other as seen from Earth
1988	Pluto passes in front of a star and the planet's atmosphere is revealed
1996	Hubble Space Telescope resolves bright and dark patches on Pluto's surface
1997	Tombaugh dies on January 17, two weeks before his 91st birthday
1999	Pluto goes outside Neptune's orbit
1999	IAU considers including Pluto in a catalogue of trans-Neptunian objects

A PLANET TOO FAR

Pluto, the ninth and outermost planet, was discovered in 1930, as the result of an extensive photographic search. But it was not the planet that astronomers had expected.

Before Pluto's discovery, two astronomers calculated that an unseen celestial body beyond Neptune—then the most distant known planet—was perturbing the orbit of Uranus. The astronomers, Bostonians William H. Pickering and Percival Lowell, estimated that the new object would be several times more massive than the Earth. When Pluto was finally revealed after an extensive search, 25 years later, its speed in the sky suggested that at the time it was not much farther away than Neptune. Yet it was so tiny—the size of a penny as seen from 3,300 miles (5,310 km)—that it had to be much smaller and less massive than the giant planet Lowell and Pickering had anticipated. It was also revealed that Pluto's orbit is more eccentric or elongated than that of any other planet: It varies its distance from the Sun between 30 and 50 times the Earth-Sun distance. Except for Mercury, all the other planets have nearly circular orbits.

Right from the beginning, it was obvious that Pluto was going to be the oddball of the solar system. But it is so far away—on average, 4.6 billion miles from the Sun—and so small that astronomers did not start uncovering any details at all of the outermost planet until the 1970s.

When they did, though, our perceptions of Pluto started to change. In 1976, spectroscopy—the analysis of light—showed that Pluto is an icy world. Its surface is covered in frozen methane, particularly at the planet's polar caps. Then, in 1978, James W. Christy took a historic photograph of Pluto. The photo showed that Pluto had a significant bump protruding from one side. This was the planet's previously unseen moon, unresolved in the photo but still unmistakable. It was named Charon, after the ferryman who brought the dead to the domain of Pluto, god of the underworld, in ancient Greek legend.

The discovery of Charon was fortunate, for it enabled astronomers to work out the

diameters and masses of both worlds quite accurately. From 1985 to 1990, Charon and Pluto were oriented in such a way that, as seen from Earth, each one passed in front of the other as they orbited. By studying how the light varied during these events, called occultations, astronomers could calculate the relative size of the two bodies, and how far apart they were. Once they could add to their equations the time it took Pluto and Charon to complete a mutual orbit—just 6.4 days—they were at last able to make an accurate calculation of the bodies' masses.

Pluto turned out to be truly tiny—hardly planet-sized at all in comparison with the rest of the planets in the solar system. Before these occultations, when astronomers had only a rough idea of the planet's mass and diameter, they had

imagined it to be a world slightly larger than Mercury. In fact, Pluto is not only smaller than Mercury, it is also smaller than the solar system's seven largest natural satellites, including our own Moon. This makes Pluto the smallest object known to have an atmosphere.

That atmosphere—revealed in 1986 when Pluto occulted a star—was perhaps the most recent major change in our perception of the ninth planet. Yet despite these refinements in our perceptions of Pluto since its discovery, the planet remains a mystery. Less is known about it than about the icy satellites of the gas giants. The reason is simple: Pluto has so far escaped the scrutiny of an interplanetary probe. We will probably have to wait until such a probe reaches this distant world before we can hope to discover anything more.

STEPS TOWARD UNDERSTANDING

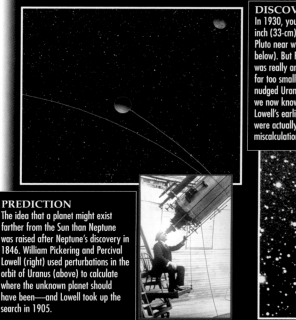

DISCOVERY
In 1930, young U.S. astronomer Clyde Tombaugh used the 13-inch (33-cm) telescope at Lowell Observatory (inset) to discover Pluto near where Lowell and Pickering had suggested (arrowed, below). But Pluto's discovery was really an accident: It is far too small to have nudged Uranus' orbit, and we now know that Percival Lowell's earlier calculations were actually miscalculations.

PREDICTION
The idea that a planet might exist farther from the Sun than Neptune was raised after Neptune's discovery in 1846. William Pickering and Percival Lowell (right) used perturbations in the orbit of Uranus (above) to calculate where the unknown planet should have been—and Lowell took up the search in 1905.

YARDSTICK
In 1978, 48 years after Pluto was found, James W. Christy discovered a small satellite with the 61-inch (1.54-m) U.S. Naval Observatory telescope (inset). The tiny moon (above, right) showed up at first merely as an odd bulge on the side of Pluto: As seen from Earth, the two objects were almost in line with each other.

YOU'RE FIRED

SOME ASTRONOMERS REGARD PLUTO NOT AS A PLANET, BUT AS ONE OF A CLASS OF OBJECTS THAT ORBIT BEYOND NEPTUNE—*TRANS-NEPTUNIAN OBJECTS*, OR TNOS. IN 1999, THE INTERNATIONAL ASTRONOMICAL UNION (ABOVE) DISCUSSED WHETHER TO ASSIGN TNOS, INCLUDING PLUTO, A NUMBER IN A TECHNICAL CATALOG. THE MEDIA TOOK THIS TO MEAN THAT PLUTO'S STATUS AS A PLANET WAS IN DANGER OF REVOCATION. BUT THAT WAS NEVER THE CASE. PLUTO IS STILL A PLANET AND NO NUMBER HAS YET BEEN ASSIGNED IT.

BEYOND PLUTO

WHAT IF...

...A PROBE IS SENT INTO THE KUIPER BELT?

By 2020 we could have close-up images of the Kuiper Belt. In 2006 or 2007, NASA plans to launch a mission to Pluto and its moon, Charon—and then beyond, into the heart of the Kuiper Belt. After its launch, New Horizons, as it has been named, will use Jupiter's gravity to whip it on its way toward Pluto.

There are many possible ways of balancing weight, speed, complexity and cost, but NASA has opted for a fast mission. The probe and its propellants together weigh around 1,000 pounds (454 kg), allowing it to reach Jupiter in just over a year. Here, it will perform a "slingshot" maneuver that will boost its speed to 47,000 miles per hour (75,640 km/h), fast enough to reach Pluto by around 2015, where it will make the first close-up record of the most distant planet.

Then New Horizons will head further into the Kuiper Belt. The mission controllers do not plan to select any Kuiper Belt targets until the probe is well on its way to Pluto—partly because more objects are being discovered there each year, so the possibilities will change as the mission goes on. In order to alter course, the spacecraft is equipped with 16 thrusters able to orient it in any direction. Since it will be blasted on its way to Jupiter by a powerful lower rocket stage, New Horizons should not have to start using its fuel reserves until it enters the Kuiper Belt.

During its close encounters, the probe will have no immediate help from home—a signal would take about 10 hours to make the round trip. Instead it will obey previously uploaded commands, and use basic artificial intelligence to make decisions.

The project offers a chance to examine the primordial stuff of the solar system. In addition, since the Kuiper Belt is thought to be the source of short-period comets, the probe might spot a comet before it begins to grow a "tail."

Once its fuel and power are exhausted, the craft will keep heading toward the edge of the solar system. When the probe is 14 times further away than Pluto—around the middle of the 22nd century—light from the Sun will grow weak and darkness will set in. By then the probe will be on its way to the Oort Cloud—and then on to follow the Voyagers and Pioneer into interstellar space.

Forty years ago, at the beginning of the space age, Pluto was routinely described as the most distant object in the solar system. But not any more: Recent discoveries and theories suggest that the planets occupy just the inner heart. Four probes have now traveled to about twice the distance of Pluto, into the realm of the comets. Another probe will follow them shortly. Eventually, all five will travel through the solar system's outer reaches, where lonely comets orbit the Sun—three-quarters of the way to the nearest star.

NEW HORIZONS SCHEDULE

January/February 2006	Primary launch window
February 2007	Secondary launch window (direct flight to Pluto)
February 2007	Jupiter fly-by (if launched in 2006)
2007–2015	Interplanetary cruise
July 2015	Pluto fly-by
2015–2020	First Kuiper Belt encounters (if launched in 2006)
2019–2020	Pluto fly-by (if launched in 2007)
2020 onwards	Possible extended mission

TO THE EDGE

Astronomers are eager to establish the boundaries to the Sun's empire. There are two possible answers. One depends on the extent of the Sun's electromagnetic influence, the other on its gravitational reach. The electromagnetic influence—the solar wind, as it is called—gradually fades as it encounters the dust, gas and radiation that make up the incredibly tenuous "atmosphere" between the stars, the interstellar medium. Pioneer 10 and Voyager 1, both at almost twice Pluto's distance, report that the intensity of cosmic rays from other stars increases by 1.3% every 100 million miles (160 million km). At that rate, like the light of a lighthouse dying away in fog, the Sun's radiation should finally fade out at about four times Pluto's distance.

Meanwhile, Pioneer 10 and Voyager 1— along with Voyager 2, now 5.5 billion miles (8.9 billion km) away—have entered the domain of the comets. Astronomers believe that the cometary sources form two diffuse regions, one that reaches to 25 times the distance of Pluto from the Sun and a second stretching out to 5,000 times that distance.

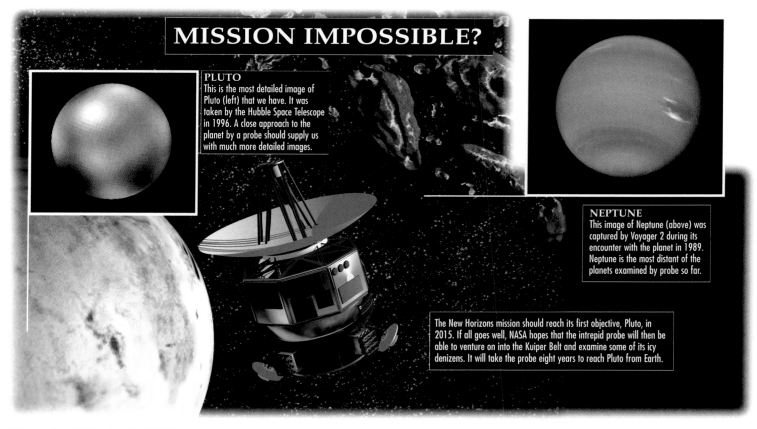

MISSION IMPOSSIBLE?

PLUTO
This is the most detailed image of Pluto (left) that we have. It was taken by the Hubble Space Telescope in 1996. A close approach to the planet by a probe should supply us with much more detailed images.

NEPTUNE
This image of Neptune (above) was captured by Voyager 2 during its encounter with the planet in 1989. Neptune is the most distant of the planets examined by probe so far.

The New Horizons mission should reach its first objective, Pluto, in 2015. If all goes well, NASA hopes that the intrepid probe will then be able to venture on into the Kuiper Belt and examine some of its icy denizens. It will take the probe eight years to reach Pluto from Earth.

ANOTHER PLANET?

In January 2005, astronomers Michael Brown, Chad Trujillo and David Rabinowitz at the Mount Palomar observatory discovered an object beyond the orbit of Neptune which seemed to have a radius of more than 1,500 miles (2,500 km). This would make it larger than Pluto. Officially named 2003 UB313, this "10th planet" was popularly dubbed Xena after the TV character. The discovery of 2003 UB313 has prompted much discussion in astronomical circles about the true definition of a planet. At the time of writing, the International Astronomical Union (IAU) is yet to deliver a conclusive ruling on the status of 2003 UB313.

INTO COMET COUNTRY

The first comet store, the source of short-period comets, is named after Dutch-American astronomer Gerard Kuiper (1905–73), who suggested its existence in 1951. He proposed that the belt stretched from Neptune's orbit to 25 times Pluto's distance from the Sun, lying in the same plane as the planets. Until 1992, the Kuiper Belt was mere theory. But then David Jewitt and Jane Luu of the University of California spotted a minute object about 200 miles (320 km) across, orbiting beyond Neptune—the first Kuiper Belt Object (KBO). Since then, some 60 have been found. Astronomers theorize that there should be anything from one billion to 6.7 billion of them.

Voyager 1, Voyager 2 and Pioneer 10 are in the Kuiper Belt now. Slight shifts in their path may allow scientists to calculate the combined gravity of the comets and estimate their numbers.

The second comet "sink" is the source of long-period comets, a cloud named for Dutch scientist Jan Oort (1900–92), who suggested its existence in the 1940s. This halo of the solar system, ranging from 6,000 to 200,000 times the Earth-Sun distance, may contain as many as 6 trillion comets. If the Voyagers and Pioneers survive to enter the Oort cloud, it will be as dead objects in 2,000 years. They will probably continue their journey unscathed, leaving the Oort Cloud—and the solar system—some 65,000 years later.

COMET FARM

The Kuiper Belt (right) may consist of up to about 6 billion objects, of which some 70,000 are planetesimals between 60 and 475 miles (97 and 764 km) across lying in a band stretching 30 to 50 AU, 200 million are in the 6- to 12-mile (10- to 19-km) size range and the rest are under a mile across. Any one of them could be nudged from its orbit and become a comet. Despite their numbers, the distances between them are immense—about one large object every 100 million miles (160 million km).

KUIPER BELT

L ate in 1992, astronomers David Jewitt and Jane Luu discovered a tiny world just 150 miles (240 km) across and over 3 billion miles (4.8 billion km) from the Sun. This lump of ice and rock was the first direct proof of the existence of the Kuiper Belt, a zone of space debris out beyond the orbit of Neptune. Until then, the Kuiper Belt had been just a theoretical possibility, but astronomers now believe that it contains thousands of small objects and is the source of many comets. Some day, it could also provide interstellar travelers with vital supplies.

WHAT IF...

...WE COULD USE KBOs AS SPACE GAS STATIONS?

O ne day, space explorers from Earth will venture out of the solar system to visit other star systems. When they do, Kuiper Belt Objects (KBOs) could prove useful by providing fuel and other supplies. A trip to another star will take decades. Even if a starship travels at a tenth of the speed of light, it will take over 40 years to reach Earth's nearest star, Proxima Centauri, and another 40 years to get back. The speed of light is 186,282 miles per second (299,792 km/sec). To travel at even a tenth of that speed means accelerating to 2,500 times the speed of an Apollo rocket.

Interstellar ships will not just be fast; they will be self-contained worlds in their own right, equipped to feed and clothe their crews and to keep themselves in good working order during voyages lasting 80 years or more. They will also have full medical facilities, and unless the crews spend most of the journey in suspended animation, these will have to include maternity units. On such long expeditions, the crews who return to Earth will be the children or grandchildren of those who set out.

To make all this possible, such ships will need to carry large supplies of raw materials and fuel. But around a quarter of the journey's time each way will be spent just traveling through the solar system. So taking on essential supplies at the last "filling station" in the system—the Kuiper Belt— would save precious weight and fuel.

The KBOs contain valuable raw materials. Their frozen water could provide drinking water and oxygen. Their organic (carbon-based) components, such as methane and carbon dioxide, could be converted into fuels and processed to make plastics or even essential nutrients such as vitamins.

Mining ships in the Kuiper Belt could extract valuable materials, such as water and methane, from the KBOs and transfer them to passing starships for use on long interstellar journeys.

MAJOR KUIPER BELT OBJECTS

Name	Discovered	Diameter	Perihelion (au)	Aphelion (au)
2003 UB313	2005	1,600 miles (2,575 km)*	38.2	97.6
Pluto	1930	1,440 miles (2,317 km)	29.5	49.3
2004 DW	2004	930 miles (1,497 km)*	30*	50*
Sedna	2003	<930 miles (<1,497 km)*	76	990
Charon	1978	795 miles (1,279 km)	29.5	49.3
2005 FY9	2005	780 miles (1,255 km)*	38.7	52.6
2003 EL61	2005	745 miles (1,199 km)*	35.2	51.5
Quaoar	2002	740 miles (1,191 km)*	40*	44*
Ixion	2001	660 miles (1,062 km)*	29*	47*
Varuna	2000	560 miles (901 km)*	39*	44*
2002 AW197	2002	550 miles (885 km)*	68.5	76.4

* APPROX

OUT ON THE EDGE

Far out in the gloom of deep space, beyond the orbit of Neptune, lies a zone of debris left over from the birth of the solar system. The debris takes the form of cold, dark chunks of ice and rock that orbit the Sun in a broad, flat ring called the Edgeworth-Kuiper Belt, or simply the Kuiper Belt.

At an average distance of around 3 billion miles (4.8 billion km) from the Sun, so-called Kuiper Belt Objects (KBOs) are so faint that astronomers didn't detect them until 1992. But occasionally, a KBO turns itself into one of the most visible objects in the night sky: The Kuiper Belt is a fertile source of comets.

The existence of the Kuiper Belt was predicted long before the first KBOs were discovered. In 1943, the Irish astronomer Kenneth Essex Edgeworth (1880–1972) suggested that certain comets must come from a region beyond the known planets. And in 1951, the same idea was put forward independently by Gerard Pieter Kuiper (1905–73), a Dutch-born American astronomer.

The Kuiper Belt came into existence shortly after the planets and their moons formed out of the swirling cloud of gas and dust that orbited the young Sun. Some leftover material settled into what is now the asteroid belt between the orbits of Mars and Jupiter, having failed to form a planet due to Jupiter's massive gravitational pull. Much of the rest was flung far beyond the orbit of Neptune by the gravity fields of the two large outer planets. There, the material formed the huge number of orbiting planetesimals—small rocks mixed with water ice and frozen gases—that comprise the Kuiper Belt.

The vast majority of the planetesimals has stayed in the belt ever since and will remain there for as long as the solar system survives. But from time to time, one of them is dragged out of its regular orbit by Neptune's gravity and propelled toward the inner solar system to become what is known as a short-period comet.

All short-period comets travel in the ecliptic, the plane in which the planets move, unlike long-period comets, which can enter the inner

solar system from any direction. It was this fact that prompted Edgeworth and Kuiper to suggest that there must be a "reservoir" of icy cometary bodies somewhere beyond Neptune.

THE SEARCH FOR KBOS

The first serious search for KBOs, which are so dimly lit by the distant Sun that they are very hard to see, got under way in 1987, when astronomers David Jewitt and Jane Luu began scanning the skies with the 7.2-inch (2.2-m) telescope of the University of Hawaii. The pair took multiple pictures of different parts of the sky, then used computers to compare the pictures and look for objects that appeared to have moved. On August 30, 1992, they found an object that had moved exactly the right distance to be in orbit beyond Neptune. This small, reddish rocky body, called 1992 QB1, is 176 miles (283 km) across. It moves in an almost circular orbit that is 44 times farther from the Sun than the

Earth and farther than the average distance of Pluto. Jewitt and Luu found a second object beyond Neptune in March 1993. Further sightings of KBOs followed rapidly, and within five years of the first discovery, 60 were known to exist.

THE KUIPER BELT

KUIPER PROBE
The Pluto-Kuiper Belt New Horizons probe (left) will arrive at Pluto in about 2015 and then travel on to explore the Kuiper Belt.

PLUTO AND CHARON
Pluto and its moon, Charon, orbit the Sun in a path that takes them within the orbit of Neptune. This unusual orbit has led some astronomers to propose that they originated in the Kuiper Belt.

KUIPER BELT OBJECTS
The Kuiper Belt lies beyond the orbits of Neptune and Pluto. It contains many thousands of small objects, including perhaps more than 35,000 that are larger than 60 miles (97 km) in diameter.

FAINT SUN
Seen from the Kuiper Belt, the Sun is 1,000 times fainter than it appears from Earth. The belt's inner edge lies about 2.8 billion miles (4.5 billion km) from our parent star; the outer edge is at least 4.4 billion miles (7 billion km), and perhaps over 90 billion miles (145 billion km), from the Sun.

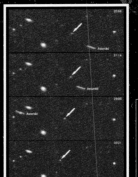

THE FIRST KBO
This sequence of pictures, taken by David Jewitt and Jane Luu, revealed the first KBO to be discovered, 1992 QB1. During the time interval covering these pictures, the faint object (arrowed) moved steadily across the sky while the background stars remained virtually stationary.

OORT CLOUD

...THE OORT CLOUD WAS SERIOUSLY DISTURBED?

Although the comets that emanate from the Oort Cloud are unpredictable, we do know that comets have hit the Earth before—and are likely to do so again. In July 1994, images from Earth-based telescopes and from the space probe Galileo gave us a chilling preview of what might happen in the event of a comet impact when comet Shoemaker-Levy 9 smashed into Jupiter with devastating force.

Some scientists are particularly concerned about the Sun's progress through the Milky Way. This exposes the outer reaches of the solar system to random gravitational effects from the neighboring stars in our galaxy, and threatens to disturb the Oort Cloud more than it has been for many millions of years.

Earth is overdue for a major comet impact. Comets are less dense than asteroids, but have a greater impact velocity. So even if the comet breaks up in mid-air, it could still inflict massive damage if it happens to strike a city area.

In 1908, the Tunguska Valley in Siberia, Russia, suffered what may have been a comet impact. Although there was no crater, suggesting that the body broke up before it hit the ground, the devastation extended over an area of 150 square miles (388 square km).

Another concern is what might happen if the solar system passes through one of the so-called Giant Molecular Clouds that inhabit the galaxy. The effect on the Oort Cloud may be minimal or it may be catastrophic. We don't really know.

Some astronomers are so worried by the threat of a comet or asteroid hitting the Earth that they have formed a group called Spacewatch. They estimate that the Earth is three times more likely to be hit by a large asteroid than by a large comet, but that major comets strike at a rate of about one per century. The last one was in 1908; when—and where—will the next one be?

Far beyond the orbit of Pluto, on the very outer limits of our solar system, lies an unseen store of dark, frozen debris that astronomers call the Oort Cloud. Occasionally, pieces of this debris—the icy bodies we know as comets—are plucked from the Oort Cloud into orbits that take them closer to Earth. Some of these comets sweep past the Sun once and disappear, never to be seen again. Others revisit the inner solar system again and again, their fiery tails lighting up the night sky in a blaze of glory.

OORT CLOUD STATISTICS

(ALL FIGURES ARE APPROXIMATE)	
RADIUS	0.5 TO 1.5 LIGHT-YEARS (3×10^{12} TO 9×10^{12} MILES) (4.8×10^{12} TO 1.4×10^{13} KM)
PROBABLE MASS	2×10^{24} TONS/1.8×10^{24} TONNES (300 X MASS OF EARTH)
POTENTIAL COMETS	AROUND 10 TRILLION
COMET COMPOSITION	ROCK AND ICES (FROZEN WATER, METHANE, AMMONIA, CARBON DIOXIDE)
AGE	FORMED 4.6 BILLION YEARS AGO
ORBITAL TIME (AROUND SUN)	16 MILLION YEARS FOR A COMET 1 LIGHT-YEAR FROM SUN

COMETS IN COLD STORAGE

The Oort Cloud is a swirling mass of debris that forms a giant halo around the solar system, possibly stretching halfway to the nearest star. This debris, which is both dark and extremely cold, consists of lumps of loosely packed rock along with various ices, including frozen water, methane, ammonia and carbon dioxide.

For billions of years, most of the debris in the Oort Cloud has been orbiting the Sun at its own leisurely pace: Whereas the Earth makes the trip in a year, the Oort Cloud debris takes much longer—several millions of years—because of its greater distance from the Sun. Occasionally, however, a piece of this debris gets "nudged" out of its regular orbit and is sent hurtling toward the inner solar system. At this point it becomes a comet—one of the most spectacular (and from our point of view potentially dangerous) sights in the night sky.

What happens to the comet next depends on its orbit. A few comets become set on a collision course with one of the planets, or even with the Sun itself. But most get "captured" by the gravity of one or more of the planets and settle into a new orbit within the inner solar system.

If the comet's orbit carries it as near to the Sun as Jupiter, the Sun's warmth begins to melt some of the frozen gases mixed up in the debris. The gases become energized and are emitted in a brightly glowing stream—the familiar "tail."

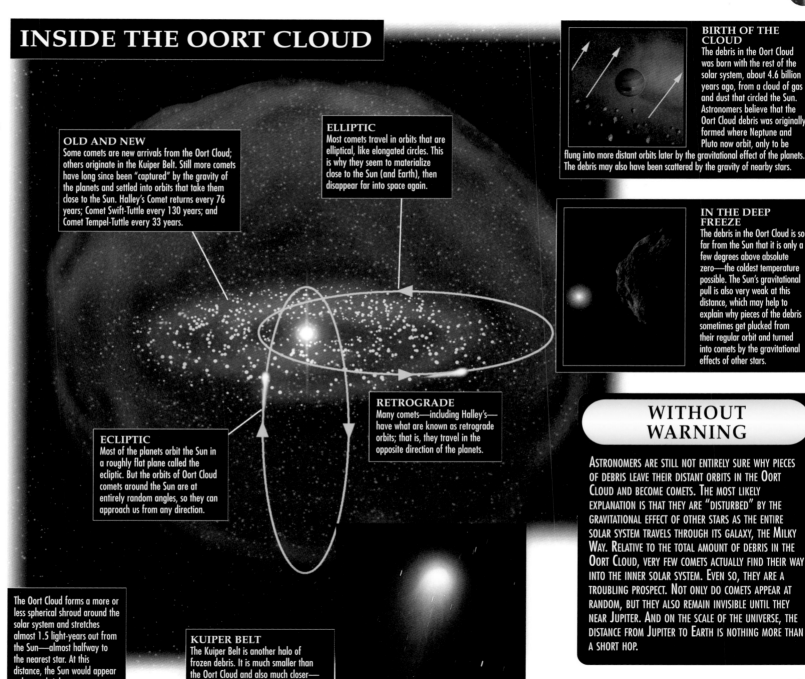

INSIDE THE OORT CLOUD

OLD AND NEW
Some comets are new arrivals from the Oort Cloud; others originate in the Kuiper Belt. Still more comets have long since been "captured" by the gravity of the planets and settled into orbits that take them close to the Sun. Halley's Comet returns every 76 years; Comet Swift-Tuttle every 130 years; and Comet Tempel-Tuttle every 33 years.

ELLIPTIC
Most comets travel in orbits that are elliptical, like elongated circles. This is why they seem to materialize close to the Sun (and Earth), then disappear far into space again.

ECLIPTIC
Most of the planets orbit the Sun in a roughly flat plane called the ecliptic. But the orbits of Oort Cloud comets around the Sun are at entirely random angles, so they can approach us from any direction.

RETROGRADE
Many comets—including Halley's—have what are known as retrograde orbits; that is, they travel in the opposite direction of the planets.

The Oort Cloud forms a more or less spherical shroud around the solar system and stretches almost 1.5 light-years out from the Sun—almost halfway to the nearest star. At this distance, the Sun would appear only as a bright star.

KUIPER BELT
The Kuiper Belt is another halo of frozen debris. It is much smaller than the Oort Cloud and also much closer—it lies just beyond the orbit of Neptune.

BIRTH OF THE CLOUD
The debris in the Oort Cloud was born with the rest of the solar system, about 4.6 billion years ago, from a cloud of gas and dust that circled the Sun. Astronomers believe that the Oort Cloud debris was originally formed where Neptune and Pluto now orbit, only to be flung into more distant orbits later by the gravitational effect of the planets. The debris may also have been scattered by the gravity of nearby stars.

IN THE DEEP FREEZE
The debris in the Oort Cloud is so far from the Sun that it is only a few degrees above absolute zero—the coldest temperature possible. The Sun's gravitational pull is also very weak at this distance, which may help to explain why pieces of the debris sometimes get plucked from their regular orbit and turned into comets by the gravitational effects of other stars.

WITHOUT WARNING

Astronomers are still not entirely sure why pieces of debris leave their distant orbits in the Oort Cloud and become comets. The most likely explanation is that they are "disturbed" by the gravitational effect of other stars as the entire solar system travels through its galaxy, the Milky Way. Relative to the total amount of debris in the Oort Cloud, very few comets actually find their way into the inner solar system. Even so, they are a troubling prospect. Not only do comets appear at random, but they also remain invisible until they near Jupiter. And on the scale of the universe, the distance from Jupiter to Earth is nothing more than a short hop.

MINOR MEMBERS OF THE SOLAR SYSTEM

Between the major planets orbit millions of smaller objects—asteroids, comets, and countless other pieces of space debris. Most of the larger objects—the asteroids or minor planets that never grew large enough to become substantial worlds in their own right—are confined to a belt between the orbits of Mars and Jupiter, held there by Jupiter's enormous gravity. A few asteroids stray out into more eccentric orbits, including some that occasionally cross Earth's own orbit, known as Near-Earth Asteroids. Comets are usually less substantial than asteroids, being composed ofa mixture of dust and ice. They originate in the icy reaches of the outer Solar System, but occasionally fall into orbits that bring them past the Sun at high speeds, burning off their icy material to form spectacular tails as they do so. Dust particles left behind by these visitors are spread throughout the inner Solar System, and when they enter Earth's atmosphere they form meteors or shooting stars. Larger chunks of material, from asteroids or even other planets, occasionally reach our planet's surface as meteorites.

Comet Hale-Bopp, photographed in March 1997 as it made its closest approach to the Sun. Hale-Bopp became visible to the naked eye in the summer of 1996, and in March 1997 a solar eclipse in Mongolia and eastern Siberia allowed observers there to see the comet in the daytime.

ASTEROID BELT

The Asteroid Belt—a broad band between the orbits of Mars and Jupiter—is home to thousands of small, rocky bodies that orbit the Sun. These asteroids, also called "minor planets," are thought to be the remains of a larger planet that tried to form in the early days of the solar system but was prevented from doing so by the powerful gravitational influence of the giant planet Jupiter. Despite their sinister reputation, the combined mass of the asteroids in the belt is still less than 1% of the mass of the Earth.

WHAT IF...

...WE COULD MAKE USE OF THE ASTEROID BELT?

There is no danger of the bodies in the Asteroid Belt ever joining up to form a planet. And asteroids almost never collide—the craters on their surfaces are nearly all the result of impacts that occurred many millions of years ago. This gives the human race plenty of time to learn more about them—and possibly to exploit them, too.

Exploratory missions of the future may follow the lead set by the NEAR (Near-Earth Asteroid Rendezvous) mission. This space probe, the first to carry out long-term studies of an asteroid, was launched in 1996 and flew past the asteroid Mathilde before going straight into orbit around Eros.

Like NEAR, future survey probes will orbit the larger asteroids in the belt, mapping their surfaces and taking measurements of their gravity and magnetic fields. Such probes will also observe the X-rays and gamma rays that are emitted from asteroids under the ceaseless bombardment of cosmic rays and charged particles from the Sun. Such observations will give scientists fairly accurate data on the composition of individual asteroids, allowing them to assess whether it is worth sending another probe to gather rock samples.

Our interest in the Asteroid Belt may well extend beyond the purely scientific: We already know that some asteroids are rich in metals, and in the future, it should be perfectly possible to mine them. Unless the mining could be done entirely by robots, such a scheme would mean establishing mining camps in the asteroid belt. Staging posts might also be needed to cope with the years-long journey to and from the belt.

Even vacation resorts are a possibility, especially if the Moon and Mars become overcrowded with tourists. Spacious living quarters could be dug into an asteroid, where they would be safe from the searing radiation of solar flares. A small asteroid could even be converted into a spacecraft for journeys to the outer solar system by hollowing it out and installing rocket engines.

The NEAR probe went into orbit around the asteroid Eros on Valentine's Day 2000, eventually landing on the rocky world a year later.

ASTEROID SIZE

ASTEROID NUMBERING

WHEN AN ASTEROID'S ORBITAL DETAILS HAVE BEEN ESTABLISHED, ASTRONOMERS GIVE IT A NUMBER THAT IS WRITTEN BEFORE THE NAME. ASTEROIDS ARE RARELY SPHERICAL. THE SIZES GIVEN BELOW ARE MAXIMUMS.

LARGE ASTEROIDS

ASTEROIDS LARGER THAN 100 MILES (160 KM) ACROSS	.26
ASTEROIDS LARGER THAN 300 FEET (90 M) ACROSS	POSSIBLY 1 MILLION
TOTAL MASS OF ASTEROIDS IN BELT	LESS THAN 15% OF THE MOON

VERMIN OF THE SKIES

At the end of the 18th century, astronomers began to search for a "missing" planet that they were convinced must exist in the extra-large gap between the orbits of Mars and Jupiter. Among these astronomers was a group that called itself the "Celestial Police."

This group scanned the skies systematically for a dot of light that could be seen to move in relation to the stars. But they were cheated of their quarry. Other astronomers established that the gap between Mars and Jupiter was filled with a number of tiny asteroids rather than the full-sized planet that the Celestial Police hoped to find. The first of these asteroids was discovered on January 1, 1801 by the Italian astronomer Giuseppe Piazzi. He named the new body Ceres, for the Roman goddess of agriculture. A second, Pallas, was found by German astronomer Heinrich Olbers in March 1802.

In the following years, more of these miniature worlds were found. Astronomers now think that there are probably tens of thousands of asteroids—one called them "vermin of the skies"—and over 8,500 have now been cataloged. Hundreds of new asteroids are added to the record every year.

MAVERICK ORBITS

The orbits of most asteroids lie between those of Mars and Jupiter, but there are some asteroids that travel in what are called maverick orbits. Some of these pass closer to the Sun than the Earth, some extend way beyond Jupiter, and others—the Trojan asteroids—travel in the same orbit as Jupiter.

Asteroids have too little gravity to support atmospheres, or even to draw themselves into a spherical shape; most resemble lumpy potatoes. The majority are also too close to the Sun to have retained any water or methane in frozen form.

Nearly all known asteroids are pitted with craters—the scars of impacts with meteoroids or other asteroids. Most of this probably happened when the solar system was forming and space was more crowded. More recent cratering could be due to a much larger asteroid that broke up.

LAGRANGE

THE TROJAN ASTEROIDS TRAVEL IN TWO GROUPS IN THE ORBIT OF JUPITER, ONE 60° AHEAD OF THE PLANET, THE OTHER 60° BEHIND IT. THESE POSITIONS ARE TWO OF WHAT ARE KNOWN AS THE LAGRANGE POINTS OF THE ORBIT. THESE POINTS, AT WHICH SMALL OBJECTS CAN SAFELY REMAIN IN A LARGER OBJECT'S ORBIT, ARE NAMED FOR THE FRENCH ASTRONOMER JOSEPH LAGRANGE (1736–1813), WHO CALCULATED THEIR EXISTENCE MATHEMATICALLY.

ASTEROID MOON
Dactyl, the moon of asteroid 243 Ida, was the first satellite of an asteroid to be discovered. Ida is about 36 miles (58 km) long; Dactyl is just a mile (1.6 km) long.

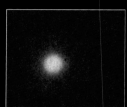

CERES
This image of 1 Ceres was taken by the Hubble Space Telescope. Ceres, the first and largest asteroid to be discovered, is almost spherical—unlike any of the other known asteroids.

MATHILDE
An image of 253 Mathilde, built up from four separate images taken by the NEAR (Near-Earth Asteroid Rendezvous) probe in 1997 from a distance of 1,500 miles (2,414 km). The deep shadow at the center is a crater estimated to be six miles (10 km) deep.

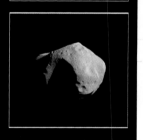

NO IMPACT

SCIENCE FICTION WRITERS, BELIEVING THAT ASTEROIDS MUST BE CROWDED TOGETHER IN SPACE, HAVE OFTEN PORTRAYED THE ASTEROID BELT AS A PLACE OF DANGER FOR SPACE TRAVELERS. SOME HAVE EVEN SPECULATED THAT SPACECRAFT ENTERING THE REGION WOULD BE BATTERED BY "ASTEROID STORMS." THE ASTEROID BELT IS CERTAINLY CROWDED, BUT ONLY BY COMPARISON WITH THE REST OF THE SOLAR SYSTEM, WHICH IS A VERY EMPTY PLACE INDEED. THE 2 PIONEER AND 2 VOYAGER PROBES THAT VISITED THE OUTER PLANETS FLEW THROUGH THE ASTEROID BELT WITHOUT ANY CLOSE ENCOUNTERS.

THE ASTEROID BELT

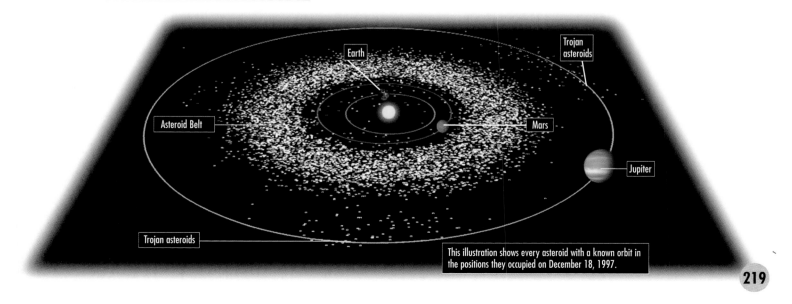

Earth

Trojan asteroids

Asteroid Belt

Mars

Jupiter

Trojan asteroids

This illustration shows every asteroid with a known orbit in the positions they occupied on December 18, 1997.

CLASSIFYING ASTEROIDS

All asteroids are bits of rock left over from the formation of planets—but they are not all alike. Though size and location help to define these rocks, it is their chemical makeup that gives us a way to categorize them—and gives asteroids their color. The tiny, distant objects are not easy to study, but fragments of them that have fallen to Earth as meteorites are more accessible. Studying their composition not only tells us more about asteroids, but may reveal secrets of the solar system's past and help us plan a future in space.

WHAT IF...

...WE HAD ACCESS TO THE MINERALS HELD IN ASTEROIDS?

The idea of sending a mining expedition to an asteroid is not new: The economic potential of these hunks of rock has been recognized for decades. In fact, scientists calculate that the mineral resources locked up in asteroids are worth trillions of dollars. One cubic mile of an M-type nickel-iron asteroid could contain over 7 billion tons (6.3 billion tonnes) of iron, 1 billion tons (0.9 billion tonnes) of nickel and enough cobalt to satisfy Earth's demand for 3,000 years.

And as our planet's resources become depleted in the centuries to come, utilizing these cosmic mineral storehouses may become a necessity. There would be an added bonus, too: Extracting minerals from asteroids instead of from the Earth's crust saves our atmosphere from the polluting effects of mining processes.

But minerals extracted by robots and humans in asteroid mining camps in the main belt would have a long and expensive delivery journey for use back on Earth. A cheaper alternative would be to put the asteroid into Earth orbit, where crews could be replaced frequently and products brought home easily. Hauling these hefty boulders close to Earth would not be difficult. Solar electric cells or a solar sail could be used to slowly propel an asteroid Earthward. Huge, lightweight solar sails work on the same principle as nautical sails—but using the gentle energy-push of sunlight instead of gusts of wind. With an asteroid in tow, the voyage to Earth orbit by solar sail would not be particularly fast—in fact, it would take several years—but the process would make effective use of a free energy source. An even more efficient strategy would be to use the asteroid's own material as fuel. Parts of the asteroid could be expelled at very high speed, using a device called a pile driver, to propel the asteroid in the opposite direction.

A ready supply of extractable minerals and metals in orbit would benefit both space-based and Earth-based projects. Construction in space would no longer entail expensive supply launches. An asteroid's mass could even provide the raw materials for the construction of satellite solar-power generators that would supply electrical power to Earth via microwaves.

And in the more distant future, when humans may be living in space, asteroids could be parked next to a space colony to provide a handy source of raw materials for constructing communications satellites or even a sister colony to house increasing numbers of space dwellers.

ASTEROID TYPES

TYPE	ALBEDO	EFLECTANCE SPECTRUM
C	3–7%	FAIRLY FLAT, FAIRLY NEUTRAL COLOR
B	4–8%	C-LIKE BUT BRIGHTER AND MORE NEUTRAL
F	3–6%	FLAT, NEUTRAL COLOUR, NO UV ABSORPTION
G	5–9%	C-LIKE BUT BRIGHTER , STRONG UV ABSORPTION
P	2–6%	SLIGHTLY REDDISH, VERY LOW ALBEDO
D	2–5%	REDDER THAN P, VERY LOW ALBEDO
T	4–11%	REDDISH (BETWEEN D AND S)

COSMIC FOSSILS

The asteroid belt that lies between Mars and Jupiter is home to billions of lumps of rock, spaced thousands of miles apart. Too small to be classified as planets, they are too big to be dismissed as mere dust. Although many have a diameter of just a few yards or less, half a million or so have a diameter of more than a mile (1.6 km), and 25 to 30 of the biggest are more than 125 miles (200 km) across.

The giant among asteroids is 1 Ceres, a ball-shaped rock about 580 miles (933 km) wide. But 1 Ceres is an exception. Virtually all asteroids are irregular, potato-shaped lumps that show up through Earth's most powerful telescopes only as dots of light. Even so, we know the positions and orbits of over 8,000 asteroids and can determine the size, rate of spin and chemical composition of some of them.

It is the chemical make-up of asteroids that forms the basis of the system by which these rocks are classified. Such chemical signatures may also provide a profile of the early chemistry of the solar system. Deep in the heart of the much more massive planets, high temperatures and pressures have altered their chemical composition over time. Since asteroids are much smaller, their composition is thought to have remained unchanged.

But how can we know the composition of a lump of rock floating in space tens of millions of miles away? One clue is its albedo—the amount of sunlight its surface reflects. Another is its reflectance spectrum—the way in which it reflects light of different wavelengths.

CLASS CODED

Different minerals reflect sunlight in characteristic ways, so an asteroid's spectral properties—its albedo and its reflectance spectrum—can reveal what its surface is made of. When scientists began to classify asteroids in the early 1970s, they came up with two types: C for carbonaceous asteroids, the surfaces of which are composed mainly of carbon and complex organic compounds, and S for stony

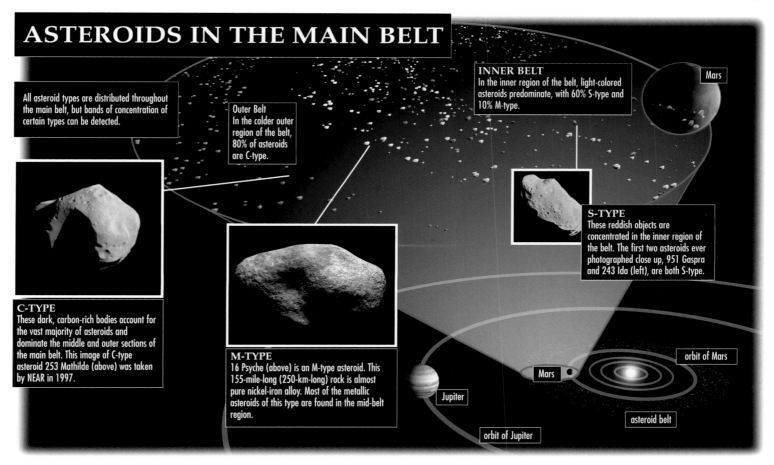

ASTEROIDS IN THE MAIN BELT

All asteroid types are distributed throughout the main belt, but bands of concentration of certain types can be detected.

Outer Belt
In the colder outer region of the belt, 80% of asteroids are C-type.

INNER BELT
In the inner region of the belt, light-colored asteroids predominate, with 60% S-type and 10% M-type.

Mars

S-TYPE
These reddish objects are concentrated in the inner region of the belt. The first two asteroids ever photographed close up, 951 Gaspra and 243 Ida (left), are both S-type.

C-TYPE
These dark, carbon-rich bodies account for the vast majority of asteroids and dominate the middle and outer sections of the main belt. This image of C-type asteroid 253 Mathilde (above) was taken by NEAR in 1997.

M-TYPE
16 Psyche (above) is an M-type asteroid. This 155-mile-long (250-km-long) rock is almost pure nickel-iron alloy. Most of the metallic asteroids of this type are found in the mid-belt region.

orbit of Mars

Mars

Jupiter

asteroid belt

orbit of Jupiter

asteroids. Today, there are a total of 14 asteroid types, all designated by a single letter.

About 75% of asteroids in the main belt are C-type; about 15% are S-type; and the remaining 10% fall into the 12 other classes, the most common of which is M-type (metallic). Certain asteroid types dominate certain parts of the belt. From the region closest to Mars out to the mid-belt, the majority of asteroids are rocky and metallic. These relatively light-colored objects are mainly M and S types. In the outer belt, where conditions are colder, carbon-rich, "sooty" C-type asteroids predominate.

Further investigation into asteroid composition is likely to increase the total of

ASTEROID SLEUTH

THE PRESENT CLASSIFICATION OF ASTEROIDS DATES FROM THE EARLY 1980S. IT IS LARGELY BASED ON THE WORK OF AMERICAN ASTRONOMER DAVID THOLEN (RIGHT). WHILE AT THE UNIVERSITY OF ARIZONA, HE OBSERVED HUNDREDS OF ASTEROIDS. HE STUDIED DATA FOR 405 OF THESE AND FOUND THAT THEY FELL INTO SEVEN MAJOR AND SEVEN MINOR CLASSES. HIS WORK WAS HONORED WHEN ASTEROID NUMBER 3255, DISCOVERED ON SEPTEMBER 2, 1980, WAS NAMED "THOLEN" AFTER HIM.

known types. And increased knowledge of the composition of these planetary remnants will help scientists to build up a picture of the early history of the solar system.

CHANGING VIEWS OF ASTEROIDS

Once called "vermin of the skies" because they were thought of as no more than cosmic trash, asteroids are now valued as fossil relics that hark back to the formation of our solar system. Astronomers have cataloged about 10,000 asteroids—most of them in a broad belt between the orbits of Mars and Jupiter. At first, scientists believed that asteroids were the remnants of an ancient, exploded planet, but now it seems more likely that they are the parts of a planet that never formed in the first place.

WHAT IF...

...OUR VIEWS OF ASTEROIDS CHANGED AGAIN?

When the Near Earth Asteroid Rendezvous spacecraft (NEAR) flew past the asteroid 253 Mathilde in June 1997, it got a surprise. Although the chemical makeup of Mathilde resembles that of coal, the asteroid is only slightly denser than water—about half the density expected. This shortfall led to the revelation that asteroids may be porous inside, like celestial styrofoam.

Despite its lightweight structure, Mathilde obviously survived a beating from other asteroids in the past. One impact crater is about half as big as Mathilde itself. Astronomers are still not sure how it survived a collision powerful enough to blast out such a large crater.

The findings highlight one of the research areas that astronomers will pursue in the 21st century. Some asteroids, such as NEAR's main target, 433 Eros, appear to be solid chunks of rock or metal. Others, like Mathilde, may be fractured inside, or could even consist of many smaller boulders held together by gravity. Astronomers hope to understand why there is such a difference in the structure of asteroids, and how these small bodies hold themselves together in violent collisions.

Another mystery that astronomers hope to solve is the relationship between asteroids and meteorites, the space rocks that survive the fiery plunge through the Earth's atmosphere and land on the surface. Scientists are certain that most meteorites are the fragments of asteroids broken off in collisions. Several classes of meteorite have been paired up with parent asteroids. But the origins of the most common type of meteorite, the ordinary chondrites, are still uncertain.

Ordinary chondrites are made of small bits of rock held together by a sort of mineral "glue." They appear to have changed little since the solar system formed.

Astronomers are trying to determine if one of the most common asteroid groups, type S, is the source of these meteorites. Studies through ground-based telescopes suggest that S-type asteroids have melted and resolidified, forming layers of successively heavier materials at greater depths. If that is so, then either S-type asteroids are not the parents of ordinary chondrites, or astronomers must explain why parents and offspring are so unlike.

NEAR's successor spacecraft may help solve these and other mysteries about some of our solar system's smallest inhabitants.

ASTEROID UNDERSTANDING

1801	Giuseppe Piazzi discovers Ceres, the first asteroid
1802	Heinrich Olbers discovers a second asteroid, Pallas, in an orbit very similar to Ceres'
1898	Gustav Witt discovers 433 Eros, the first "near-Earth" asteroid
1980	Luis and Walter Alvarez suggest that an asteroid impact wiped out the dinosaurs 65 million years ago
1991	Galileo probe makes the first close approach to an asteroid, flying just 1,000 miles (1,600 km) past Gaspra
1994	Galileo probe discovers a small moon, later named Dactyl, orbiting the asteroid Ida
1997	NEAR finds that about half of the interior of 253 Mathilde may be empty space
2000	NEAR arrives at 433 Eros, orbiting it for a year before landing on its surface

BRIGHT SIGHTS

Science was missing a few planets at the start of the 19th century. At the time, only seven were known, and astronomers were convinced that there must be more. So in 1801, when Italian astronomer Giuseppe Piazzi (1746–1826) spotted a point of light between the orbits of Mars and Jupiter, many assumed that the elusive eighth planet had been found. Piazzi named the new "planet" Ceres.

But other astronomers were not satisfied. They had expected a large new world, and tiny Ceres did not fit the bill. A year after the discovery of Ceres, German astronomer Heinrich Olbers (1758–1840) found a second object in a similar orbit. Olbers thought that the bodies—called asteroids, from the Latin word for "starlike"—were "fragments of a trans-Martian planet, blown to pieces in the past." Most asteroids logged in the 19th century were found between Jupiter and Mars, and scientists assumed that they were indeed pieces of an ancient planet.

By the end of the 19th century, nearly 500 "minor planets" were known, all with different orbits, and the explosion idea was in doubt. The view accepted by the turn of the 20th century still holds today. The asteroids started off as planetesimals, the precursors to planets, in the early solar system. Anywhere else, these planetesimals might have merged together to become a planet. But the pull of Jupiter's powerful gravity accelerated them so that they simply shattered on collision instead of growing into a true planet.

ASTEROID INTRIGUE

The exploded planet idea persisted into the 20th century. Before it was finally abandoned, several mechanisms for the planet's demise were suggested—one involving a nuclear war among its inhabitants.

Meanwhile, asteroids had provided early observers with another mystery. They were perplexed by the brightness of one particular body, Vesta, compared with other known

asteroids. This suggested that Vesta must be cloud-covered—yet its small size meant that it could not hold an atmosphere. We now know that Vesta is made of different rock than many of the other asteroids.

Our views of asteroids have recently been complicated by the discovery of hazes around some, as if they were comets. These are probably bodies that came from farther out in the solar system and were dragged into their present orbits by the influence of Jupiter.

To begin with, asteroids were regarded as a mere curiosity. But the discovery of bodies with Earth-crossing orbits bought asteroids—and their potential strikes—uncomfortably close to home. Although asteroids may be our undoing, they are also our best clues to the origins of the solar system. Some have remained unchanged since the birth of the Sun. Now, space agencies hope that their probes will provide us with a new angle on asteroids—and reveal much about the cloud of gas and dust from which the planets formed.

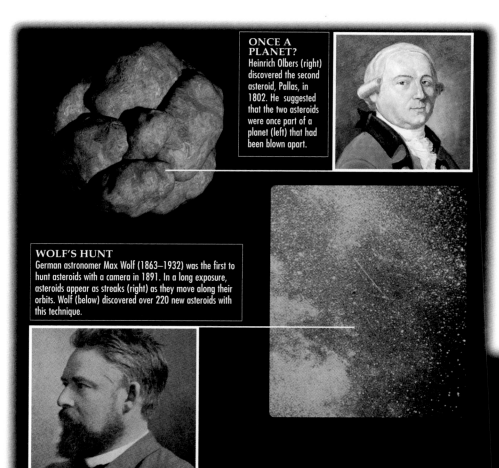

ONCE A PLANET?
Heinrich Olbers (right) discovered the second asteroid, Pallas, in 1802. He suggested that the two asteroids were once part of a planet (left) that had been blown apart.

WOLF'S HUNT
German astronomer Max Wolf (1863–1932) was the first to hunt asteroids with a camera in 1891. In a long exposure, asteroids appear as streaks (right) as they move along their orbits. Wolf (below) discovered over 220 new asteroids with this technique.

ASTEROIDS THROUGH THE AGES

From fragments of an exploded planet to the raw material of a planet that never formed—our views of asteroids have not stood still for long. In the future, asteroids may tell us what the solar system is really made of.

CLOSE-UP
The "exploded planet" idea, although wrong, found echoes in the first close-up images of an asteroid, taken by the Galileo probe (left) in 1991. Gaspra (above) lacks impact craters, and probably formed recently from the breakup of a larger body.

CERES

Most of the many thousands of asteroids already cataloged are tiny. But a handful of these space rocks are the size of cities or even small countries. None deserves the label "minor planet" more than the biggest of them all—Ceres, with a diameter of around 600 miles (965 km), the first asteroid to be discovered. Ceres' dark surface may not have changed significantly in billions of years. It may retain the scars and craters it acquired when the solar system was young. We could learn much from this mysterious giant.

WHAT IF...

...WE WENT TO CERES?

Only a handful of asteroids have ever been seen close up—Ida and Dactyl, Gaspra, Mathilde, Toutatis and Eros—and these have all been small and irregular. Ceres has so far eluded close scrutiny—although it is easily the solar system's largest main-belt asteroid and was the first to be discovered—so little is known of this world. There are currently no missions planned to Ceres. If there were, what could be gained by them?

For a start, Ceres' surface is rich in solar system history. Despite being in the asteroid belt, Ceres probably does not suffer frequent large-scale impacts. The asteroid belt is not as densely packed as is sometimes imagined—an observer on Ceres would be extremely lucky to see any other asteroids with the naked eye. Also, although the asteroid is relatively large, it is too small to be active volcanically.

In consequence, Ceres may not have changed significantly in billions of years, since the solar system was in its infancy. Thus the asteroid's surface should provide a record of the early days of the solar system and give us an insight into the "teen" years of a growing planet. A mission to Ceres could therefore yield fascinating new data on the process by which the planets, including Earth, and the asteroids were created.

Another reason to go to Ceres would be to mine it. Other asteroids could serve the same purpose, of course, but Ceres offers a truly massive reserve of useful raw materials. The asteroid has a carbon top layer, but deeper below the surface Ceres could be bountiful in metal-rich ores—raw material for ships and space stations—and precious minerals. Meanwhile, spectroscopic studies, which analyze light reflected from the asteroid, have revealed that, as well as the carbon typical of a C-class asteroid, Ceres contains a lot of water—trapped in the minerals on the asteroid's surface. Such water could easily be released by heating the minerals. So perhaps another compelling reason for going to Ceres in the future would be to use this water. The asteroid is too small to provide much gravity for an astronaut—its surface gravity is only about 3% that of Earth's—so a permanent base would not be comfortable. But Ceres would make an ideal "watering hole," a place where spacecraft could refuel mid-mission—using the hydrogen in water—on their way elsewhere in the solar system.

HEAVYWEIGHT STATS

FULL DESIGNATION	1 CERES
YEAR OF DISCOVERY	1801
DISCOVERER	GIUSEPPE PIAZZI (1746–1826)
CLASSIFICATION	C OR CARBONACEOUS
ALBEDO (REFLECTIVITY)	5.7%
MASS	1.29×10^{18} TONS (1.17×10^{18} TONNES)
DENSITY	0.097 POUNDS PER CUBIC INCH (2,684 KG PER CUBIC METER)
DIAMETER	ABOUT 600 MILES (965 KM)
ROTATION PERIOD	9.075 HOURS
ORBITAL PERIOD	4.6 YEARS
MAXIMUM DISTANCE FROM SUN	2.98 AU

WOULD-BE PLANET

The first asteroid to be discovered, Ceres—or 1 Ceres, to give its proper designation—is 600 miles (965 km) across. It is almost twice the size of the next two largest asteroids, 2 Pallas and 4 Vesta. Though Ceres has never been observed close up, pictures taken by the Hubble Space Telescope show that Ceres is round like a planet, while virtually all other asteroids are irregular.

Astronomers have also been able to make an estimate of Ceres' general shape thanks to the fact that the asteroid rotates. As the giant spins through its 9-hour day, the light it reflects varies very little, suggesting a fairly regular shape. Another clue lies in its size. The bigger an asteroid, the more massive it is, and thus the stronger its gravitational pull. Those around 250 to 400 miles (400 to 644 km) across are sufficiently massive that their gravity can compress them into a rough sphere—though the exact form depends on density and strength.

Ceres grew to its great size in the same way that all the planets did: by mopping up debris left behind after the Sun formed. As it grew larger at the expense of smaller objects, its gravity increased and allowed it to draw in even more material. In fact, Ceres would have swept up all the smaller asteroids to form a single small planet, but Jupiter's gravity probably prevented the process from reaching its natural conclusion.

C-CLASS ASTEROIDS

Ceres orbits at an average distance of just under 260 million miles (420 million km) from the Sun. The orbit places Ceres neatly in the middle of the asteroid belt, in a transition zone between two different types of asteroid. Nearer to the Sun, asteroids tend to be light in color, and are similar in composition to rocks on Earth. But beyond Ceres, where lower temperatures allowed carbon to condense in the early solar system, asteroids are much darker, with dark grey or soot-black surfaces that reflect very little sunlight. These are called carbonaceous or C-class asteroids; Ceres itself is one of them.

Despite the passing of two centuries since its

discovery, Ceres is still shrouded in mystery. Its dark surface and its distance put it at the limits of the Hubble Space Telescope's powers. If we want

to learn more, we will have to send a spacecraft— and no such mission has yet been planned, or even given a budget.

HEAVYWEIGHT CONTENDER

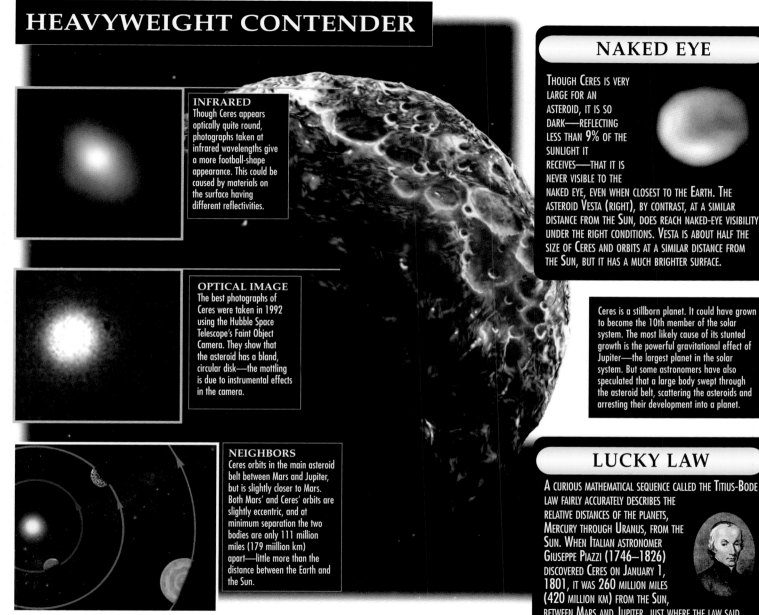

INFRARED
Though Ceres appears optically quite round, photographs taken at infrared wavelengths give a more football-shape appearance. This could be caused by materials on the surface having different reflectivities.

OPTICAL IMAGE
The best photographs of Ceres were taken in 1992 using the Hubble Space Telescope's Faint Object Camera. They show that the asteroid has a bland, circular disk—the mottling is due to instrumental effects in the camera.

NEIGHBORS
Ceres orbits in the main asteroid belt between Mars and Jupiter, but is slightly closer to Mars. Both Mars' and Ceres' orbits are slightly eccentric, and at minimum separation the two bodies are only 111 million miles (179 miillion km) apart—little more than the distance between the Earth and the Sun.

NAKED EYE

THOUGH CERES IS VERY LARGE FOR AN ASTEROID, IT IS SO DARK—REFLECTING LESS THAN 9% OF THE SUNLIGHT IT RECEIVES—THAT IT IS NEVER VISIBLE TO THE NAKED EYE, EVEN WHEN CLOSEST TO THE EARTH. THE ASTEROID VESTA (RIGHT), BY CONTRAST, AT A SIMILAR DISTANCE FROM THE SUN, DOES REACH NAKED-EYE VISIBILITY UNDER THE RIGHT CONDITIONS. VESTA IS ABOUT HALF THE SIZE OF CERES AND ORBITS AT A SIMILAR DISTANCE FROM THE SUN, BUT IT HAS A MUCH BRIGHTER SURFACE.

Ceres is a stillborn planet. It could have grown to become the 10th member of the solar system. The most likely cause of its stunted growth is the powerful gravitational effect of Jupiter—the largest planet in the solar system. But some astronomers have also speculated that a large body swept through the asteroid belt, scattering the asteroids and arresting their development into a planet.

LUCKY LAW

A CURIOUS MATHEMATICAL SEQUENCE CALLED THE TITIUS-BODE LAW FAIRLY ACCURATELY DESCRIBES THE RELATIVE DISTANCES OF THE PLANETS, MERCURY THROUGH URANUS, FROM THE SUN. WHEN ITALIAN ASTRONOMER GIUSEPPE PIAZZI (1746–1826) DISCOVERED CERES ON JANUARY 1, 1801, IT WAS 260 MILLION MILES (420 MILLION KM) FROM THE SUN, BETWEEN MARS AND JUPITER, JUST WHERE THE LAW SAID ANOTHER PLANET SHOULD BE. ASTRONOMERS ARE STILL DEBATING THE VALIDITY OF THE TITIUS-BODE LAW.

BINARY ASTEROIDS

En route to Jupiter in 1993, the Galileo spacecraft made an unexpected discovery. As the probe flew past the asteroid Ida, it photographed a tiny, rocky body in orbit around it. The moonlet made Ida the first known binary asteroid. Scientists began to wonder if such doublets were common—a possibility that became a probability when an Earth-based telescope found an ice moon around the asteroid Eugenia in 1998. Binaries may be the result of gentle collisions between asteroids that are more floating rubble piles than rocks.

WHAT IF...

...EARTH'S GRAVITY BROKE UP A PASSING ASTEROID?

The year is 2012, and a newly discovered asteroid is approaching Earth at 11 miles per second (18 km/sec). But astronomers have been tracking the 500-yard-wide (460-m-wide) object for weeks with radar and powerful telescopes. They have an accurate plot of its orbit, and the news seems good. The asteroid will pass very close—but it will not hit our planet. Reassured, people get on with their daily lives, and a nervous stock market settles down.

Real asteroid experts, though, are seriously worried. They believe the asteroid is little more than a large rubble pile, held together only by its own feeble gravity. They fear that tidal strains caused by the Earth's much stronger gravity will be enough to rip it apart. That could mean sizeable chunks pulled from their original orbit—and many of them could fall to Earth. If a large piece struck a city, the loss of life would be terrible.

Meanwhile, the U.S. and Russian governments jointly launch a spacecraft with a load of nuclear missiles. Most scientists are appalled. If the asteroid really is a rubble pile, a nuclear explosion will break it up as surely as the Earth's gravity—and with equally devastating effects. Debate rages in the U.N.: To shoot or not to shoot? Opinions are divided, but in the end the wait-and-see lobby wins the argument. The safety interlocks on the orbiting missiles stay on.

As the asteroid hurtles by at its closest approach, astronomers watch its appearance change from an irregular fist shape to an elongated teardrop. Then the teardrop splits, and splits again. The asteroid, with no more internal cohesion than a pile of gravel, has been broken up by Earth's gravity into a long string of 20- to 50-yard (18- to 50-m) blocks, linked by a haze of smaller fragments.

Some of these fragments come streaking down into the atmosphere, where they create the most spectacular meteor shower in living memory. But the big pieces skim past the Earth and continue on a changed orbit into interplanetary space.

All is well this time. But astronomers quickly calculate the new orbit—and warn their governments. Next time around, all of the fragments will hit the Earth. And governments have exactly 3 years, 12 days and 43 minutes to figure out what to do.

BINARY MATERIAL

Asteroid	Imaged By	Size	Features
243 Ida	Galileo spacecraft (1993)	36 x 14 miles (58 x 23 km)	Possible rubble pile; has moon, Dactyl
45 Eugenia	CFHT telescope, Hawaii	135 miles (217 km) in diameter	Has ice moon, 8 miles (13 km) diameter
4179 Toutatis	radar telescope	about 3 miles (5 km) long	Possibly two loosely bound boulders (Goldstone and Arecibo)
4769 Castalia	radar telescope (Arecibo)	1.1 x 0.5 miles (1.8 x 0.8 km)	Possibly two or more loosely bound boulders
951 Gaspra	Galileo spacecraft (1991)	12.5 x 7.5 miles (20 x 12 km)	Possible rubble pile
1620 Geographos	radar telescope	3.2 x 1.1 miles (5.1 x 1.8 km)	Possible rubble pile (Goldstone and Arecibo)

SPACE COUPLES

We know relatively little about asteroids. Most are too small to be seen as anything more than points of light even in powerful telescopes, so a few passing encounters with spacecraft and some radar surveys of a few near-Earth objects are our best sources of information. It was a spacecraft encounter that showed the tiny moon Dactyl in orbit around Ida—a binary arrangement that may be turn out to be common.

These binaries were probably created by impacts or by tidal forces. And they suggest that small asteroids may be structurally much weaker than was previously thought. Instead of these space rocks having the solidity of miniature planets, many could be little more than piles of dust and stony fragments, glued together by nothing stronger than their own gravity. Struck by a large, fast-moving object, such rubble-pile asteroids would disintegrate—but given time, the fragments could reform into one or two "reshuffled" descendants.

Planetary gravity, and the tidal forces that it can induce, would also account for some binary asteroids. If a rubble-pile asteroid strayed too close to a planet, the tides raised by gravity in its fragile body would easily be strong enough to pull it apart. Some of the fragments might impact the nearby planet, but others would escape into space, orbiting in a double, triple or even more complex "family." With luck, the spacecraft of the 21st century may find some of these—and bring back some family photographs.

ASTEROIDS ENTANGLED

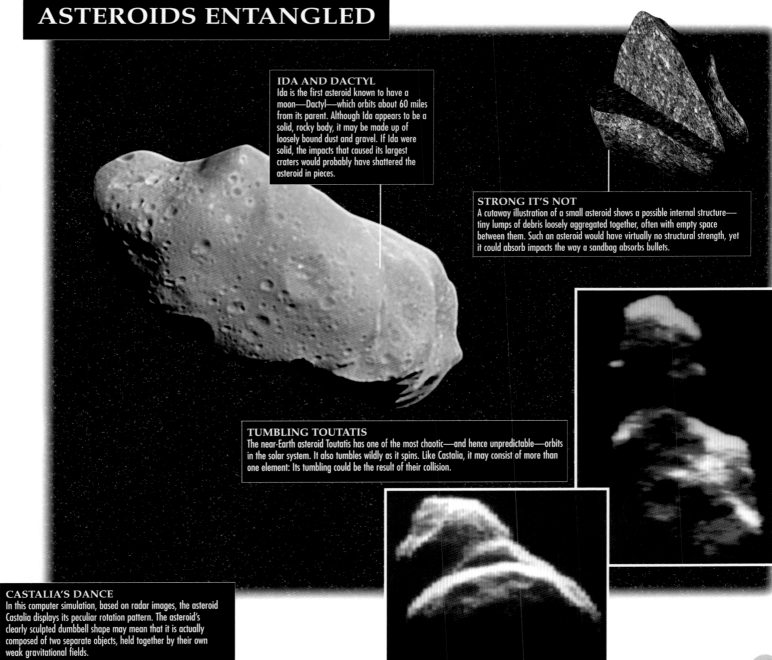

IDA AND DACTYL
Ida is the first asteroid known to have a moon—Dactyl—which orbits about 60 miles from its parent. Although Ida appears to be a solid, rocky body, it may be made up of loosely bound dust and gravel. If Ida were solid, the impacts that caused its largest craters would probably have shattered the asteroid in pieces.

STRONG IT'S NOT
A cutaway illustration of a small asteroid shows a possible internal structure—tiny lumps of debris loosely aggregated together, often with empty space between them. Such an asteroid would have virtually no structural strength, yet it could absorb impacts the way a sandbag absorbs bullets.

TUMBLING TOUTATIS
The near-Earth asteroid Toutatis has one of the most chaotic—and hence unpredictable—orbits in the solar system. It also tumbles wildly as it spins. Like Castalia, it may consist of more than one element: Its tumbling could be the result of their collision.

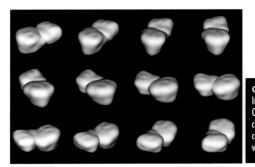

CASTALIA'S DANCE
In this computer simulation, based on radar images, the asteroid Castalia displays its peculiar rotation pattern. The asteroid's clearly sculpted dumbbell shape may mean that it is actually composed of two separate objects, held together by their own weak gravitational fields.

NEAR-EARTH ASTEROIDS

Most of the thousands of asteroids in the solar system orbit the Sun between Mars and Jupiter. But in the last 70 years, astronomers have discovered groups of asteroids that move in orbits far from the principle asteroid belt. Named after the first of their families to be discovered, the Apollo and Aten asteroids jostle for space with Earth in the solar system, each crossing our planet's orbit in two places. So if a large object does smash into Earth, it will most probably be an Apollo or an Aten asteroid.

WHAT IF...

...WE USED CHAOS THEORY TO EXPLAIN THE EARTH-CROSSERS?

Ever since the discovery of the Apollo and Aten asteroids, astronomers have pondered how they came to be in such unusual orbits. What forces could be responsible for shifting them from their original places in the asteroid belt between Mars and Jupiter and turning them into dangerous Earth-crossers?

The latest theories have concentrated on the effect of Jupiter on the asteroid belt. Jupiter, the largest planet in our solar system, played a crucial part in the creation of the asteroid belt. In the early millennia of the solar system, the sheer mass of Jupiter prevented another planet forming between it and Mars. Instead, the components for the would-be planet were left in pieces—pieces that eventually formed the asteroid belt.

The most promising theories on how asteroids come to leave the asteroid belt center around the so-called Kirkwood gaps, named for English astronomer Daniel Kirkwood (1814–95). These gaps are bands within the asteroid belt where there are no, or relatively few, asteroids. Only recently, astrophysicists have used chaos theory to show that these empty lanes are created by orbital resonance with Jupiter.

Orbital resonance works much like a child pushed on a swing. If the person pushing the child pushes at the just the right moment during each swing, the child will climb higher and higher. But if they push at the wrong intervals, the child does not swing high at all: The mistimed pushes actually cancel each other out. The same happens to the asteroids as they orbit the Sun. As they pass inside Jupiter, the giant planet's gravitational field gives them a nudge. In certain bands within the asteroid belt this achieves a resonance—just like the child being pushed on the swing—and the asteroids within the bands are gradually shifted. The asteroids that once orbited in the empty areas may have moved vast distances.

The 3:1 gap—where objects complete three orbits for Jupiter's one—may be the principal source of Earth-crossers. American astrophysicist Jack Wisdom suggested in 1983 that this 3:1 gap is particularly "chaotic." Asteroids in this zone suffer strong resonant nudges from Jupiter, with unpredictable consequences.

Such asteroids are gradually tugged into a more and more elongated orbit. Eventually they either crash into the Sun or encounter another planet. In the case of Apollo and Aten asteroids, the objects have been pulled by Earth's gravity into new, Earth-crossing orbits that sooner or later could intercept our planet.

ROGUES GALLERY

Name	H	Diam	P	Type	Discovered	Discoverer
(2062) Aten	16.8	0.9–1.7 (1.4–2.7)	0.95	Aten	Jan 07, 1976	E. F. Helin
(2100) Ra-Shalom	16.05	1.2–2.4 (1.9–3.9)	0.76	Aten	Sep 10, 1978	E. F. Helin
(1862) Apollo	16.25	1.1–2.2 (1.8–3.5)	1.78	Apollo	Apr 24, 1932	K. Reinmuth
(1566) Icarus	16.9	0.8–1.6 (1.3–2.6)	1.12	Apollo	Jun 27, 1949	W. Baade
(1864) Daedalus	14.85	2.5–4.3 (4–6.9)	1.77	Apollo	Mar 24, 1971	T. Gehrels
(1866) Sisyphus	13.0	5.0–9.3 (8–15)	2.60	Apollo	Dec 05, 1972	P. Wild
(3200) Phaethon	14.6	2.5–4.3 (4–6.9)	1.43	Apollo	Oct 11, 1983	IRAS

THE MAJOR APOLLO AND ATEN ASTEROIDS. H IS THE ABSOLUTE MAGNITUDE (BRIGHTNESS), DIAM IS THE RANGE OF DIAMETER OF THE OBJECT IN MILES (KM), AND P IS THE ORBITAL PERIOD IN YEARS.

COSMIC ROULETTE

Since 1801, when Italian Giuseppe Piazzi (1746–1826) discovered the first asteroid, astronomers' knowledge of "minor planets" has steadily grown. As the 19th century progressed, observations showed that these objects are mostly concentrated between the orbits of Mars and Jupiter in a wide band called the asteroid belt. It was not until 1932 that a more sinister aspect of these "vermin of the skies" was stumbled upon. An asteroid named 1862 Apollo was discovered frighteningly close to Earth. The object, nearly a mile across, came within 6.5 million miles (10.5 million km) of Earth—a close shave in astronomical terms.

More bad news followed. Apollo was not alone. By 1999, over 400 Apollo asteroids had been found. Most of these objects were discovered to be Earth-crossing asteroids: Their orbits cut across that of Earth. And in 1976 the first of another family of Earth-crossers was discovered: 2062 Aten. Since then, over 50 Aten asteroids have been spotted.

The two families are distinguished by the general shape of their orbits. Apollo asteroids spend most of their orbit outside that of Earth, but at their closest point to the Sun can pass inside Earth's orbit. Aten asteroids, on the other hand, orbit mostly closer to the Sun than Earth, but at their farthest point from the Sun they can pass outside Earth's orbit. Individual Apollo and Aten orbits can vary in orientation—but all their orbits follow these basic shapes.

FRIEND OR FOE

The Apollo and Aten families are relatively small asteroids: Most of them are no bigger than three miles (5 km) across. But this is little comfort to the inhabitants of Earth. Small asteroids are harder to spot in space, and most are not noticed by astronomers until they are already close to Earth—leaving little time for defensive action. And although astronomers call these objects "small," they are still quite big enough to have a cataclysmic effect if they were to it hit our planet.

But Earth is not completely unprepared. Observatories around the world have set up asteroid watches. These include the Spacewatch project at the University of Arizona's Kitt Peak Observatory. Here a team of astronomers hunts down dangerous asteroids. Using CCD (charge-coupled device) cameras and a 36-inch (91-cm) telescope, Spacewatch can see one-mile (1.6-km) asteroids up to 93 million miles (150 million km) away. But we are still a long way off from a comprehensive and reliable database of so-called NEOs (near-Earth objects).

Apollo and Aten asteroids' proximity to Earth can also be seen in a more positive light. As Earth's closest solar system neighbors, the two families of Earth-crossers could prove beneficial to mankind. They could, for example, serve as stepping stones for space exploration missions. Some of the closest asteroids are already easily within the reach of today's rockets. A well-chosen Apollo or Aten asteroid could act as a staging post between Earth and the planets. Astronauts could even mine the asteroids for water, organic compounds and metals, reducing the need for expensive supply runs.

So the Apollo and Aten asteroids represent a mixed blessing for Earth dwellers. Quite probably, one of their number was responsible for the extinction of the dinosaurs, and they may produce future cataclysms. But they could prove to be invaluable as we move out into the solar system—and beyond.

DISCOVERER

GERMAN ASTRONOMER KARL REINMUTH DISCOVERED THE FIRST EARTH-CROSSING ASTEROID, APOLLO, IN 1932. BASED IN HEIDELBERG, GERMANY, REINMUTH SPENT NEARLY 40 YEARS AS AN ASTEROID OBSERVER. HE FOLLOWED UP HIS DISCOVERY BY FINDING ANOTHER EARTH-CROSSER, HERMES, IN 1937. BOTH ASTEROIDS WERE LOST SOON AFTER THEIR DISCOVERY. APOLLO TURNED UP IN 1973, BUT HERMES HAS YET TO BE LOCATED AGAIN. IT IS THE ONLY NAMED ASTEROID OFFICIALLY "LOST."

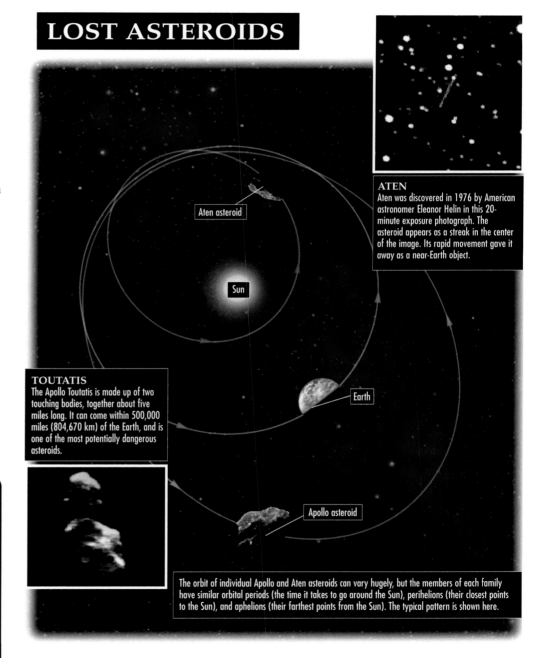

LOST ASTEROIDS

Aten asteroid

Sun

Earth

Apollo asteroid

ATEN
Aten was discovered in 1976 by American astronomer Eleanor Helin in this 20-minute exposure photograph. The asteroid appears as a streak in the center of the image. Its rapid movement gave it away as a near-Earth object.

TOUTATIS
The Apollo Toutatis is made up of two touching bodies, together about five miles long. It can come within 500,000 miles (804,670 km) of the Earth, and is one of the most potentially dangerous asteroids.

The orbit of individual Apollo and Aten asteroids can vary hugely, but the members of each family have similar orbital periods (the time it takes to go around the Sun), perihelions (their closest points to the Sun), and aphelions (their farthest points from the Sun). The typical pattern is shown here.

KILLER ASTEROIDS

Earth-crossing asteroids—lumps of stray matter whose orbits cross that of the Earth—are not just the invention of disaster movies: Asteroid strikes are a real possibility, and have actually happened in the past. The chances of a collision in our lifetime are small. But if an asteroid were to hit us, there is a strong possibility that it would occur without warning. And if the asteroid were large enough, it could cause enough damage to signal the end of life on Earth as we know it.

WHAT IF...

...WE COLLIDED WITH AN EARTH-CROSSER?

Imagine a large asteroid was on a collision course for Earth. How much notice would we get? Duncan Steele is an astronomer who has been trying to alert the world's governments to the danger: "If a half-mile asteroid hits us, we can expect six seconds' warning. When it hits, the sky will light up like a thousand suns. By the time you've turned to look at it, it will have struck the ground, releasing energy equivalent to 10,000 times the Hiroshima bomb. Then it's goodbye."

Although asteroids this size are known to exist, most are only spotted when they are close to Earth, and on their way past us. Our closest known encounter so far was with 1994 XL1, which was located only 14 hours before its nearest

approach. It was 100,000 times fainter than can be seen with the naked eye, and was about 30 feet (9 m) across. Scientists have since worked out that 1994 XL1 missed us by just 52 minutes.

The discoverer of 1994 XL1, Jim Scotti, said afterward: "While I did find this little guy, I missed the 40 or 50 objects of similar size that passed within the Moon's distance that day."

...WE HAD SOME WARNING?

Even if we had warning of an Earth-crosser, what could we do? Small ones could be blown up with missiles, but with larger objects, there is still a risk from the debris. A better plan would be to alter the asteroid's orbit. Johndale Solem of Los Alamos National Laboratory has worked out that exploding a nuclear bomb in front of the oncoming body would be enough to nudge it off course without shattering it—but we would need at least five months' warning.

Another option would be to land rocket boosters on the asteroid to blast it off-course. Far into the future, we might even be able to make icy comets self-propelling by melting their ice core with nuclear power so that the resulting steam acted like a giant rocket.

Firing a nuclear missile at the body risks exposing the Earth to the debris. Detonation near the asteroid, to shatter it, is a safer bet.

CLOSE CALLS

NAME	DATE	CLOSEST TO EARTH
2004 YDS	DEC 19, 2004	20,000 MILES (32,100 KM)
2004 FH	MARCH 18, 2004	26,500 MILES (42,650 KM)
2003 SQ222	SEPT 27, 2003	54,700 MILES (88,000 KM)
1994 XL1 (FORMERLY XM1)	DEC 9, 1994	65,000 MILES (104,600 KM)
1993 KA2	MAY 20, 1993	93,000 MILES (149,700 KM)
1991 BA	JAN 18, 1991	93,000 MILES (149,700 KM)
4581 ASCLEPIUS	MAR 1, 1989	372,000 MILES (598,600 KM)

DOOMSDAY STRIKE

Every night, thousands of shooting stars appear in the night sky, evidence of the huge quantity of space matter that is constantly bombarding the Earth's atmosphere. Most of these objects are small meteors that burn up harmlessly before they reach the surface. However, out in space there are some very large objects, some a mile or two (up to 3.2 km) across. And between the small rocks and the really large ones lie many more, too small to spot, yet big enough to cause considerable damage if they hit the Earth.

Such impacts have occurred in uninhabited areas several times this century. Scientists are warning that it is only a matter of time before there is a major disaster.

CLOSE SHAVE

Asteroids exist mostly in orbit around the Sun between Mars and Jupiter, in a vast ring known as the Asteroid Belt. Occasionally, though, an asteroid will leave its secluded orbit, and veer off into the inner solar system. This sudden change may be caused by an impact with another asteroid, or by the gravitational pulls of Jupiter or Mars.

In recent years, astronomers have chanced on rocks a few hundred yards in diameter hurtling near to Earth. Records for the closest recorded approach are constantly changing as astronomers detect even smaller objects—often when they have already passed the Earth.

Each year, other asteroids are spotted coming almost as close, and it seems that the harder astronomers look, the more they find. It is now estimated that there are over 1,000 near-Earth asteroids roughly a mile (1.6 km) in size. And although there is no known body due to hit the Earth, there are many asteroids out there that we don't know about.

Perhaps it is time to start putting more effort into searching for Earth-crossers. One may not strike for a thousand years, or it may hit us tomorrow.

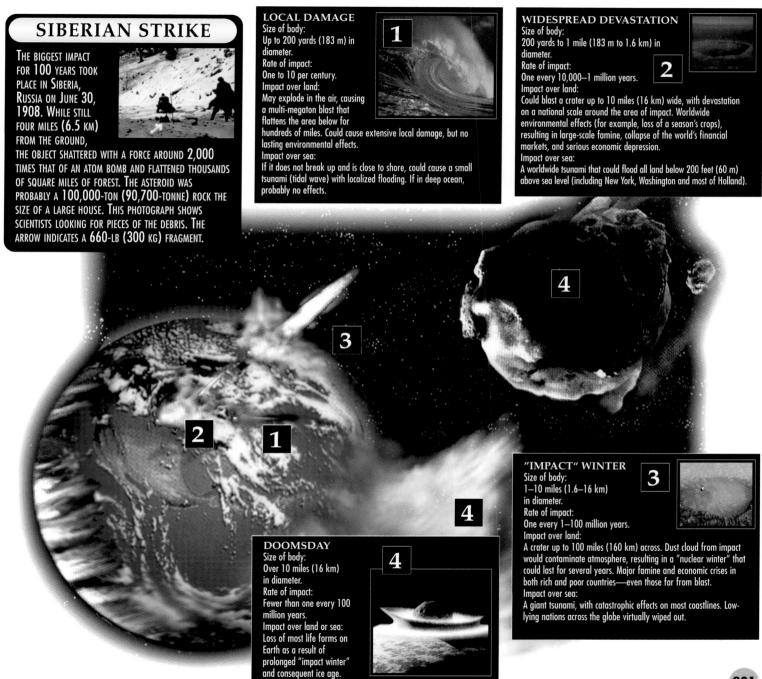

SIBERIAN STRIKE

THE BIGGEST IMPACT FOR 100 YEARS TOOK PLACE IN SIBERIA, RUSSIA ON JUNE 30, 1908. WHILE STILL FOUR MILES (6.5 KM) FROM THE GROUND, THE OBJECT SHATTERED WITH A FORCE AROUND 2,000 TIMES THAT OF AN ATOM BOMB AND FLATTENED THOUSANDS OF SQUARE MILES OF FOREST. THE ASTEROID WAS PROBABLY A 100,000-TON (90,700-TONNE) ROCK THE SIZE OF A LARGE HOUSE. THIS PHOTOGRAPH SHOWS SCIENTISTS LOOKING FOR PIECES OF THE DEBRIS. THE ARROW INDICATES A 660-LB (300 KG) FRAGMENT.

LOCAL DAMAGE
1
Size of body:
Up to 200 yards (183 m) in diameter.
Rate of impact:
One to 10 per century.
Impact over land:
May explode in the air, causing a multi-megaton blast that flattens the area below for hundreds of miles. Could cause extensive local damage, but no lasting environmental effects.
Impact over sea:
If it does not break up and is close to shore, could cause a small tsunami (tidal wave) with localized flooding. If in deep ocean, probably no effects.

WIDESPREAD DEVASTATION
2
Size of body:
200 yards to 1 mile (183 m to 1.6 km) in diameter.
Rate of impact:
One every 10,000–1 million years.
Impact over land:
Could blast a crater up to 10 miles (16 km) wide, with devastation on a national scale around the area of impact. Worldwide environmental effects (for example, loss of a season's crops), resulting in large-scale famine, collapse of the world's financial markets, and serious economic depression.
Impact over sea:
A worldwide tsunami that could flood all land below 200 feet (60 m) above sea level (including New York, Washington and most of Holland).

DOOMSDAY
4
Size of body:
Over 10 miles (16 km) in diameter.
Rate of impact:
Fewer than one every 100 million years.
Impact over land or sea:
Loss of most life forms on Earth as a result of prolonged "impact winter" and consequent ice age.

"IMPACT" WINTER
3
Size of body:
1–10 miles (1.6–16 km) in diameter.
Rate of impact:
One every 1–100 million years.
Impact over land:
A crater up to 100 miles (160 km) across. Dust cloud from impact would contaminate atmosphere, resulting in a "nuclear winter" that could last for several years. Major famine and economic crises in both rich and poor countries—even those far from blast.
Impact over sea:
A giant tsunami, with catastrophic effects on most coastlines. Low-lying nations across the globe virtually wiped out.

COMETS

omets are really nothing more than small, dirty balls of ice, dust and rock that swoop around the solar system in elongated, elliptical orbits. In deep space, they are inert; but as they near the inner solar system, the Sun's heat brings them to life. The ice melts, and the dust and gas boil off to form great glowing tails that can stretch tens of millions of miles through space. For thousands of years, comets were thought to herald great events—or doom. Now, some scientists believe they harbor the seeds of life itself.

WHAT IF...

...COMETS CARRIED VIRUSES?

Back in the 1960s, British astronomer Fred Hoyle suggested that the unexpected appearance of new diseases was linked to comets: When Earth passed through the debris trail of a comet, the planet collected a dose of viruses from space. Other scientists disagreed. But in 1986, the Giotto space probe found traces of complex organic substances during its close approach to the nucleus of Halley's comet. The crust around the nucleus seems to protect these big molecules from cosmic radiation that would otherwise break them up. However, we

Astronomer Fred Hoyle proposed that organic material in cometary tails could reach Earth and bring sickness to its people. His theory has never been proved.

still do not know if any life form, even one as simple as a virus, could exist in a comet. And no one has yet explained how it would get from the interior of the nucleus down to the surface of Earth, or how it would survive the journey.

...COMETS BROUGHT LIFE TO EARTH?

Around 4 billion years ago, comets hit the Earth a thousand times more often than they do today. So if comets could contain the chemical building blocks of life, is it possible that in the distant past they could have brought life to Earth?

The problem with this theory is that comets enter Earth's atmosphere so fast that they reach temperatures of well over 5,000°F (2,760°C)—enough to disintegrate most large molecules. But NASA geochemist Chris McKay has found that

although complex molecules would indeed break up, they would later recombine as they cooled.

McKay's team at NASA's Ames Research Center in Moffett Field, California, focused a laser beam on a jar filled with comet-like gases, flash-heating them to 15,000°F (8,316°C) to simulate impact with the atmosphere. Amazingly, a dozen new compounds were formed, including water, carbon dioxide, methane, nitrogen and hydrogen sulfide. If reactions such as these can happen in a lab, perhaps they could happen elsewhere. Comet impacts on Earth and other planets may be destructive, but they might also have a positive, life-creating effect.

SOME RECENT COMETS

Name	Closest Approach to Sun (million miles/million km)	Date of Closest Approach	Period
Bennett	49.9/80.3	March 3, 1970	1,678
Kohoutek	13.1/21	December 28, 1973	6.67
Kobayashi-Berger-Milon	39.6/63.7	September 9, 1975	Not periodic
West	18.3/29.4	February 25, 1976	Not periodic
Encke	31.6/50.9	August 17, 1977	3.3
Halley	54.8/88	February 9, 1986	76.0
Hyakutake	21.4/34.4	May 1, 1996	Not periodic
Hale-Bopp	84.9/136.6	April 1, 1997	2,380

DIRTY SNOWBALLS

Comets were formed along with the rest of the solar system around 4.5 billion years ago. These fragile lumps of ice and frozen gas could never have coalesced anywhere close to the Sun or they would have been quickly evaporated by its heat. Instead, they were probably born much farther out, in the cold vicinity of the orbits of Uranus or Neptune. There, the comets would have been strongly affected by the gravitation of these two planets. Many of them would have been slung far from the planetary realm to form the Oort Cloud—a spherical region of inert comets that surrounds our solar system and reaches perhaps halfway to the nearest star.

Even at that distance, comets remain in orbit around the Sun—although they may take millions of years to complete their long, elliptical journey. A few—classified by astronomers as short-period comets—make a full orbit in less than 200 years. All the rest are grouped together as long-period comets. Often, they approach from the Oort Cloud at a steep angle to the plane of the planets, and many of them may be on their first visit to the inner solar system. The human race could well be extinct before some of these voyagers return—if they ever do so.

Whatever its orbital period, though, every comet spends most of its existence as an insignificant speck of deep-frozen cosmic debris that is invisible even to the most powerful of telescopes. But as the comet begins its approach to the Sun, the growing warmth triggers a miraculous transformation.

COMET STRUCTURE

At the heart of a comet is a nucleus, at most only tens of miles across. Often called a "dirty snowball," it is largely made of ice, sometimes with a rocky core perhaps a few miles across. The ice consists of layers of frozen gas—hydrogen, nitrogen, oxygen, carbon dioxide and carbon monoxide—as well as water. Dust and pockets of unfrozen gas are mixed in with them, and everything is encased in a thick crust of frozen dust that can be as dark as coal.

As the comet approaches the Sun, it begins to warm up. Near the orbit of Jupiter, the Sun's heat is enough to turn its ices into gas, which leaks through cracks in the crust and surrounds the nucleus in a vast halo called the coma. As the comet plunges still closer to the Sun, this material spreads out behind it in the two enormous tails that make comets so spectacular. One is slightly curved and made of dust, which glows yellow-white as it reflects sunlight. The other tail—straight, often blue and always pointing away from the Sun—is made of a thin scattering of gas pushed out millions of miles by the pressure of solar radiation.

HALLEY'S COMET

EDMUND HALLEY WAS THE FIRST PERSON TO REALIZE THAT COMETS ORBIT THE SUN AND SO ARE PERIODIC VISITORS TO OUR SKIES. HE ANALYZED THE ORBITS OF COMETS SEEN IN 1531, 1607 AND 1682, CONCLUDED THAT THEY WERE THE SAME OBJECT, AND PREDICTED THAT IT WOULD REAPPEAR IN 1758. SADLY, HE DID NOT LIVE TO SEE THE REAPPEARANCE OF THE COMET THAT NOW BEARS HIS NAME (RIGHT), BECAUSE HE DIED IN 1742.

INSIDE A COMET
In deep space, a comet is an inert lump of dust and frozen gas. But as the comet approaches the Sun, the gas begins to boil off. It forms a thick "coma" that obscures the original lump, which now becomes the nucleus of the developing comet. Sometimes, the Sun heats pockets of gas trapped beneath the still-frozen surface. Eventually, they burst through, releasing jets of hazy debris that will form the comet's tails. Repeated passes of a comet around the Sun will in time rob it of anything that can still evaporate. Only a rocky core—if the comet has one—will remain, quietly orbiting the Sun as a tiny asteroid.

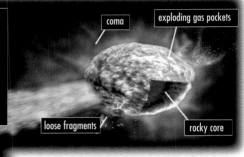

coma

exploding gas pockets

loose fragments

rocky core

Comet Hale-Bopp, one of the brightest comets of the 20th century, was prominent in the night sky in the spring of 1997.

HALLEY'S NUCLEUS

WHEN HALLEY'S COMET APPROACHED THE SUN IN 1986, THE SPACE PROBE GIOTTO FLEW THROUGH ITS COMA. SCIENTISTS WERE EXPECTING TO FIND THAT THE COMET WAS MADE OF MINERAL ELEMENTS SUCH AS IRON, CARBON, CALCIUM AND SILICON. INSTEAD, GIOTTO FOUND HYDROGEN, NITROGEN AND ORGANIC MOLECULES. THE NUCLEUS ITSELF WAS MAINLY ICE.

inert comet nucleus

tails die away

ices begin to melt

Jupiter

Sun

gas tail

LIFE OF A COMET
Far out in its orbit, the comet is nothing but a frozen nucleus. When it nears the orbit of Jupiter, it starts to thaw. Ice melts, and tails of gas and dust begin to form. As the comet continues inward to the Sun, the tails may expand to a spectacular length. They gradually die away after the comet swings around the Sun and heads off again to the cold fringes of the solar system.

fully formed comet

dust tail

COMETARY NUCLEI

Of all the astronomical objects visible to the naked eye, few are more spectacular than comets. A comet's tail can stretch for millions of miles, its microscopic particles gleaming in the sunlight. Yet the body of the comet itself, called the nucleus, may be no more than a few miles in diameter, and darker than coal. It is a mixture of rock and frozen gases left over from the formation of the solar system. Many trillions of comet nuclei probably merged to form the outer planets, while a trillion more still orbit the Sun.

WHAT IF...

...COMETS WERE FRAGILE?

In the summer of 1994, the splintered remains of Comet Shoemaker-Levy 9 pounded the planet Jupiter. As almost two dozen fragments slammed into Jupiter's dense atmosphere, they exploded with the force of hundreds of hydrogen bombs, in some cases creating scars as big as Earth. Shoemaker-Levy 9 confirmed that although a cometary nucleus is a single, solid body, it falls apart as easily as a multi-scoop ice cream cone.

Astronomers had already seen several comets split apart as they flew close to the Sun. The combination of intense heat and tidal gravity—a difference in the gravitational pull on the side of the comet facing the Sun and the side opposite the Sun—split them apart. But Shoemaker-Levy 9 demonstrated that gravity alone can tear a comet apart, indicating that a nucleus may be a pile of snowballs glued together by ice or by their own weak gravity. If so, then destroying a comet on a collision course with Earth could be difficult.

The impact of a large comet or asteroid could blast enough debris into the atmosphere to block out the Sun and kill most of the life on Earth. Such an impact 65 million years ago may have exterminated the dinosaurs.

If astronomers discover a comet on a collision course months before impact, it might be possible to blast it with a nuclear missile. But if the comet splits apart, scientists say we could be in even worse shape: Several sizeable comet chunks, striking different regions of Earth a few minutes apart, could inflict even more damage than a single larger body.

The solution may be to deflect the comet instead of destroying it. Nuclear bombs detonated to one side of the comet might alter its course just enough for it to miss Earth completely. But the results of a recent a study question the viability of using nuclear weapons to avert an incoming comet. In 1998, Erik Asphaug of the University of California created a complex computer simulation of the effects of a mid-space blast on an asteroid. He used various hypothetical structures for the model asteroids. One structure, the so-called "rubble-pile" asteroid, survived the blast almost unscathed. The loose structure allowed the asteroid to absorb the energy of the blast, and its trajectory was not affected. The spectacular ejection of material created by NASA's Deep Impact probe as it smashed into Comet Tempel 1 in 2005 suggests that comets do indeed have remarkably loose internal structures.

HALLEY'S NUCLEUS

SIZE	10 MILES (16 KM) LONG, 5 MILES (8 KM) WIDE
COMPOSITION	WATER, CARBON, NITROGEN, SULFUR
SURFACE COLOR	BLACK
BRIGHTNESS	REFLECTS 4% OF SUNLIGHT
VOLUME	120 CUBIC MILES (500 CUBIC KM)
AVERAGE DENSITY	10–25% DENSITY OF WATER
CLOSEST APPROACH TO EARTH	ABOUT 3 MILLION MILES (4.8 MILLION KM) (APRIL 837 A.D.)
LAST APPEARANCE	1986
NEXT APPEARANCE	2061

SOFT CELL

Each time a comet flies through the inner solar system, the Sun's warmth vaporizes ice just beneath its surface, releasing gas and particles of dust into space. This material forms a bright cloud hundreds of thousands of miles wide and a tail millions of miles long.

All the known comets are so small and dark, though, that without the Sun's heat, we would never see them with our own eyes. The body of a comet, called the nucleus, is a fluffy agglomeration of frozen water and gases mixed with rock, generally no more than a few miles across. Carbon-based compounds at the surface make the nucleus darker than coal. Comet Halley, for example, reflects just 4% of the sunlight that strikes it.

For decades, astronomers described comets as "flying hailstorms"—clouds of small chunks of ice and rock orbiting the Sun in formation. But in 1950, astronomer Fred Whipple came up with the description "dirty snowballs." He theorized that a comet is a single, solid object that sheds material as it orbits the Sun. Spacecraft that flew past Halley in the mid 1980s found that the truth may actually be somewhere between these two models: Nuclei may consist of large clumps of ice and rock held together by an icy "glue," gravity, or both.

Cometary nuclei probably formed during the birth of our solar system 4.5 billion years ago, when small grains of ice and rock in the early dust and gas coalesced into ever-larger bodies. Many nuclei then clumped together to form the outer gas giant planets.

NUCLEUS ENCOUNTERS

We know much less about the nuclei of comets than we would like. Not only are they very minor members of the solar system, but on their rare travels close to the Earth, they are wreathed in dust and gas that makes their true nature hard to fathom. What appears to be the nucleus through even the most powerful telescope is in reality a much larger region of glowing gas surrounding it.

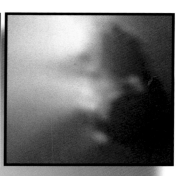

NEAR MISS
One of six comets discovered by the U.S.-British-Dutch probe IRAS (InfraRed Astronomical Satellite), IRAS-Araki-Alcock was spotted just 16 days before it passed less than 3 million miles (4.8 million km) from Earth.

CLOSE-UP
The most detailed picture of a cometary nucleus that we have, this image is actually a composite of 60 photographs taken by the European Space Agency's probe Giotto during its close encounter with Comet Halley in March 1986.

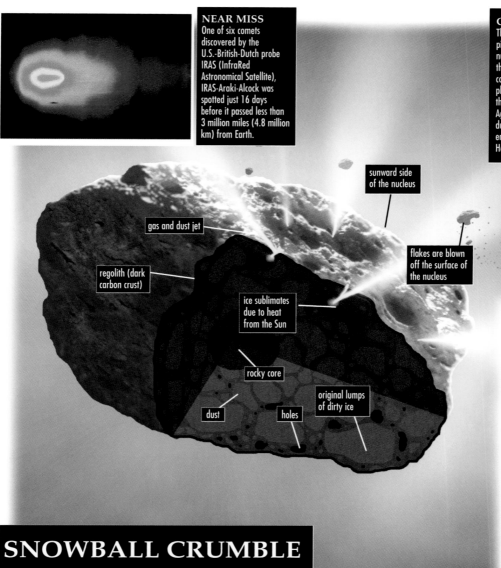

- sunward side of the nucleus
- gas and dust jet
- flakes are blown off the surface of the nucleus
- regolith (dark carbon crust)
- ice sublimates due to heat from the Sun
- rocky core
- dust
- holes
- original lumps of dirty ice

SNOWBALL CRUMBLE

A comet comes to life in sunlight. Heat from the Sun sublimates the ice (turns it straight from a solid to a gas) just below the surface. The resulting jets of gas and dust form the comet's coma (the bright cloud around the nucleus) and tail. Some astronomers theorize that nuclei have a rocky core.

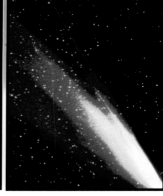

BREAK-UP
The appearance of Comet West in 1976 gave astronomers a chance to study a comet's nucleus. The comet broke up in March 1976, allowing unprecedented study of its internal structure. It was revealed to be homogenous: the same throughout.

The first cometary nucleus to be studied in detail was that of Halley's Comet, during its 1986 passage around the Sun. An armada of space probes went to meet the visitor, but the closest approach was made by the European Giotto spacecraft, which plunged into the coma and passed within about 300 miles (483 km) of the nucleus. It sent back pictures of a peanut-shaped, 5- by 9-mile (8- by 15-km) object.

More recently, NASA's Stardust probe made a close approach to Comet Wild-2 while collecting particles from its tail for an eventual return to Earth, and in 2005 the Deep Impact spacecraft launched a high-speed projectile into the surface of Comet Tempel-1, studying the resulting ejecta. An even more ambitious project is Rosetta, a European-led mission to put a probe in orbit around a comet nucleus and place a robot lander on its surface in 2014.

235

COMETS IN HISTORY

Throughout history, even to the present day, comets have provoked strange reactions in human observers. Often thought to foretell the deaths of rulers, comets have always been seen as bad news. Ignorance has made them history's most feared omen. It was not until the Renaissance that astronomers discovered that comets are not atmospheric phenomena—an idea that had survived since before the birth of Christ. And despite our present-day knowledge of comets, they can still surprise and alarm us.

WHAT IF...

...A COMET CHANGED HISTORY?

Our ancestors were deeply fearful of comets. Modern humans like to think of themselves as beyond such superstitious dread. But how would our world cope with the appearance of a really big new comet?

Imagine that this comet—named McCoy-Alvarez after its discoverers—is first observed in 2020 A.D., a little beyond the orbit of Jupiter. Its own orbit, perturbed by a passing star, has brought it on a million-year sweep from the Oort cloud at the edge of the solar system: McCoy-Alvarez is making its first close approach to the Sun.

Astronomers are excited. Calculations show that McCoy-Alvarez will pass within 4 million miles of the Earth, and since the frozen gases of its nucleus have never before been exposed to the Sun, it should put on as spectacular a show as any comet in history. The world's space agencies are working feverishly to launch what amounts to a small armada of space probes: A crewed mission is ruled out as too dangerous.

A large proportion of the human race starts to feel nervous. Church attendance soars. Religious cults announce the imminent end of the world—and they are not the only ones. Practically every TV channel on the planet runs talk shows that discuss the possibility that the comet will hit the Earth. Astronomers deny that there is any risk, but there are plenty of conspiracy theorists convinced that the establishment is lying in order to prevent worldwide panic.

The panic happens anyway. No one is buying stocks and shares any more—although everyone with a credit card is using it recklessly. The market crashes and the world economy teeters on the brink of collapse. And McCoy-Alvarez continues to swell.

The first serious riots break out in a northern Indian city, where Hindus and Muslims decide to settle old scores before the world ends—and loot a few consumer goods while they are it. The rioting soon spreads. A million people converge on Cape Canaveral to prevent NASA's comet probe launches. Some believe that the spacecraft will accelerate the end, others that the launches are a getaway plan for government officials. It takes regular Army units to hold them back.

In the general confusion, hardly anyone notices that McCoy-Alvarez is receding. The comet has done no harm to the Earth—at least, not directly. As rioters come to their senses in their ruined cities, they are glad to know it will not be back for a million years.

GREAT COMETS IN HISTORY

COMET	YEAR	MAGNITUDE	CLOSEST APPROACH TO EARTH (MILLION MILES/MILLION KM)
(NO NAME)	1577	−4	94/151
DE CHESEAUX	1743	−7	125/201
DONATI	1858	−1	80/129
GREAT SOUTHERN COMET	1882	−10	148/238
DAYLIGHT COMET	1910	−4	130/209
IKEYA-SEKI	1965	−10	21/34
HALE-BOPP	1997	−1.5	193/311

BAD OMENS

The history of comets—as far as humanity is concerned—begins with the Chinese. Between about 1400 B.C. and 1600 A.D., they noted at least 338 cometary appearances, and kept remarkably accurate records of what they saw. Most impressive of all was the Mawangdui textbook, excavated in the 1970s from an ancient burial chamber. Dating from about 300 B.C., the book contains information on 29 comets, illustrated and classified with great care. Not until the 15th century would the Western world make comparable observations.

Until the Renaissance, though, not even the Chinese had any real idea what comets were. The view of the 4th-century-B.C. Greek philosopher Aristotle that they were atmospheric phenomena was the most commonly held opinion among experts. Ordinary people saw them as evil omens. The first person to contradict Aristotle and suggest that comets were actually celestial bodies was the Roman philosopher Seneca. His famous book *Natural Questions*, written in the first century A.D., rejects Aristotle's version of the cosmos. Unfortunately, Seneca's ideas were not widely accepted.

Aristotle's version was only really overturned by the Danish astronomer Tycho Brahe (1546–1601). In 1577, Brahe used the parallax method of measuring stellar distances, to measure the distance of a comet. He found that it was moving far beyond the Moon, certainly not in Earth's upper atmosphere. With Aristotle's universe dismissed, astronomers needed a new comet theory.

NEW THEORIES

English astronomer Edmond Halley (1656–1742) took up the challenge. He worked out the orbits of 24 comets that had been observed between 1337 and 1698. With one he found a regular pattern. The comets of 1531, 1607 and 1682 appeared to be the same object, returning at approximately 76-year intervals. Halley predicted that it would return again in 1758.

Although he died before then, the comet did indeed return, and was named after Halley in recognition of his achievement.

The next great leaps in comet science were not taken until the 1950s. It was then that American astronomer Fred Whipple worked out that comets are lumps of dirty ice. More recently, researchers have delved even deeper by sending space probes to the dirty snowballs to study them at close range.

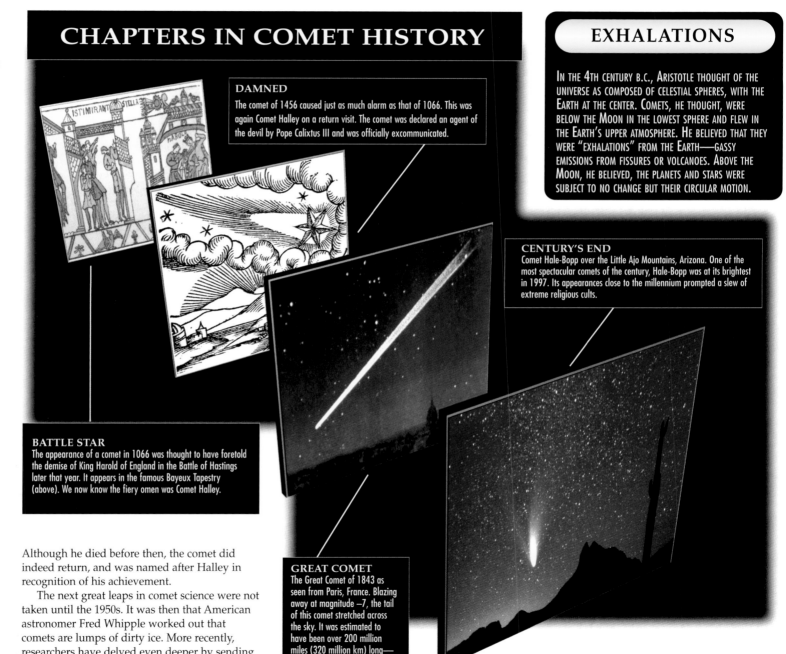

CHAPTERS IN COMET HISTORY

DAMNED
The comet of 1456 caused just as much alarm as that of 1066. This was again Comet Halley on a return visit. The comet was declared an agent of the devil by Pope Calixtus III and was officially excommunicated.

BATTLE STAR
The appearance of a comet in 1066 was thought to have foretold the demise of King Harold of England in the Battle of Hastings later that year. It appears in the famous Bayeux Tapestry (above). We now know the fiery omen was Comet Halley.

GREAT COMET
The Great Comet of 1843 as seen from Paris, France. Blazing away at magnitude –7, the tail of this comet stretched across the sky. It was estimated to have been over 200 million miles (320 million km) long—the longest comet tail ever recorded.

CENTURY'S END
Comet Hale-Bopp over the Little Ajo Mountains, Arizona. One of the most spectacular comets of the century, Hale-Bopp was at its brightest in 1997. Its appearances close to the millennium prompted a slew of extreme religious cults.

EXHALATIONS

In the 4th century B.C., Aristotle thought of the universe as composed of celestial spheres, with the Earth at the center. Comets, he thought, were below the Moon in the lowest sphere and flew in the Earth's upper atmosphere. He believed that they were "exhalations" from the Earth—gassy emissions from fissures or volcanoes. Above the Moon, he believed, the planets and stars were subject to no change but their circular motion.

SUN GRAZING COMETS

A few times a year, a new comet appears in the inner solar system. Most come no nearer to the Sun than the orbit of Venus. But occasionally, a comet will skim so close to the Sun's surface that it may not survive its passage. In the course of its destruction, the Sun-grazer may lose so much material that it puts on a spectacular show before it vanishes. Strangely, many of these Sun-grazing comets seem to belong to the same family—and recent space images show that there are many more of them than astronomers thought.

WHAT IF...

...ANOTHER PARENT COMET APPEARED?

So far as we know, the Kreutz Sun-grazers are the most spectacular group of Sun-grazing comets to visit the solar system. But they are almost certainly not the only cometary group to have originated from the earlier breakup of a giant parent body. Some astronomers think that comets Liller of 1988 and Tabur of 1995, though not in Sun-grazing orbits, are both fragments of a larger body that broke up a few thousand years ago. If this is true, then there may be other potential parent bodies out there—icy planetoids whose orbits have not carried them near the vicinity of the inner solar system since human beings first walked on the Earth.

If such a body—say a few hundred miles across—did appear, we would have very little warning. The nucleus of the spectacular 1997 comet, Hale-Bopp, was 25–30 miles (40–50 km) in diameter, yet it was only detected a couple of years before perihelion (its closest approach to the Sun), by which time it was well inside the orbit of Saturn. And at around this point, the body would begin to fall under the massive gravitational sway of Jupiter, which has a reputation for snaring comets and either smashing them or catapulting them toward the Sun.

A large, unstable body might fragment as it nears Jupiter—but because of its size, the result could be very different from Comet Shoemaker-Levy 9's demise in July 1994. S-L9 smashed into pieces so small that they were all captured by the giant planet's gravity. If the same thing happened on a larger scale, the fragments might escape Jupiter's gravity and be sent careening toward the Sun—and the Earth.

We would then have a nervous wait of 12 to 18 months before the approaching shower of Sun-grazers reached their perihelion. Once inside Jupiter's orbit, their proximity to the Sun would cause them to glow like a swarm of fireflies. But each night, the dots of light would grow steadily bigger and brighter. Scientists would spend long hours at their computers, calculating the random gravitational effects that might send one of the fragments on a collision course with Earth. At the height of the shower, the largest of the comets would appear as large as the Moon, and the sky itself would seem to be on fire. People would remember it as the Greatest Show on Earth—if they lived to tell about it.

FAMOUS SUN GRAZING COMETS

YEAR OF DISCOVERY	NAME	TAIL LENGTH	NOTES
371 B.C.	Unnamed	60°	Described by Aristotle
1668 A.D.	Unnamed	30°	Observed by Cassini
1680	Unnamed	80°	May have sparked Edmond Halley's interest in comets
1843	Great March Comet	70°	At peak, as bright as first-quarter Moon
1882	Great September Comet	30°	At peak, brighter than full Moon
1887	Great Southern Comet	50°	Discovered by South African farmer in twilight
1965	Ikeya-Seki	60°	Nucleus fragmented; as bright as full Moon
1998	1998J1	20°	One of many sungrazers discovered by SOHO

CLOSE ENCOUNTERS

Comets are fragile objects—lumps of ice and rock a few miles across, held together only by their own feeble gravity. Yet some of these delicate agglomerations regularly survive passages so close to the Sun that the comets actually pass through the outer layers of the solar atmosphere.

Apart from the ferocious conditions they must endure, Sun-grazers are no different from any other comet. Their central nucleus contains water ice, and frozen gases such as carbon dioxide, methane and ammonia, mixed with small dusty particles. Like all comets, they grow a huge tail as they near the Sun and solar radiation boils off dust and gas from the surface of their nuclei. Most comets swing away from the Sun before heat can penetrate through miles of insulating ice to the interior of the nucleus. But Sun-grazers come so close that their frozen cores are thawed and vaporized. The result is a spectacular show in the night sky—and in the case of a few Sun-grazers, the daytime sky, too.

For many Sun-grazers, their first perihelion—their closest approach to the Sun—is also their last. Weakened by heat, their nuclear structure tears apart. In 1976, for example, the nucleus of the spectacular Comet West broke into four pieces a few days after perihelion.

Even if a Sun-grazer survives its close encounter, we are unlikely to see it again. Its orbit will take it far from the center of the solar system, out into the deep dark in an orbit that takes 500 or even 1,000 years to complete.

FAMILY RESEMBLANCE

The details of a comet's orbit—its duration, its shape and its angle with respect to the Earth's own orbit—amount to a distinguishing "fingerprint." Astronomers have noted that many Sun-grazers have orbits that share some key characteristics. The length of their "year" is similar, they are all tilted at 141° to the Earth's orbit and they are retrograde—that is, they travel around the Sun in the opposite direction of the planets.

Such common features make it likely that all, or most, of the Sun-grazing comets are the children of an original giant parent, perhaps hundreds of miles across, that fragmented at its perihelion—a theory first suggested by the German astronomer Heinrich Kreutz in 1888. Kreutz gave his name to the principal group of Sun-grazers. Since then, as astronomers find new comets that match the general orbital characteristics that Kreutz identified, they add those comets to the Kreutz list.

Usually, a Sun-grazer's performance ends with its own breakup. A few Sun-grazers, though, come so close to the Sun that they actually hit it. The cometary wreckage smashes into the Sun's outer layers at several hundred miles per second, providing a substantial burst of energy. Although the impact itself is usually hidden in the glare of the Sun's disk, the outcome may be another spectacle.

In June 1998, astronomers discovered two small Sun-grazing comets on an collision course with the Sun. Hours later, from that very spot, a giant bubble of hot gas erupted out of the Sun's external atmosphere. Although at the time the sequence was dismissed by cautious scientists as coincidence, some astronomers consider it possible that the comet impacts may have triggered the solar burst.

SUN-GRAZERS CLOSE UP

1 INCOMING COMETS
An image from the observation satellite SOHO shows two small comets approaching the Sun. The picture uses a device called a coronograph that masks the Sun's own light behind a central disk.

2 CLOSE APPROACH
The first comet crosses a bright stream of gas. The stream reaches the edge of the dark blue disk—about 2 million miles (3.2 million km) from the Sun (indicated as a white circle, top left).

3 SOLAR IMPACT?
The first comet has passed the prominence and is just visible at its upper edge. Its trajectory suggests that it is about to impact the Sun—the possible fate of many Sun-grazing comets.

ARRIVAL
The comet falls into the inner solar system. Its nucleus warms in the increasing sunlight and boils off dust and vapor. This material streams out behind the comet, directly away from the Sun.

PERIHELION
The intense radiation of the sun peaks at perihelion—the point that marks its closest approach.

DESTRUCTION
The Sun's heat and gravity tear the comet apart. The pieces may take many centuries to return—if they ever do so.

SATELLITE OBSERVATORIES

SOME RARE SUN-GRAZERS, SUCH AS 1965's IKEYA-SEKI, ARE SPECTACULAR. BUT MANY LESSER COMETS ARE SO SMALL THAT THEY ARE COMPLETELY INVISIBLE FROM THE EARTH. WITH THE HELP OF INSTRUMENTS CALLED CORONAGRAPHS (RIGHT), WHICH BLOCK OUT THE BRIGHT DISK OF THE SUN, SOLAR RESEARCH SPACECRAFT HAVE REVEALED THAT SUN-GRAZERS ARE FAR MORE COMMON THAN WE ONCE THOUGHT. ASTRONOMERS NOW USE CORONAGRAPHS TO STUDY THE SUN'S ATMOSPHERE. ONE OF THE FIRST ORBITING CORONAGRAPHS TO RECORD SUN-GRAZING COMETS WAS THE SOLWIND INSTRUMENT, CARRIED ABOARD THE U.S. MILITARY SATELLITE P78-1, WHICH OPERATED FROM THE LATE 1970s UNTIL IT WAS DESTROYED DURING TESTS FOR THE "STAR WARS" PROGRAM. THE 1980s SOLAR MAXIMUM MISSION SATELLITE RECORDED NINE SUN-GRAZERS. THE TOP COMET-SPOTTER, HOWEVER, IS THE SOHO SATELLITE, WHICH DISCOVERED 55 SUN-GRAZERS IN ITS FIRST THREE YEARS. ONE OF SOHO's DISCOVERIES WAS BRIGHT ENOUGH TO BE SEEN IN THE SKY OVER THE SOUTHERN HEMISPHERE IN JULY 1998.

COMET SHOEMAKER-LEVY 9

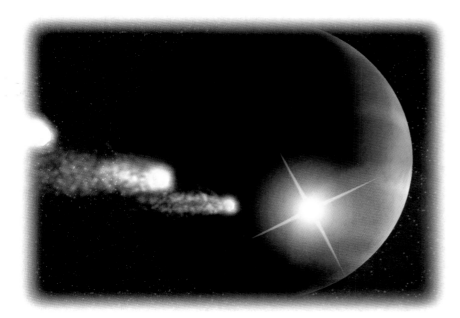

...JUPITER OR SATURN COULD BE DAMAGED BY COMETS?

As the largest planet in the solar system, Jupiter is by far the biggest influence on comets apart from the Sun. When comets are diverted from the farthest reaches of the solar system, it is often Jupiter's gravity that draws them into a closer orbit around the Sun.

We know of no other comets that have become trapped into orbiting Jupiter itself. Yet there is some evidence that comets like SL9 have collided with the giant planet before, or at least with its moons. Both Ganymede and Callisto have chains of craters on their surfaces that are thought to be the result of split comet impacts. So could a large comet do any really serious damage?

The answer is no. Although the SL9 impacts were spectacular, and traces of them were visible for months afterward, the energy released in the collision was insignificant compared with the vast amounts of energy tied up in the rotating planet. Comets are not even powerful enough to disrupt Jupiter's wind systems. And even if Jupiter has a solid core beneath the visible clouds, a fragile comet would be destroyed long before it reached it.

Saturn, the second-largest planet in the solar system, is another gas giant like Jupiter. If a comet were to hit it, we could expect a similar display to the SL9 impact. But what would happen to its famous rings? Could they be warped by a comet, or might they even disintegrate after an impact?

Saturn's main rings are just over twice the diameter of the planet, while the fainter outer rings extend to five times the diameter, or 380,000 miles (611,550 km). They represent a sizable cosmic net—but also a loosely woven one. Because the rings consist of bits of ice and rock, ranging in size from snow-flakes to boulders, an impacting comet would simply punch a hole in them, after which they would close up again as if nothing had happened.

Saturn and its rings (left) would survive even a large comet impact. Crater chains on Jupiter's moon Callisto suggest that comets have struck it in the past.

Way back in 1929, a comet strayed too close to Jupiter and was captured by the giant planet's gravity. For 63 years it swung around Jupiter in an unstable orbit until 1992, when it broke up into a string of pieces. On its next and final approach, in the summer of 1994, the string of debris known as Comet Shoemaker-Levy 9 (SL9) plunged deep into Jupiter's thick, gaseous atmosphere. The comet's fiery demise astonished the world; it was the most spectacular fireworks display ever seen in the solar system.

SL9 COMET STATISTICS

NUMBER OF FRAGMENTS IDENTIFIED	.23
NUMBER OF IMPACTS RECORDED	.20
IMPACT VELOCITY	.37 MILES PER SECOND (60 KM/SEC)
ANGLE OF IMPACT	.45°
TIME OF FIRST IMPACT (FRAGMENT A)	.JULY 16 20:11:00 GMT
TIME OF LAST IMPACT (FRAGMENT W)	.JULY 22 08:05:30 GMT
SIZE OF FRAGMENTS	.0.5 TO 2 MILES (0.8 TO 3.2 KM)
SIZE OF BODY BEFORE FRAGMENTATION	.5 MILES (8 KM)

FIRE ON JUPITER

When it was first spotted in March 1993, Comet Shoemaker-Levy 9 (SL9) had only 16 months to live. Unlike other comets, SL9 was not in orbit around the Sun, but in an unstable orbit around Jupiter. Calculations showed that the comet must have passed unnoticed within 13,000 miles (20,920 km) of Jupiter's cloud-tops on July 7, 1992. During that unseen flyby, the gravity of our solar system's largest planet pulled it apart. On its next passage, in July 1994, the fragments were destined to crash into Jupiter.

Early predictions suggested that the fragments of SL9 would impact the far side of Jupiter and would not be visible from Earth. But as collision time grew nearer and more observations were made, revised calculations showed that we would only narrowly miss the action: Jupiter's 10-hour rotation period would bring the crash sites into view just a few minutes after each impact.

Observatories around the world mounted a coordinated campaign to monitor the impacts. Spacecraft such as the Hubble Space Telescope and the Galileo probe (then on its way to Jupiter) were also pressed into service. Jupiter has no solid surface, so it was hard to predict what would happen. Some astronomers thought that the pieces would simply be swallowed by the planet's thick atmosphere. Others hoped to see some more spectacular sights.

As time went by, the pieces of the comet spread out into a chain over 2 million miles (3.2 million km) in length. This chain crashed into Jupiter over a six-day period in July 1994, with the first of many sightings of the aftermath coming from the Calar Alto Observatory in Spain on the evening of July 16.

CRUNCH TIME

Each impact began with a faint glow, as minor debris hurtled into Jupiter's atmosphere just ahead of the main fragment. Then came the fragment itself. Traveling at 37 miles per second (60 km/sec), and to the delight of astronomers, it struck with a brilliant flash that sent a giant fireball hurtling 2,000 miles (3,220 km) into space.

The impact was visible from Earth as a plume of gas on the edge of Jupiter's disk. Within minutes, debris falling from this plume had scorched a bright spot in the atmosphere. When the impact site itself rotated into view, 11 minutes after the collision, astronomers saw a dark spot surrounded by an expanding wave like ripples in a lake. A crescent-shaped scar marked the point at which the plume fell back onto Jupiter's atmosphere.

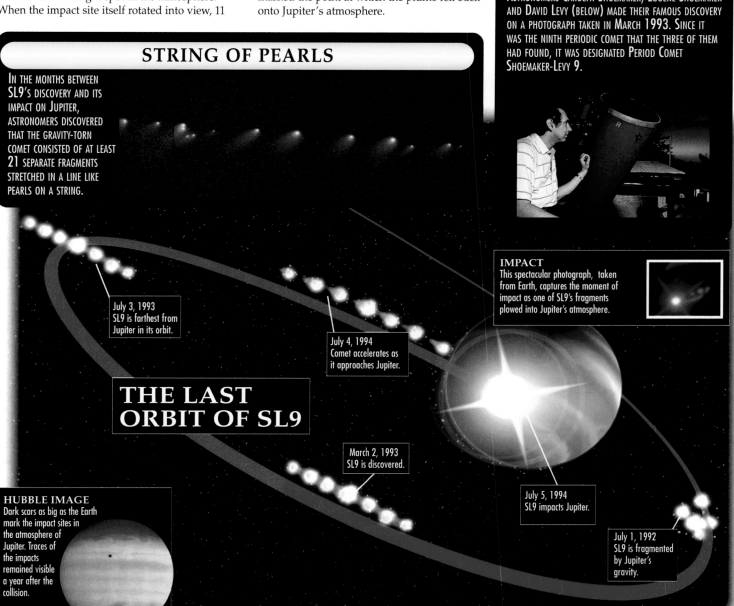

STRING OF PEARLS

IN THE MONTHS BETWEEN SL9'S DISCOVERY AND ITS IMPACT ON JUPITER, ASTRONOMERS DISCOVERED THAT THE GRAVITY-TORN COMET CONSISTED OF AT LEAST 21 SEPARATE FRAGMENTS STRETCHED IN A LINE LIKE PEARLS ON A STRING.

THE LAST ORBIT OF SL9

July 3, 1993
SL9 is farthest from Jupiter in its orbit.

July 4, 1994
Comet accelerates as it approaches Jupiter.

March 2, 1993
SL9 is discovered.

July 5, 1994
SL9 impacts Jupiter.

July 1, 1992
SL9 is fragmented by Jupiter's gravity.

IMPACT
This spectacular photograph, taken from Earth, captures the moment of impact as one of SL9's fragments plowed into Jupiter's atmosphere.

HUBBLE IMAGE
Dark scars as big as the Earth mark the impact sites in the atmosphere of Jupiter. Traces of the impacts remained visible a year after the collision.

241

CENTAURS

P art asteroid and part comet, the strange objects that astronomers call Centaurs originate in the Kuiper Belt, a zone of ancient debris beyond the orbit of Neptune. Far from the Sun's warmth, Kuiper objects are rich in frozen organic chemicals, gases and water ice. Occasionally, an encounter with Neptune's gravity sends one of them into an eccentric orbit that brings it closer to the Sun. This can cause the volatile ices to boil off—briefly transforming the asteroid-like object into a comet.

WHAT IF...

...ALL ASTEROIDS WERE ONCE COMETS OR CENTAURS?

T he orbits of Centaurs are not stable. These objects were ejected from the Kuiper belt by Neptune's gravity, and pulled into trajectories that often take them perilously close to the outer giant planets. Eventually, all of them will crash into a gas giant, careen out of the solar system, or plummet toward the inner planets.

If a Centaur happens to fall toward the Sun, it may develop a coma—a visible head of gas—just as Chiron does. And if Chiron or another Centaur of about the same size were nudged into a typical cometary orbit that passed very close to the Sun, it might develop into the most spectacular comet ever seen. But it seems more likely that Centaurs tumble through the inner solar system for a few million years while the Sun's rays slowly melt away their volatile chemicals—methane, for example, or water ice. Eventually they will be baked dry—toasted into arid, rocky hulks. Many will be shattered in impacts with other asteroids, and some of the debris could even end up in the solar system's asteroid belt, between the orbits of Mars and Jupiter.

At present, we can study asteroids only by examining meteorites—space rocks that fall to the Earth's surface. These meteorites are completely waterless. It is possible that they have never contained water. But many of them are riddled with internal fractures and cavities, where water might once have been. And when scientists cut through such meteorites, they find deposits of chemicals called carbonates, an almost certain indication that the dry crevices once contained ice. Perhaps many of the asteroids between Mars and Jupiter were once rich in water ice. If so, then some asteroids could be the dehydrated remains of prehistoric Centaurs and comets.

Created in the freezing depths of the solar system, for billions of years they orbited the Sun at a safe distance, preserving within them their frozen cargo of water. But then they flew closer to the solar flame, and for a few millennia, they became comets, squandering their volatiles into space in successive visits to the Sun, and developing spectacular cometary tails in the process. Their orbits must have been in some way perturbed so that they no longer ventured near to the Sun, and eventually they became mere cinders. The asteroid belt is now a graveyard of Centaurs and comets, more ancient than the dinosaurs.

FIRST CENTAURS

Name	Distances from Sun in AU*			Name	Distances from Sun in AU*		
	Perihelion	Aphelion	Mean		Perihelion	Aphelion	Mean
1999 OX3	17.42	28.60	23.01	1994 TA	11.77	21.95	16.86
1999 JV127	10.72	22.73	16.72	1995 DW2	18.85	30.90	24.88
1999 HD12	8.90	33.75	21.32	1997 CU26	13.07	18.41	15.74
1998 TF35	17.88	37.30	27.59	1995 GO	6.83	29.14	17.99
1998 SG35	5.85	10.99	8.37	7066 Nessus	11.81	37.10	24.46
1998 QM107	17.32	22.88	20.10	5145 Pholus	8.65	31.80	20.23
1998 BU48	18.47	32.43	25.45	2060 Chiron	8.45	18.77	13.6
1995 SN55	7.94	39.19	23.56				

*AU = ASTRONOMICAL UNITS

STRANGE HYBRID

On October 18, 1977, U.S. astronomer Charles Kowal was studying a photographic plate when he noticed a point of light that did not appear on any star chart. The next evening he turned a telescope to the same region of the sky. The speck of light had moved—and Kowal had discovered a minor planet.

Kowal named the object Chiron. Subsequent research found that Chiron had been photographed as long ago as 1895—but never noticed. It circles the Sun in an eccentric, 51-year orbit between Saturn and Uranus. A mere 125 miles (200 km) across, Chiron was classified as an asteroid. But in 1988, as Chiron neared perihelion—its closest approach to the Sun—astronomers discovered a faint shroud of glowing matter around it. It seemed that the Sun's heat was blasting volatile material away from the planetoid's surface.

Chiron was behaving less like an asteroid and more like a comet. But if it were a comet, it would be by far the most massive ever discovered. So for some years Chiron was considered an oddity—too large to be a comet, too active to be an asteroid. But 1992 brought the discovery of Pholus, roughly the same size as Chiron, and in a similar orbit. The hunt was on: Astronomers around the world scanned the sky for more of these curious objects. Chiron now has more than 150 known companions, and the number grows yearly. Together, they are known as the Centaurs.

LURED AWAY

All of the Centaurs almost certainly originate in the Kuiper belt, a band of asteroids that orbits the Sun beyond Neptune. It contains approximately 70,000 objects more than 50 miles (80 km) across. The Kuiper belt was formed 4.5 billion years ago and is as old as the Sun itself. The Kuiper asteroids consist of dust and rubble frozen into bergs of ice, methane and other solid organic chemicals. Occasionally one of them passes close to the planet Neptune. The gravitational tug of the giant planet wrenches the little asteroid into a new, more elliptical orbit, which periodically brings it much closer to the Sun—and a new Centaur is born.

Since Kuiper belt objects are too far away for easy observation, the Centaurs give astronomers somewhat more accessible samples of those outer-system bodies. The Centaurs are hardly near neighbors. At perihelion, Chiron is still more than 780 million miles (1,255 million km) from the Sun. And even at that distance, the solar heat is enough to boil off some of its outer material. This cometary behavior makes Chiron especially interesting—and certainly changes its surface from the pristine state in which it left the Kuiper belt.

Chiron is the only Centaur that has so far been seen to develop a bright, comet-like haze of gas and dust around it—perhaps because it is the only Centaur to have been observed near its perihelion. It is also a different color from other members of the Centaur group. Most of the Centaurs—such as Pholus and Nessus—are very red, probably due to the mixture of frozen chemicals that makes up their surface. But Chiron, is a neutral, dark color, possibly because the Sun has already vaporized much of its icy exterior. Another theory is that solar radiation actually changes the chemistry of Chiron's surface. But unless a spacecraft visits one of the Centaurs—and no such mission is planned as yet—we may never know for sure.

HORSE SENSE

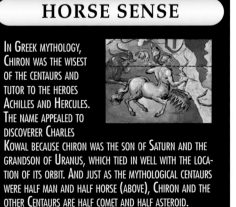

In Greek mythology, Chiron was the wisest of the Centaurs and tutor to the heroes Achilles and Hercules. The name appealed to discoverer Charles Kowal because Chiron was the son of Saturn and the grandson of Uranus, which tied in well with the location of its orbit. And just as the mythological Centaurs were half man and half horse (above), Chiron and the other Centaurs are half comet and half asteroid.

THE WAYWARD JOURNEY OF A CENTAUR

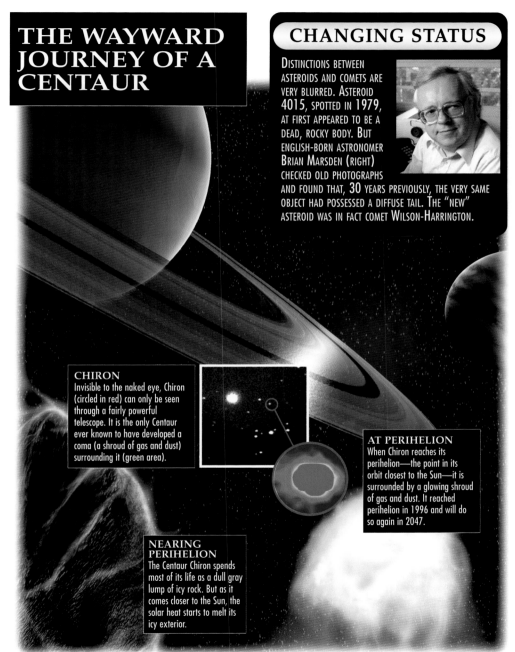

CHIRON
Invisible to the naked eye, Chiron (circled in red) can only be seen through a fairly powerful telescope. It is the only Centaur ever known to have developed a coma (a shroud of gas and dust) surrounding it (green area).

NEARING PERIHELION
The Centaur Chiron spends most of its life as a dull gray lump of icy rock. But as it comes closer to the Sun, the solar heat starts to melt its icy exterior.

AT PERIHELION
When Chiron reaches its perihelion—the point in its orbit closest to the Sun—it is surrounded by a glowing shroud of gas and dust. It reached perihelion in 1996 and will do so again in 2047.

CHANGING STATUS

Distinctions between asteroids and comets are very blurred. Asteroid 4015, spotted in 1979, at first appeared to be a dead, rocky body. But english-born astronomer Brian Marsden (right) checked old photographs and found that, 30 years previously, the very same object had possessed a diffuse tail. The "new" asteroid was in fact comet Wilson-Harrington.

METEORS

...THE EARTH PASSED THROUGH A COMET'S TAIL?

Almost all of the regular meteor showers that happen at various times of the year take place when the Earth crosses the orbit of a comet and sweeps up the debris that the comet has left behind it. Most of this debris tends to be scattered evenly around the comet's orbit. But some of the debris may be concentrated at one point in the orbit, and when the Earth crosses such a point, the resulting meteor shower can be spectacular. A good example is the Leonid shower, which revisits once every 33 years.

If the Earth were to pass through the actual tail of a comet, which contains still greater concentrations of debris (freshly swept off the comet by the effects of radiation from the Sun at close range), the result would be not a shower but a storm. Countless billions of cometary meteoroids would be swept up by the atmosphere, and the particles that create noticeable meteor trails—sand grain-sized and bigger—would be present in colossal numbers.

Scientists can only speculate what kind of display this huge concentration of meteoroids would put on, but it would make the average fireworks display seem tame by comparison.

The chances of any of these cometary meteoroids reaching ground level would be very slim, so the storm, although dramatic, would not present any physical danger to the Earth's inhabitants. The danger would be if the Earth were hit by the comet's head—or even by just a part of it. The impact of separate fragments of comet Shoemaker-Levy 9 on Jupiter in 1994 clearly demonstrated the potential destructive power of a comet impact on a planet.

The Earth is thought to have been hit by just such a cometary fragment on June 30, 1908, when an object devastated some 850 square miles (2200 square km) of forest in the Tunguska region of central Siberia in Russia. It is not known whether the object actually hit the ground, but it is thought to have delivered the same energy to the atmosphere as a 10-megaton hydrogen bomb. As it flamed through the sky, it was reported to be as bright as the Sun—it is the brightest meteor ever documented.

Gaze upward on a clear night and sooner or later you will see streaks of light that race across the sky for up to a few seconds. Sometimes the streaks give off what look like sparks, or leave a glowing trail. You might even see a storm of streaks, all of which seem to radiate from a fixed point. The objects in question are meteors—natural fireworks displays that are put on by small particles of cometary debris called meteoroids as they burn to a cinder high up in the Earth's atmosphere.

METEOR SHOWER CALENDAR

Shower Name	Maximum Intensity	Duration (min.)	Visual Strength	Associated Comet
Quadrantids	January 3	5	Medium	Not Known
Lyrids	April 22	5	Irregular	Thatcher
Eta Aquarids	May 4	5	Weak	Halley
Delta Aquarids	July 29	8	Medium	Not Known
Capricornids	July 30	3	Medium	Not Known
Perseids	August 12	6	Strong	Swift-Tuttle
Andromedids	October 3	11	Weak	Biela
Draconids	October 9	1	Irregular	Giacobini-Zinner
Orionids	October 21	2	Medium	Halley
Taurids	November 3	30	Weak	Encke
Leonids	November 17	2	Irregular	Tempel-Tuttle
Geminids	December 14	4	Strong	Asteroid 3200 Phaethon
Ursids	December 22	7	Weak	Tuttle

SPACE TRAILS

To the ancients, the fiery trails of meteors must have looked very much as though a star had come adrift from the heavens and blazed its way across the sky. But the term "shooting star" for a meteor is misleading: Meteors are not stars, but the glowing tracks left by small particles called meteoroids that enter the Earth's atmosphere at high speed and burn up as they do so.

Meteoroids are mostly particles of dust from the tails of comets that crossed our part of the solar system up to several thousands of years ago. They orbit the Sun at speeds of many miles per second, and if they happen to encounter Earth's atmosphere, their closing speed increases from between seven and 45 miles per second (11–72 km/sec).

At such speeds, the particles rapidly find themselves plunging through increasingly dense layers of the atmosphere. Although they weigh no more than a tiny fraction of an ounce, they carry more kinetic energy than a bullet from a gun. Within a fraction of a second, friction heats them up and their outer layers begin to vaporize, throwing atoms off into the atmosphere.

The atoms cause the surrounding air molecules to become ionized and they begin to glow brightly. The meteoroid itself soon breaks up, but not before it has left a trail of glowing air several yards wide and maybe 20 miles (32 km) long. On the ground, 50 miles (80 km) below, anyone looking upward during the night hours will see this trail of destruction as the brief appearance of a shooting star.

Almost as soon as it appears, the meteor is gone and the trail fades. Sometimes, however, the passage of the meteoroid releases so much energy that a train of glowing gas is left for several minutes.

SAME TIME

The Earth's part of the solar system is full of interplanetary debris that results in sporadic meteors. But far more interesting are the regular streams of meteoroids that give rise to what are called meteor showers. These appear without fail on certain dates of the year when the Earth plows through a cloud of debris that has spread out in the wake of a comet's orbit probably within the past few hundred years.

From the ground, the meteors give the impression that they are radiating away from a particular point in the sky, known as the radiant. This is simply an effect of perspective, like the way the parallel lanes of a highway seen from a bridge seem to radiate away from a single point on the horizon.

These meteor showers are named after the constellation in which the radiant appears. The Perseids, for example, which regularly produce 60 to 100 meteors an hour around August 12, appear to come from a point in Perseus. Some showers can give rise to true meteor storms, in which hundreds of meteors appear every second.

IN THE WAKE OF A COMET

FIERY TRAIL
During its brief dash through the Earth's atmosphere, a meteor may create a trail of glowing air molecules up to 20 miles (32 km) long. But the trail is seldom more than a few yards wide and usually fades in seconds.

PARALLEL TRACKS
Although the paths of meteors from a particular source are parallel to one another, from the ground they appear to radiate away from a point in the sky, called the radiant.

COMETARY DEBRIS
The dust from a comet's tail spreads out far beyond the comet's original orbit, though it continues to orbit the Sun. Eventually the dust may encounter Earth at high velocity, resulting in a meteor shower.

METEOR TRAIN
Occasionally a meteoroid creates a long-lasting "train" in the upper atmosphere that can last for minutes or even hours and is sometimes distorted by high-level winds.

COMET DUST
The dust from comets that causes meteors is usually light and crumbly, rather than sandy like the particles that split from asteroids. So-called Brownlee particles (right), gathered from the atmosphere by high-altitude airplanes, are probably the closest we will ever get to seeing real meteoroids.

APPROACH SPEED
Meteoroids enter the atmosphere at speeds of between 25,000 and 160,000 mph (40,200 an 257,500 km). They start to glow at an altitude of about 70 miles (110 km).

FIERY END
Most meteoroid bodies burn up completely about 50 miles (80 km) above the ground, and the meteor trail ends.

METEOR STREAMS

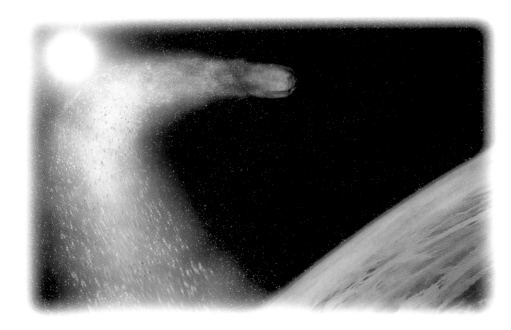

...WE WERE SUDDENLY DELUGED BY METEORS?

Imagine yourself outside on a dark, moonless night, gazing up at a cloudless sky filled with stars. Just as that strange feeling of awe and serenity begins to wear off, something catches your eye. You know instinctively that it is a shooting star, blazing across the heavens as it burns up in the Earth's atmosphere. Suddenly, two more meteors appear, then another three. Before you know it, the sky is ablaze with celestial streakers, showering down in the hundreds from what appears to be a point above your head.

Taken aback by the sudden extraterrestrial lightshow, you wonder why no one had clued you in on it. Surely astronomers could have predicted such a dramatic event? Next morning, as you scan the news channels, you discover that the entire world was caught unaware: The Earth had passed through an especially dense part of an undiscovered meteor stream.

A scenario like this could take place at any time, any day of the year, because encounters with meteor streams cannot be predicted until they have first been recorded crossing the Earth's orbit. There are probably millions of such streams out there, with widely varying particle densities at different points along their paths. But until their presence is confirmed by a meteor shower, we on Earth are left effectively in the dark.

A new stream could pass unnoticed—or it could produce a meteor storm the likes of which the world has never seen. Current records catalog around 150 short-period comets whose orbits around the Sun take less than 200 years. But that's not to say that new short-period comets won't turn up—and longer-period comets can take millions of years to orbit the Sun, which means that nobody knows when they last passed by. Even so, you can rest easy: If a meteor storm does occur, only the tiniest particles are ever likely to reach the ground.

An unexpected meteor storm is statistically unlikely—but far from impossible. Only after a stream's first encounter with Earth can its return be predicted.

Meteor streams, the icy trails of debris left in the wake of comets as they pass close to the Sun, are among the most elusive phenomena in the solar system. For most of the time, apart from the odd dent they leave in spacecraft, these clouds of tiny, dark particles might as well not exist—after all, we cannot see them. But when the Earth's orbit crosses their path and they rain down through the atmosphere, burning up as they go, the result is a meteor shower—one of the most dazzling spectacles in the night sky.

ANNUAL METEOR SHOWER

NAME	DATE AT MAXIMUM INTENSITY	NUMBER OF METEORS PER HOUR	PARENT COMET
QUADRANTIDS	JAN. 4	110	UNKNOWN
PERSEIDS	AUG. 12	68	COMET 1862 III
ORIONIDS	OCT. 21	30	COMET HALLEY
LEONIDS	NOV. 17	10	COMET P/TEMPEL-TUTTLE
GEMINIDS	DEC. 14	58	3200 PHAETHON

TRAIL BLAZERS

As a comet nears the Sun, its icy outer layers begins to evaporate and it sheds tiny particles, or meteoroids, in its wake. This trail of debris is a known as a meteor stream. The particles in the stream move in roughly the same direction as the comet, but at different speeds. In time, many of them come to lag far behind their parent body, while others race ahead.

Each time a comet reaches perihelion, its closest point to the Sun, it adds to the trail of meteoroids left behind on its previous visit—for example, Halley's Comet loses around 300 million tons (272 million tonnes) of material on each pass. In this way, the comet steadily shrinks while the meteor stream behind it builds up.

In time, the meteor stream becomes strewn across the comet's path—denser in some places than in others, because the comet does not shed material at an even rate. Eventually, the stream may spread all the way around, and slightly beyond, the comet's orbit. By this stage, the comet that created it may well have "died"—its ice having evaporated to leave a dark, dusty core that no longer produces a fiery tail. With no fresh dust to replenish it, the meteor stream, too, will gradually melt away to nothing.

SHOWER FORECASTING

Comets can be one of the brightest and most beautiful objects in the night sky, but the meteor streams that they leave behind are completely invisible from Earth. The particles within them are simply too small to be detected, even through the most powerful telescopes. We only become aware of a meteor stream's existence when its path crosses the Earth's orbit around the Sun. At this point, some of the particles pass through the atmosphere, where they reveal themselves as a meteor shower. By comparing the intensity of the shower with the orbit of the stream's parent comet, we can begin to picture how it is distributed across outer space.

The Earth collides with around 10 significant meteor streams each year, producing roughly the same number of regular meteor showers. Not all

are spectacular, or even visible. But, on very rare occasions, the Earth runs into an especially dense part of a meteor stream. When this happens, the result is a spectacular meteor "storm" in which the sky becomes ablaze with thousands of fiery streaks.

We may not be able to see a meteor shower approaching, but its return can usually be predicted. Only a close encounter with a giant planet like Jupiter will drag the stream away from its regular path and cause it to miss the Earth completely. But even if the stream returns, there is still no guarantee of a fireworks display because many comets leave little or no stream beyond their immediate vicinity. The Leonids are a good example.

Around November 17 every year, the Earth crosses the Leonid meteor stream. But only once every 33 years can we expect a really great display—that is, when the Earth crosses the stream just ahead of, or behind, its parent comet, Comet P/Tempel-Tuttle. The 1833, 1866, and 1966 returns of Tempel-Tuttle resulted in some of the most spectacular meteor showers ever seen. The 1999 return also saw spectacular showers in some parts of the world.

PULLED OFF THE COURSE

THE QUADRANTID METEOR SHOWER PEAKS AROUND JANUARY 3 EVERY YEAR, WITH A MAXIMUM INTENSITY OF 45–200 METEORS PER HOUR. AS WITH ALL SUCH SHOWERS, THE FIERY TRAILS SPRING FROM WHAT APPEARS TO BE A SINGLE POINT IN THE SKY, KNOWN AS THE *RADIANT* (RIGHT). BUT IN 1918, OBSERVERS NOTICED THAT THE QUADRANTIDS HAD SHIFTED TO A POSITION 8° NORTH OF THEIR USUAL SPOT. ASTRONOMERS NOW BELIEVE THAT JUPITER'S POWERFUL GRAVITY WAS RESPONSIBLE. IT SEEMS LIKELY THAT THE GIANT PLANET "CAPTURED" THE QUADRANTID METEOR STREAM'S PARENT COMET THOUSANDS OF YEARS AGO AND STILL DRAGS THE COMET'S REMAINS AROUND ITS ELLIPTICAL ORBIT TO THIS DAY.

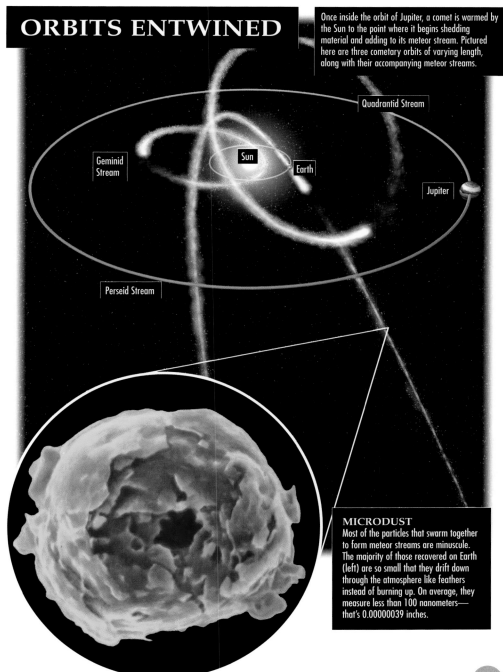

ORBITS ENTWINED

Once inside the orbit of Jupiter, a comet is warmed by the Sun to the point where it begins shedding material and adding to its meteor stream. Pictured here are three cometary orbits of varying length, along with their accompanying meteor streams.

Quadrantid Stream

Geminid Stream

Sun

Earth

Jupiter

Perseid Stream

MICRODUST
Most of the particles that swarm together to form meteor streams are minuscule. The majority of those recovered on Earth (left) are so small that they drift down through the atmosphere like feathers instead of burning up. On average, they measure less than 100 nanometers—that's 0.00000039 inches.

METEOR MYSTERIES

Most meteors—the regular, annual Perseid and Leonid showers, for example—sleet through the atmosphere as straight streaks of light against the background of the stars. But every so often, meteor observers report arrivals that are distinctly unusual—curved or strongly colored light tracks, and sometimes even meteors that appear darker than the night sky. Some of these meteor mysteries are optical illusions; others are caused by the very nature of the small, fast-moving fragments that appear as shooting stars.

WHAT IF...

...METEORS PRESENTED A HAZARD TO EARTH?

The vast majority of meteoroids that enter Earth's atmosphere are burned to a fiery vapor at high altitudes. Occasionally, larger pieces of material—fragments thrown off by collisions between asteroids—survive atmospheric passage, making it to ground as meteorites.

Most meteors are produced by debris shed from comets. Spectroscopic analysis has shown the presence of poisonous gases including cyanogen and carbon monoxide in comet tails, boiled from a comet's nucleus by the heat of the Sun. Could meteoroids recently released from a comet have pockets of these gases embedded in their fluffy structure? It seems unlikely—solar radiation quickly strips away these gases, and after a few decades in space, only insignificant quantities will remain. Even major meteor storms from recently ejected material—from Comet Tempel-Tuttle, for example—do not appear to present a chemical threat.

Some comet nuclei have been observed to break up into smaller, but still sizeable chunks. Could recently separated iceberg-sized cometary fragments pose a threat to life on Earth? Even quite large pieces of cometary debris are completely destroyed on entering the atmosphere: Their low mechanical strength results in rapid disintegration, sometimes in a spectacularly bright explosion. Material from these airbursts disperses rapidly at high altitudes, and large-scale meteors of this kind are no more of a threat than the relative small-fry that produce fainter meteors.

There was a time, during the Cold War, when the arrival of such substantial meteorites was a very real threat—not because of the effects of impact, but as a trigger that might have set off World War III. Satellites and early warning systems, programmed to watch out for missile launches or weapons tests, frequently recorded these super-meteors. Fortunately, neither the Soviet nor the U.S. military was ever quite jumpy enough to launch a wave of nuclear missiles in retaliation for a cometary fragment.

Today, the main hazard presented by meteors is probably to communications and other satellites during intense shower activity. Engineers worry that major Leonid storms, for example, could damage satellites either by direct collisions of small high-speed particles with the spacecraft, or by producing clouds of ionization which may affect delicate electrical circuits. And the forthcoming International Space Station—a permanent human presence in space—could also be vulnerable.

METEOR MYSTERIES

REPORTED PHENOMENON	REAL OR ILLUSION?	LIKELY EXPLANATION
Curved meteors	Illusion.	Observer disorientation
Zigzag meteors	Open to debate.	Release of trapped gas
Corkscrew meteors	Real.	Rotation of incoming meteoroid
Nebulous meteors	Real.	Fragmentation of incoming meteoroid and/or gas release
Colored meteors	Real.	Excitation of atmospheric oxygen and metallic ion species in the meteoroid
Dark meteors	Illusion.	Physiology of dark-adapted eye; wildlife, and wind-blown terrestrial debris
Meteor sounds	Open to debate.	Psychological, or remission of electromagnetic radiation as sound by charged objects at ground level

FLASHER'S TRAIL

Meteoroids hurtle into the upper atmosphere at speeds between 7 and 45 miles per second (11–72 km/sec). At such velocities, friction with the air usually burns them up before they can come within 50 miles (80 km) of the Earth's surface. As seen from the ground, their demise amounts to a streak of light across the sky that lasts a few tenths of a second.

Sometimes, though, meteoroids meet their ends in surprising ways. Long-lasting meteor trails—those that cross most of the visible sky—are occasionally described as following a curved trajectory. One explanation is that the curve is an illusion, caused by disorientation when the observer's head turns to follow what is in reality a long, straight path. Certainly, no one has yet produced a convincing photograph of a curved meteor.

It is harder to explain reported zigzag meteors, which may abruptly change direction—though again, hard photographic evidence is lacking. One possibility is that in such cases the incoming particle has retained small pockets of gas—frozen as ice residues inside its fluffy lattice grain structure. The heat of atmospheric entry turns the gas into rocket propellant that boosts the meteor into a strange flight pattern.

While most meteors are sharp, thin streaks, observers sometimes report diffuse objects—not so much a shooting star as a "shooting nebula." These fuzzy streaks may also be a consequence of gas embedded in an incoming meteoroid particle.

NEBULOUS TRACKS

Almost all meteors were once part of an icy comet nucleus: a mixture of dust and frozen gas. Once a meteoroid particle has been separated from its parent comet, it should quite quickly become depleted of any residual gas by solar radiation. Only material released very recently from a comet should retain significant amounts of gas, so nebulous meteor tracks—reported too often by experienced observers to be illusory—remain a real mystery.

Observers sometimes report seeing dark

streaks, apparently traveling at the same speeds as regular meteors across the sky, quite often in contrast against the faint illumination of the Milky Way. These dark meteors are quite probably another illusion, produced by physiological effects in the eye: Most reports come from observers whose eyes are settling into night-vision state, or from tired watchers nearing the end of a long stint. Night flying insects and bats, and wind-blown terrestrial debris might also explain the alleged sightings.

Meteors travelling on tight corkscrew paths do seem to be real—a few photographs exist—and are probably an indication that the incoming meteoroid is spinning. And strongly colored meteor events are also no illusion. Most meteors emit too little light to stimulate the color receptors of the human eye. But rarer, bigger and brighter incomers do display pronounced colors. Bright meteors in the Perseid meteors shower are usually yellowish; Geminids often appear blue or green. Red meteors are the rarest. The colors come from metal ions in the meteoroid fragment, but atmospheric ions also contribute: Some green emissions come from oxygen.

There are many reports of meteors that make a simultaneous hissing or crackling noise. But the air 50 miles (80 km) up is too thin to carry sound, which in any case travels very much more slowly than light. Probably, the effect is purely psychological.

SKIPPING STONE

On August 10, 1972, a bright fireball blazed across the early afternoon skies of the western United States at an altitude of about 40 miles (64 km). The object, probably a comet fragment 10 yards (9 m) across, skipped off the atmosphere and back into space, continuing its orbit around the Sun. It is expected to return in the future, when it may completely burn up or fall to Earth as a meteorite.

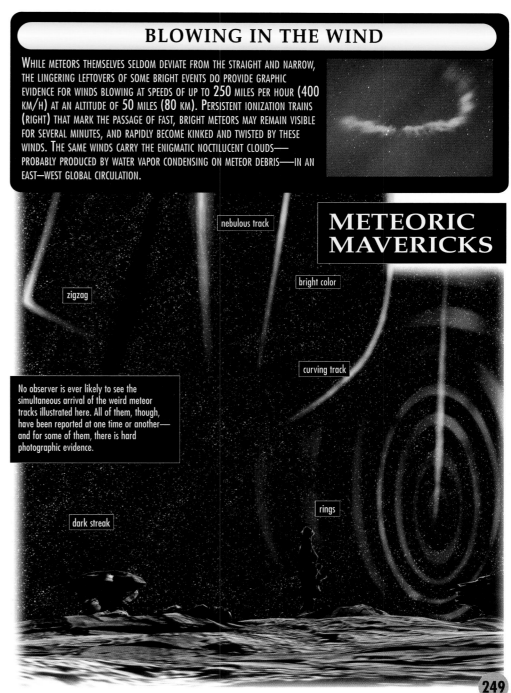

BLOWING IN THE WIND

While meteors themselves seldom deviate from the straight and narrow, the lingering leftovers of some bright events do provide graphic evidence for winds blowing at speeds of up to 250 miles per hour (400 km/h) at an altitude of 50 miles (80 km). Persistent ionization trains (right) that mark the passage of fast, bright meteors may remain visible for several minutes, and rapidly become kinked and twisted by these winds. The same winds carry the enigmatic noctilucent clouds—probably produced by water vapor condensing on meteor debris—in an east–west global circulation.

nebulous track

METEORIC MAVERICKS

bright color

zigzag

curving track

No observer is ever likely to see the simultaneous arrival of the weird meteor tracks illustrated here. All of them, though, have been reported at one time or another—and for some of them, there is hard photographic evidence.

rings

dark streak

METEORITES

Every year the Earth receives thousands of visitors from space. They are meteorites—pieces of rocky debris that have crossed the Earth's path through space to produce spectacular fireballs as they plunge into the atmosphere. Most meteorites simply fall into the sea and are lost. But some are recovered on land, especially in Antarctica, where they lie preserved in the ice. And meteorites are more than just lumps of rock: Many are as old as the solar system itself and offer important clues to the formation of the Earth.

WHAT IF...

...YOU WERE HIT BY A METEORITE?

At the top end of the meteorite scale, impacts are devastating. The Earth still bears scars from house-sized rocks that released far more energy than any atom bomb. But are modestly sized meteorites dangerous?

Small meteorites lose almost all of their colossal velocity as they pass through the atmosphere. Even so, they land about as fast as a stone falling from a high building and are capable of doing serious harm. Larger meteorites, weighing a few pounds, strike at much higher speeds, and anybody hit by one would probably be killed instantly.

Fortunately, the risk is very low, and no deaths have been reported for centuries. But old Chinese records do mention fatalities caused by rocks falling from the sky, and more recently, there have been a few lucky escapes. In 1992, a young Ugandan was playing soccer with his friends when he suddenly fell to the ground, stunned. A meteorite, which lay nearby, had hit him on the arm. The boy escaped with no broken bones. The only American meteorite target so far is Elizabeth Hodges, of Sylacauga, Alabama.

In 1954, she survived a meteorite impact with nothing worse than a bruised hip. In 1911, a shower of 48 space stones, which probably originated on Mars, lashed the village of Nakhla, Egypt. There was only one fatality—a dog—the Earth's only known victim of a "Martian attack."

More often, meteorites damage property. English gardener Arthur Pettifor was tending his onions one spring day in 1991, when a 4-inch by 2-inch (10-cm by 5-cm) meteorite plowed through his carefully trimmed hedge. And on October 9, 1992, Michelle Knapp of Peekskill, New York, was watching TV when she heard a loud bang. A 27-pound (12-kg) space rock had just wrecked her car.

An irate Michael Aponte shows what a space rock did to his girlfriend's car. The offending meteorite (inset) was later sold— for enough cash to buy a new car.

METEORITE CRATERS

Location	Diameter	Age (MILLIONS OF YEARS)
1 Aouelloul, Mauritania	0.23 miles (0.37 km)	3.1
2 Brent, Ontario, Canada	2.4 miles (3.9 km)	450
3 Clearwater Lake, Quebec, Canada	14 miles (23 km)	290
4 Kara, Russia	31 miles (50 km)	57
5 Meteor Crater, Arizona, U.S.	0.75 miles (1.2 km)	0.1 OR LESS
6 Popigay, Russia	62 miles (100 km)	39
7 Ries, Germany	15 miles (24 km)	15
8 Taban Khara Obo, Mongolia	0.8 miles (1.3 km)	30 OR LESS

HARD RAIN FALLING

Most of the pieces of space debris known as meteoroids that streak across the sky are dust grains from comet tails. When these tiny specks of matter hit the atmosphere, the heat generated by friction vaporizes them. But sometimes a much larger and more solid object makes it through the Earth's protective shroud and hits the ground. No longer a meteoroid, its arrival on Earth has earned it the name of meteorite.

Some meteorites originate on the Moon or Mars, where they were once dislodged by ancient asteroid impacts, and a few even began life in the heads of comets. But the majority appear to come from the Asteroid Belt—the ring of debris, possibly the remains of a planet that failed to form, that orbits the Sun between Mars and Jupiter.

When two asteroids collide within the belt, the energy of the impact sends pieces flying. Sooner or later, some of these fragments find themselves on a collision course for Earth. And if a fragment is big enough—at least four ounces (100 g) in weight, but more often a pound (500 g) or more—it creates a spectacular fireball as it arrives.

After its scorching descent through the atmosphere, the meteorite reaches the surface as a blackened lump of rock. Yet despite its atmospheric roasting, a meteorite found soon after landing is often coated with frost. This is because, like a cosmic Baked Alaska, its interior remains at the freezing cold temperature of outer space.

HISTORY LESSONS

Meteorites come in two distinct classes. Iron meteorites are composed of a near-pure nickel-iron alloy. Stony meteorites are made from a range of minerals, including some iron. Many stony meteorites have a complex, grainy structure that is quite different from any other known rocks. Scientists believe that these so-called chondrites may have condensed directly from the original dusty nebula that formed our solar system.

Older than any rocks on Earth, chondrites are time capsules from an age before our planet existed and, as such, provide a unique source of information from an era of which we know little. One subgroup, the carbonaceous chondrites, may even yield insights into the origins of life. Formed at the same time as other chondrites, they contain carbon-based organic matter, including amino acids—the fundamental building blocks from which all living things are made. People have even suggested that carbonaceous chondrites were the seeds from which life on Earth first sprang.

Iron meteorites usually survive weathering on Earth better than their stony companions, which in any case are often mistaken for ordinary Earth rocks. An iron meteorite is noticeably heavier than a stone or rock of the same size. But any recently fallen meteorite is characterized by a black, sooty crust—sometimes covered with ripples—that signifies its fiery arrival in the vicinity of our planet.

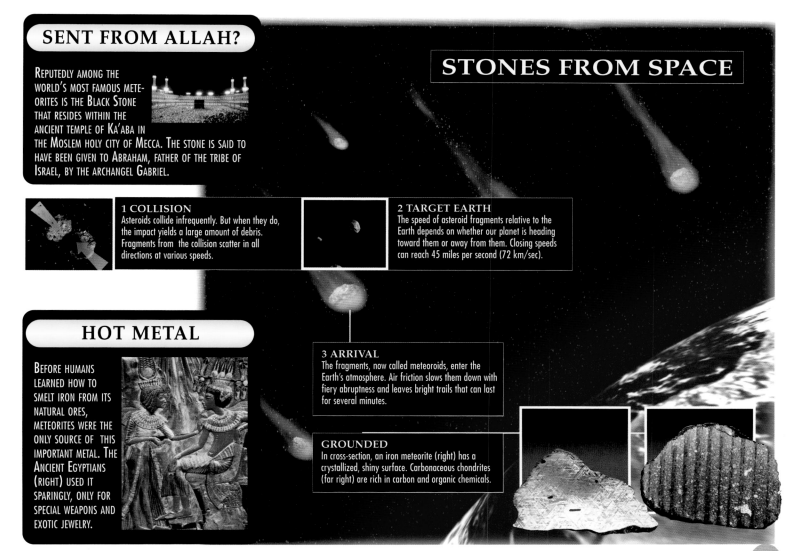

SENT FROM ALLAH?

REPUTEDLY AMONG THE WORLD'S MOST FAMOUS METEORITES IS THE BLACK STONE THAT RESIDES WITHIN THE ANCIENT TEMPLE OF KA'ABA IN THE MOSLEM HOLY CITY OF MECCA. THE STONE IS SAID TO HAVE BEEN GIVEN TO ABRAHAM, FATHER OF THE TRIBE OF ISRAEL, BY THE ARCHANGEL GABRIEL.

STONES FROM SPACE

1 COLLISION
Asteroids collide infrequently. But when they do, the impact yields a large amount of debris. Fragments from the collision scatter in all directions at various speeds.

2 TARGET EARTH
The speed of asteroid fragments relative to the Earth depends on whether our planet is heading toward them or away from them. Closing speeds can reach 45 miles per second (72 km/sec).

HOT METAL

BEFORE HUMANS LEARNED HOW TO SMELT IRON FROM ITS NATURAL ORES, METEORITES WERE THE ONLY SOURCE OF THIS IMPORTANT METAL. THE ANCIENT EGYPTIANS (RIGHT) USED IT SPARINGLY, ONLY FOR SPECIAL WEAPONS AND EXOTIC JEWELRY.

3 ARRIVAL
The fragments, now called meteoroids, enter the Earth's atmosphere. Air friction slows them down with fiery abruptness and leaves bright trails that can last for several minutes.

GROUNDED
In cross-section, an iron meteorite (right) has a crystallized, shiny surface. Carbonaceous chondrites (far right) are rich in carbon and organic chemicals.

CHANGING VIEWS OF METEORITES

Meteorites are chunks of rock or metal big enough to endure a fiery passage through the atmosphere and reach the surface more or less intact. In ancient times, they were associated with the wrath of the gods and inspired tales of dragons. Early scientists, eager to distance themselves from such superstition, refused to believe that they fell from the sky at all. Now we know that meteorites are the legacy of countless solar system impacts—mostly collisions between asteroids, but sometimes huge strikes on planets or moons.

WHAT IF...

...THE IMPACT THAT KILLED THE DINOSAURS BLASTED PIECES OF EARTH TO MARS?

NASA researchers announced in 1996 that they had found what appeared to be fossilized bacteria inside a meteorite from Mars—a lump of rock that had been flung from the Red Planet's surface by the violent impact of a comet or asteroid. Could a similar giant impact on Earth—such as the asteroid strike that is thought to have eradicated the dinosaurs 65 million years ago—have sent Earth rocks to Mars? And if so, could they have carried with them a cargo of Earthy microorganisms?

The dinosaur-killing impact was certainly powerful enough to have hurled millions of rock fragments, some over 30 feet (9 m) across, into interplanetary space. Scientists estimate that around 1 in 500 of these fragments would have maintained internal temperatures below the boiling point of water, and so thousands of rocks containing live terrestrial microbes could have escaped into space. The chances of larger creatures in the rocks surviving, though, are infinitesimal. They would have had to endure crushing accelerations equivalent to 10,000 times the pull of Earth's gravity.

Once ejected, each fragment would have gone into orbit around the Sun. Sometimes a fragment would return to the Earth's vicinity, and a kick from the planet's gravity would nudge it into a new orbit that might intercept Mars after a few million years. But microbes can live for a long time—on Earth, live bacteria have been extracted from salt deposits that formed some 255 million years ago. The surrounding rock would shield the microbes from the harsh vacuum and high radiation levels in space.

Impact on Mars would be the next hazard for the organisms to endure. Luckily, the planet has an atmosphere. Rocks striking bodies without an atmosphere—such as the Moon—hit so fast that they often vaporize on impact. But Mars' atmosphere would slow down the incoming rock fragment, softening the blow. If the conditions on Mars were conducive to the development of life, as some scientists think they may have been long ago, the organisms could then have thrived on the planet's surface.

Of course, it is possible that the microbes were only returning to their ancestral home. If meteorite-carried Martian life had reached Earth first, Mars could turn out to be the wellspring of solar system life.

METEORITE TIMELINE

Date	What happened
2000 B.C.	Phrygian stone falls to Earth in ancient Rome
861 A.D.	The earliest meteorite impact from which material still exists takes place, in Nogata, Japan
1492	A meteorite falling near Ensisheim, France, is declared to be a sign of God's anger
1794	Ernst Friedrich Chladni publishes a book arguing that rocks from the sky come from space
1803	A hail of meteorites striking L'Aigle, Normandy, convinces many to accept Chladni's theory
1807	U.S. President Thomas Jefferson dismisses the notion that meteorites landing in Connecticut could have come from space
1940s–50s	The idea of volcanism as the cause of cratering on Earth is gradually overthrown in favor of meteorite impacts
1962	The Zagami meteorite—the largest known meteorite from Mars—is found; its Martian identity is established in the 1980s
1982	The first meteorite fragment from the Moon is identified
1996	NASA researchers announce that ALH84001, a meteorite from Mars, appears to contain fossilized bacteria
1999	Scientists find similar fossil-like structures in four other Martian meteorites

WRATH OF THE GODS

A meteorite is any stony or metallic object from space that survives passage through the atmosphere to reach the surface of the Earth. Almost all meteorites are fragments chipped from colliding asteroids. The remainder are debris from massive crater-forming impacts on the planets and moons of the solar system.

But this understanding is a modern one. Long ago, rocks falling from the sky were a dangerous mystery, often taken as a sign of divine displeasure. For example, when a large meteorite fell near the French town of Ensisheim in 1492, King Maximilian of Germany interpreted it as an expression of God's anger at the French for waging war against the Holy Roman Empire.

Scientists of the 17th and 18th centuries were inclined to be skeptical about such reports. They had gone to much trouble to build up a rational view of the world, and they preferred to flatly disbelieve tales of rocks from the sky. But in 1794, German physicist Ernst Friedrich Chladni published a book in which he argued that lumps of stone and metal really do fall to Earth from space, leaving bright, fiery trails behind them as they plow through the planet's atmosphere. Initially, the idea was met with mockery, but, backed up by a growing number of respectable eyewitness reports, Chladni's theory was gradually accepted.

Today, meteorites are so well studied that dozens of different types are known and cataloged. And modern astronomers are more inclined than their predecessors to credit some supposedly tall stories from the past.

Some scientists argue that meteorites may have been the cause of a sudden change in climate around the year 540 A.D., as evidenced by studies of the ring structure of very old trees. Dust and smoke from meteorite impacts would account for the climatic disturbances. Mike Baillie, a paleontologist at Queen's University, Belfast, Northern Ireland, points out that the event roughly coincides with the writing of the Anglo-Saxon poem *Beowulf*, which describes a fire-breathing dragon that was blighting the land of the Geat people. If the tree-ring anomaly was indeed due to meteorites, then the fire in the sky

created by their passage through the Earth's atmosphere could have been mistaken for a dragon's breath.

Most meteorite rocks are very small. Rarely, a much larger chunk will strike the Earth, leaving behind a crater to mark its arrival. Such craters were once thought to be of volcanic origin. But when geologists investigated the sites, they found them to be devoid of basalt or any other volcanic rocks. Instead, there were many small rock fragments that matched the composition of known meteorite types. The volcanic theory was dropped.

In 1990, scientists identified the outline of a 110-mile (177-km) wide crater on the Yucatán peninsula of Mexico. The crater was probably formed 65 million years ago by a 6-mile (9.5-km) wide rock—just when the dinosaurs became extinct, and the most likely reason for their demise.

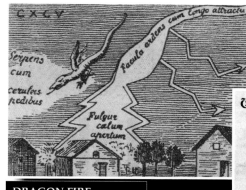

DRAGON FIRE
A sketch of a meteorite falling on a European town in 1000 A.D. combines imagination with observation. The meteoric fireball is represented as a fire-breathing dragon. But the track of the meteorite's fall seems to depict the object breaking up and falling in several pieces.

DIVINE DISAPPROVAL
A 16th-century woodcut depicts the arrival of a "thunderstone" near the town of Ensisheim on the Franco-German border in 1492. Ensisheim was French, and France had invaded German princedoms: It was obvious to German rulers that the meteorite was God's warning. Later, scientists refused to believe that meteorites fell from the sky.

INTERPRETING METEORITES

JEFFERSON'S YANKEES

IN 1807, TWO PROFESSORS FROM YALE UNIVERSITY CONCLUDED THAT A NUMBER OF ROCKS THAT HAD LANDED IN THE TOWN OF WESTON, CONNECTICUT, MUST HAVE FALLEN FROM THE SKY. IT WAS TOO MUCH FOR U.S. PRESIDENT THOMAS JEFFERSON (RIGHT), HIMSELF SOMETHING OF A SCIENTIST. LESS THAN CONVINCED BY WHAT HE SAW AS AN ABSURD EXPLANATION, HE REPORTEDLY SAID, "GENTLEMEN, I WOULD RATHER BELIEVE THAT TWO YANKEE PROFESSORS WOULD LIE THAN BELIEVE THAT STONES FALL FROM HEAVEN."

LIFE FROM SPACE

METEORITES MAY BE RESPONSIBLE FOR INTRODUCING NEW LIFE TO THE EARTH. THE THEORY OF PANSPERMIA SUPPOSES THAT MICROORGANISMS, SUCH AS VIRUSES AND BACTERIA, CAN TRAVEL THROUGH SPACE INSIDE ROCKS, BLASTED FROM THE SURFACE OF THEIR HOME WORLD BY LARGE IMPACTS. IN 1996, NASA SCIENTISTS DISCOVERED WHAT MIGHT HAVE BEEN FOSSIL BACTERIA IN A METEORITE FROM MARS. THE FIND—SHOWN HERE AT A PRESS CONFERENCE WITH THE JOHNSON SPACE CENTER'S DAVID MCKAY—IS STILL CONTROVERSIAL.

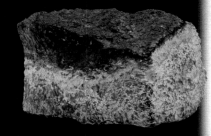

ROCKS FROM THE SKY
Fire leaps from a French farmhouse after a meteorite impact in 1846. A spate of such impacts—observed by reliable witnesses—forced scientists to accept the theories of German physicist Ernst Chladni that meteorites really did fall from space through the atmosphere.

CLASSIFICATION
By the 20th century, meteorites were a well-understood phenomenon. Scientists had learned to classify them into different types—for example, stony or nickel-iron—and often to deduce their origin. Most began as part of an asteroid. This 1-pound (450 g) example, discovered in California's Mojave Desert, was once part of the planet Mars. The dark brown "fusion crust" at the top of the meteorite was caused by intense atmospheric heating.

INDEX

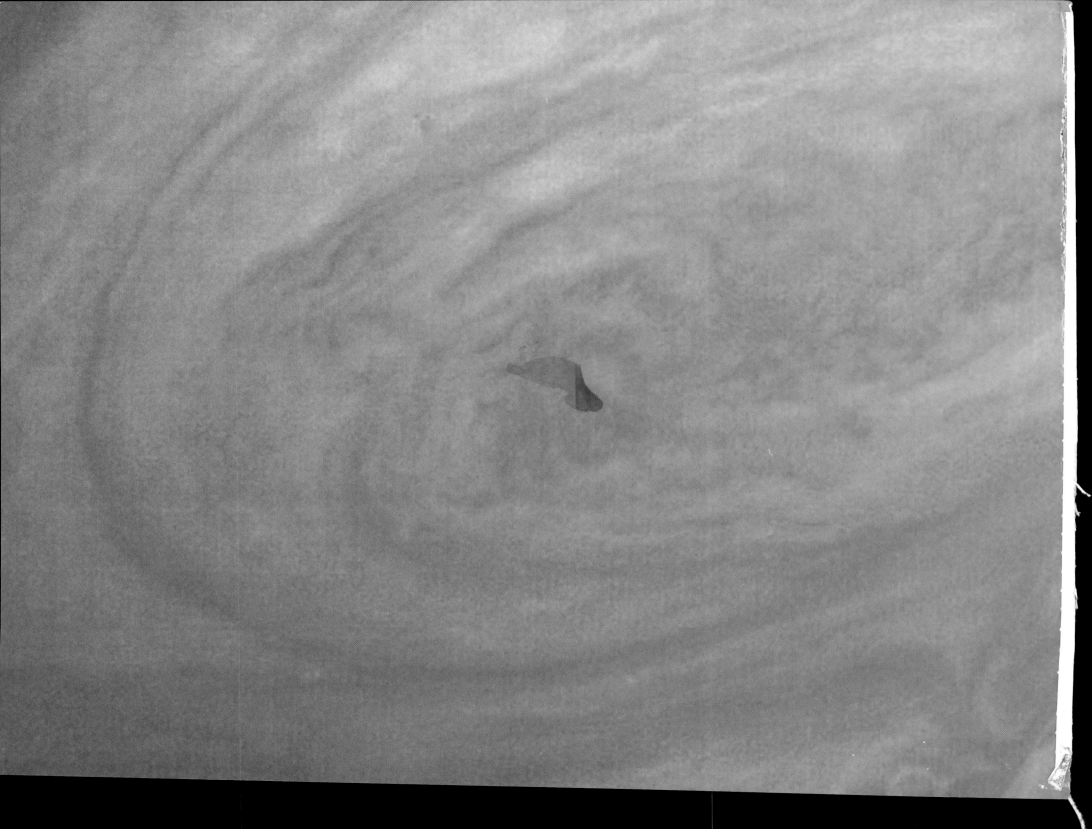